Lecture Notes in Computer Science 13349

More information about this series at https://link.springer.com/bookseries/558

Nicolas Behr · Daniel Strüber (Eds.)

Graph Transformation

15th International Conference, ICGT 2022
Held as Part of STAF 2022
Nantes, France, July 7–8, 2022
Proceedings

Editors
Nicolas Behr
Université Paris Cité
Paris, France

Daniel Strüber
Computer Science and Engineering
Chalmers | Gothenburg University
Gothenburg, Sweden

ISSN 0302-9743 ISSN 1611-3349 (electronic)
Lecture Notes in Computer Science
ISBN 978-3-031-09842-0 ISBN 978-3-031-09843-7 (eBook)
https://doi.org/10.1007/978-3-031-09843-7

This Springer imprint is published by the registered company Springer Nature Switzerland AG
The registered company address is: Gewerbestrasse 11, 6330 Cham, Switzerland

Preface

This volume contains the proceedings of ICGT 2022, the 15th International Conference on Graph Transformation, held during July 7–8, 2022, in Nantes, France. ICGT 2022 was affiliated with STAF (Software Technologies: Applications and Foundations), a federation of leading conferences on software technologies. ICGT 2022 took place under the auspices of the European Association of Theoretical Computer Science (EATCS), the European Association of Software Science and Technology (EASST), and the IFIP Working Group 1.3, Foundations of Systems Specification.

The ICGT series aims at fostering exchange and the collaboration of researchers from different backgrounds working with graphs and graph transformation, either by contributing to their theoretical foundations or by applying established formalisms to classic or novel areas. The series not only serves as a well-established scientific publication outlet but also as a platform to boost inter- and intra-disciplinary research and to stimulate for new ideas. The use of graphs and graph-like structures as a formalism for specification and modeling is widespread in all areas of computer science as well as in many fields of computational research and engineering. Relevant examples include software architectures, pointer structures, state-space and control/data flow graphs, UML and other domain-specific models, network layouts, topologies of cyber-physical environments, quantum computing, and molecular structures. Often, these graphs undergo dynamic change, ranging from reconfiguration and evolution to various kinds of behavior, all of which may be captured by rule-based graph manipulation. Thus, graphs and graph transformation form a fundamental universal modeling paradigm that serves as a means for formal reasoning and analysis, ranging from the verification of certain properties of interest to the discovery of fundamentally new insights.

ICGT 2022 continued the series of conferences previously held in Barcelona (Spain) in 2002, Rome (Italy) in 2004, Natal (Brazil) in 2006, Leicester (UK) in 2008, Enschede (The Netherlands) in 2010, Bremen (Germany) in 2012, York (UK) in 2014, L'Aquila (Italy) in 2015, Vienna (Austria) in 2016, Marburg (Germany) in 2017, Toulouse (France) in 2018, Eindhoven (The Netherlands) in 2019, and online in 2020 and 2021, following a series of six International Workshops on Graph Grammars and Their Application to Computer Science from 1978 to 1998 in Europe and in the USA.

This year, the conference solicited research papers describing new unpublished contributions in the theory and applications of graph transformation as well as tool presentation papers that demonstrate main new features and functionalities of graph-based tools. All papers were reviewed thoroughly by at least three Program Committee members and additional reviewers. We received 19 submissions, and the Program Committee selected 10 research papers and one tool presentation paper for publication in these proceedings, after careful reviewing and extensive discussions. The topics of the accepted papers cover a wide spectrum, including theoretical approaches to graph transformation, logic and verification for graph transformation, and model transformation, as well as the application of graph transformation in some areas. In addition to these paper presentations, we were delighted to host an invited talk by Christian Doczkal (Max Planck Institute for Security and Privacy, Bochum, Germany).

A special focus of ICGT 2022 consisted of new approaches to formalizing the knowledge in the research field of graph transformation theory via proof assistants such as Coq. A long-term goal of this kind of approach consists of establishing a Coq-enriched wiki for this research field akin to the nLab. This platform will serve as a sustainable mechanism for curating applied and mathematical knowledge in graph transformation research, and eventually as a research tool in its own right, notably through the provision of interactive database-supported proof construction. Another avenue of research concerns executable applied category theory (ExACT), i.e., code extraction from formalized categorical structures, with the perspective of curating a database of correct-by-construction reference prototype algorithms for various forms of graph transformation semantics and graph-like data structures. To introduce the initiative, facilitate the broad involvement of the ICGT community, and collect feedback from participants regarding the scope and format of such a wiki project, a peer-reviewed brainstorming session was conducted as one of the events at the conference. Results of this session as well as further information on this initiative are available via the GReTA ExACT working group website (https://www.irif.fr/%7Egreta/gretaexact/).

We would like to thank all who contributed to the success of ICGT 2022, the invited speaker Christian Doczkal, the authors of all submitted papers, and the members of the Program Committee, as well as the additional reviewers, for their valuable contributions to the selection process. We are grateful to Reiko Heckel, the chair of the Steering Committee of ICGT, for his valuable suggestions; to Massimo Tisi and Gerson Sunye, the general chair and the local chair, respectively, of STAF 2022; and to the STAF federation of conferences for hosting ICGT 2022. We would also like to thank EasyChair for providing support for the review process.

May 2022

Nicolas Behr
Daniel Strüber

Organization

Program Committee

Paolo Baldan	Università di Padova, Italy
Nicolas Behr (Chair)	Université Paris Cité, CNRS, IRIF, France
Paolo Bottoni	Sapienza University of Rome, Italy
Andrea Corradini	Università di Pisa, Italy
Juan De Lara	Universidad Autonoma de Madrid, Spain
Juergen Dingel	Queen's University, Canada
Fabio Gadducci	Università di Pisa, Italy
Holger Giese	Hasso Plattner Institute at the University of Potsdam, Germany
Russ Harmer	CNRS and ENS Lyon, France
Reiko Heckel	University of Leicester, UK
Thomas Hildebrandt	University of Copenhagen, Denmark
Wolfram Kahl	McMaster University, Canada
Timo Kehrer	University of Bern, Switzerland
Aleks Kissinger	University of Oxford, UK
Barbara König	University of Duisburg-Essen, Germany
Leen Lambers	Brandenburgische Technische Universität Cottbus-Senftenberg, Germany
Yngve Lamo	Western Norway University of Applied Sciences, Norway
Koko Muroya	Kyoto University, Japan
Fernando Orejas	Universitat Politècnica de Catalunya, Spain
Detlef Plump	University of York, UK
Arend Rensink	University of Twente, The Netherlands
Leila Ribeiro	Universidade Federal do Rio Grande do Sul, Brazil
Andy Schürr	TU Darmstadt, Germany
Daniel Strüber (Chair)	Chalmers University of Technology and University of Gothenburg, Sweden/Radboud University Nijmegen, The Netherlands
Gabriele Taentzer	Philipps-Universität Marburg, Germany
Matthias Tichy	Ulm University, Germany
Uwe Wolter	University of Bergen, Norway
Steffen Zschaler	King's College London, UK

Additional Reviewers

Barkowsky, Matthias
Courtehoute, Brian
Fritsche, Lars
Kosiol, Jens
Righetti, Francesca
Sakizloglou, Lucas
Schaffeld, Matthias
Schneider, Sven
Söldner, Robert
Trotta, Davide

Graph Theory in Coq: Axiomatizing Isomorphism of Treewidth-Two Graphs (Abstract of Invited Talk)

Christian Doczkal

Max Planck Institute for Security and Privacy (MPI-SP), Bochum, Germany

Despite the importance of graph theory in mathematics and computer science, there are relatively few machine-checked proofs of graph theory results and even fewer general purpose libraries. After the formalization of some basic concepts in HOL [1] and a formalization of Euler's theorem in Mizar [12] during the 90s, the 2000s saw several results on planar graphs: Gonthier's celebrated formal proof of the four-color theorem [9] in Coq, the formalization of tame graphs as part of the Flyspeck project [13] in Isabelle/HOL, and a study on Delaunay triangulations [8] by Dufourd and Bertot. More recently, Noschinski developed a library for both simple and multigraphs in Isabelle/HOL [14].

Over the past couple of years, Damien Pous and I have developed a graph theory library[1,2] for the interactive theorem prover Coq[3] based on the Mathematical Components Library[4].

The initial goal was to formalize soundness and completeness of a finite axiomatization of isomorphism for the class of labeled treewidth-two multigraphs [10], a new result answering positively – for this particular class of graphs – a question posed by Courcelle [2, p. 118]. Since none of the available libraries suited our needs, we started to develop a new graph theory library [3, 6, 7]. Since then, there has been some renewed interest in the formalization of graph theory, both in Coq [15, 16] and in other systems [11].

The development of our library and the aforementioned axiomatizability result that guided the development process highlight the fruitful interplay between the development of pen-and-paper proofs and the development of machine-checked mathematical libraries. While the initial design of the library allowed us to formally verify [4] parts of the original proofs, formalizing the full proof seemed out of reach. This prompted us to develop a new pen-and-paper proof [5] that was significantly simpler and written with formalization in mind. In addition, we revised and extended the library [7], allowing us to overcome those difficulties that could not be sidestepped using the new proof. This allowed us to finally verify [6] the completeness result we initially set out to prove.

Despite the use of a graph rewrite system, the completeness proof and its formalization in Coq are elementary in the sense that we do not employ results from graph transformation theory. This is due, at least in part, to us not being familiar with these

1 https://coq-community.org/graph-theory/.

2 With contributions from Daniel Severín, Guillaume Combette and Guillaume Ambal.

3 https://coq.inria.fr.

4 https://math-comp.github.io.

techniques and there not being any preexisting graph transformation libraries, in particular none that would interface well with the Mathematical Components library. Further, our simple 4-rule rewrite system could still be reasoned about directly, However, for reasoning about more complex graph rewrite systems, using a more abstract approach appears necessary.

References

1. Chou, C.T.: A formal theory of undirected graphs in higher-order logc. In: Melham, T.F., Camilleri, J. (eds.) HUG 1994. LNCS, vol. 859, pp. 144–157. Springer, Berlin, Heidelberg. https://doi.org/10.1007/3-540-58450-1_40
2. Courcelle, B., Engelfriet, J.: Graph Structure and Monadic Second-Order Logic - A Language-Theoretic Approach, Encyclopedia of mathematics and Its Applications, vol. 138. Cambridge University Press (2012)
3. Doczkal, C.: A variant of Wagner's theorem based on combinatorial hypermaps. In: ITP. LIPIcs, vol. 193, pp. 17:1–17:17. Dagstuhl (2021). https://doi.org/10.4230/LIPIcs.ITP.2021.17
4. Doczkal, C., Combette, G., Pous, D.: A formal proof of the minor-exclusion property for treewidth-two graphs. In: Avigad, J., Mahboubi, A. (eds.) ITP 2018. LNCS, vol. 10895, pp. 178–195. Springer, Cham (2018). https://doi.org/10.1007/978-3-319-94821-8_11
5. Doczkal, C., Pous, D.: Treewidth-two graphs as a free algebra. In: MFCS. LIPIcs, vol. 117, pp. 60:1–60:15. Dagstuhl (2018). https://doi.org/10.4230/LIPIcs.MFCS.2018.60
6. Doczkal, C., Pous, D.: Completeness of an axiomatization of graph isomorphism via graph rewriting in Coq. In: CPP, pp. 325–33. ACM (2020). https://doi.org/10.1145/3372885.3373831
7. Doczkal, C., Pous, D.: Graph theory in Coq: minors, treewidth, and isomorphisms. J. Autom. Reason. **64**(5), 795–825 (2020). https://doi.org/10.1007/s10817-020-09543-2
8. Dufourd, JF., Bertot, Y.: Formal study of plane Delaunay triangulation. In: Kaufmann, M., Paulson, L.C. (eds.) ITP 2010. LNCS, vol. 6172, pp. 211–226. Springer, Heidelberg (2010). https://doi.org/10.1007/978-3-642-14052-5_16
9. Gonthier, G.: Formal proof—the four-color theorem. Notices Amer. Math. Soc. **55**(11), 1382–1393 (2008)
10. Llópez, E.C., Pous, D.: K4-free graphs as a free algebra. In: MFCS. LIPIcs, vol. 83, pp. 76:1–76:14. Dagstuhl (2017). https://doi.org/10.4230/LIPIcs.MFCS.2017.76
11. Lochbihler, A.: A mechanized proof of the max-flow min-cut theorem for countable networks. In: ITP. LIPIcs, vol. 193, pp. 25:1–25:18. Dagstuhl (2021). https://doi.org/10.4230/LIPIcs.ITP.2021.25
12. Nakamura, Y., Rudnicki, P.: Euler circuits and paths. Formaliz. Math. **6**(3), 417–425 (1997)
13. Nipkow, T., Bauer, G., Schultz, P.: Flyspeck I: tame graphs. In: Furbach, U., Shankar, N. (eds.) IJCAR 2006. LNCS, vol. 4130, pp. 21–35. Springer, Heidelberg. https://doi.org/10.1007/11814771_4
14. Noschinski, L.: A graph library for Isabelle. Math. Comput. Sci. **9**(1), 23–39 (2015). https://doi.org/10.1007/s11786-014-0183-z
15. Severín, D.E.: Formalization of the domination chain with weighted parameters (short paper). In: ITP. LIPIcs, vol. 141, pp. 36:1–36:7. Dagstuhl (2019). https://doi.org/10.4230/LIPIcs.ITP.2019.36
16. Singh, A.K., Natarajan, R.: A constructive formalization of the weak perfect graph theorem. In: CPP. ACM (2020). https://doi.org/10.1145/3372885.3373819

Contents

Tool Presentation

Theoretical Advances

Acyclic Contextual Hyperedge Replacement: Decidability of Acyclicity and Generative Power

Frank Drewes[1(✉)], Berthold Hoffmann[2(✉)], and Mark Minas[3(✉)]

[1] Umeå Universitet, Umeå, Sweden
drewes@cs.umu.se

[2] Universität Bremen, Bremen, Germany
hof@uni-bremen.de

[3] Universität der Bundeswehr München, Neubiberg, Germany
mark.minas@unibw.de

Abstract. Graph grammars based on contextual hyperedge replacement (CHR) extend the generative power of the well-known hyperedge replacement (HR) grammars to an extent that makes them useful for practical modeling. Recent work has shown that acyclicity is a key condition for parsing CHR grammars efficiently. In this paper we show that acyclicity of CHR grammars is decidable and that the generative power of acyclic CHR grammars lies strictly between that of HR grammars and unrestricted CHR grammars.

Keywords: Graph grammar · Hyperedge replacement · Contextual hyperedge replacement · Acyclicity · Decidability · Generative power

1 Introduction

Contextual hyperedge replacement (CHR, [3,4]) strengthens the generative power of hyperedge replacement (HR, [11]) significantly, e.g., to languages of unbounded treewidth. This is achieved by a moderate extension: productions may glue a graph not only to nodes attached to the nonterminal hyperedge being replaced, but also to nodes in the context. The applicability of such productions thus depends on the presence of context nodes created by other derivation steps.

In previous work, we have devised efficient parsing algorithms for subclasses of HR grammars, which rely on canonical orders for replacing nonterminals [5,6]. When these algorithms are extended to CHR, the canonical orders may be in conflict with dependencies arising from the creation and use of context nodes. Recently [9] we have shown that a CHR grammar Γ can be turned into an HR grammar generating graphs where the context nodes of Γ are "borrowed", i.e., generated like ordinary nodes. From these graphs, those generated by Γ can be obtained by "contraction", i.e., merging borrowed nodes with other nodes. This is correct provided that contractions cannot create cyclic chains of dependencies.

N. Behr and D. Strüber (Eds.): ICGT 2022, LNCS 13349, pp. 3–19, 2022.
https://doi.org/10.1007/978-3-031-09843-7_1

In this paper, we establish two important properties of acyclic CHR grammars that have been left open in [9]. (1) It is decidable whether a CHR grammar is acyclic or not (Sect. 3). (2) Acyclicity reduces the generative power of CHR grammars by limiting the possibility to exploit node dependencies between productions (Sect. 4).

We start by recapitulating CHR grammars, borrowing grammars and contractions as well as acyclicity taken from [9] before we present the results of this paper in Sect. 3 and Sect. 4. Finally, we conclude the paper by discussing related and future work in Sect. 5.

2 Contextual Hyperedge Replacement

We let $\mathbb{N}$ denote the set of non-negative integers, and $[n]$ the set $\{1, \ldots, n\}$ for all $n \in \mathbb{N}$. A^* denotes the set of all finite sequences over a set A; the empty sequence is denoted by ε, and the length of a sequence α by $|\alpha|$. For a function $f\colon A \to B$, its extension $f^*\colon A^* \to B^*$ to sequences is defined by $f^*(a_1 \cdots a_n) = f(a_1) \cdots f(a_n)$, for all $n \in \mathbb{N}$ and $a_1, \ldots, a_n \in A$. As usual, $\to^+$ and $\to^*$ denote the transitive and the transitive reflexive closure of a binary relation $\to$.

Graphs. Let $\Sigma = \dot{\Sigma} \uplus \bar{\Sigma}$ be an alphabet of *labels* for nodes and edges respectively, where edge labels come with a *rank function* $rank\colon \bar{\Sigma} \to \mathbb{N}$.

Then a (*hyper-*) *graph* over Σ is a tuple $G = (\dot{G}, \bar{G}, att_G, lab_G)$, where $\dot{G}$ and $\bar{G}$ are disjoint finite sets of *nodes* and (*hyper-*) *edges*, respectively, the function $att_G\colon \bar{G} \to \dot{G}^*$ attaches sequences of nodes to edges, and the *labeling function* $lab_G\colon \dot{G} \cup \bar{G} \to \Sigma$ maps $\dot{G}$ to $\dot{\Sigma}$ and $\bar{G}$ to $\bar{\Sigma}$ in such a way that $|att_G(e)| = rank(lab_G(e))$ for every edge $e \in \bar{G}$. We assume that the attachment sequences are free of repetitions. $\mathcal{G}_\Sigma$ denotes the class of graphs over Σ. An edge carrying a label $\sigma \in \bar{\Sigma}$ is called *σ-edge*. G° denotes the discrete subgraph of a graph G, which is obtained by removing all edges.

A graph $G \in \mathcal{G}_\Sigma$ is called a *σ-handle* (or just a *handle*) if G has a single σ-edge e with $\sigma \in \bar{\Sigma}$, and each node of G is attached to e. $\mathcal{H}_\Sigma$ shall denote the set of *handles* of Σ. If $rank(\sigma) = 0$, a σ-handle is unique (up to isomorphism); we denote such a handle by $\sigma^\bullet$.

$G - x$ shall denote the graph G without the edge $x \in \bar{G}$. A set of edges $E \subseteq \bar{G}$ *induces* the subgraph consisting of these edges and their attached nodes. Given graphs $G_1, G_2 \in \mathcal{G}_\Sigma$ with disjoint edge sets, a graph $G = G_1 \cup G_2$ is called the *union* of G_1 and G_2 if G_1 and G_2 are subgraphs of G, $\dot{G} = \dot{G}_1 \cup \dot{G}_2$, and $\bar{G} = \bar{G}_1 \cup \bar{G}_2$. Note that $G_1 \cup G_2$ exists only if common nodes are consistently labeled, i.e., $lab_{G_1}(v) = lab_{G_2}(v)$ for $v \in \dot{G}_1 \cap \dot{G}_2$.

For graphs G and H, a *morphism* $m\colon G \to H$ is a pair $m = (\dot{m}, \bar{m})$ of functions $\dot{m}\colon \dot{G} \to \dot{H}$ and $\bar{m}\colon \bar{G} \to \bar{H}$ that preserve attachments and labels, i.e., $att_H(\bar{m}(v)) = \dot{m}^*(att_G(v))$, $lab_H(\dot{m}(v)) = lab_G(v)$, and $lab_H(\bar{m}(e)) = lab_G(e)$ for all $v \in \dot{G}$ and $e \in \bar{G}$.

The morphism is *injective* or *surjective* if both $\dot{m}$ and $\bar{m}$ have this property, and a *subgraph inclusion* of G in H if $m(x) = x$ for every node or edge x in G; then we write $G \subseteq H$. If m is surjective and injective, we say that G and H are *isomorphic*, written as $G \cong H$.

Contextual Hyperedge Replacement (CHR). The set $\bar{\Sigma}$ of edge labels is assumed to contain a subset $\mathcal{N}$ of *nonterminal labels*; edges with labels in $\mathcal{N}$ are *nonterminal* while all others are *terminal*.

A *production* $p = (L, R)$ consists of graphs L and R over Σ such that (1) the *left-hand side* L contains exactly one edge x, which is a nonterminal, and (2) the *right-hand side* R is an arbitrary supergraph of $L - x$. Nodes in L that are not attached to x are the *context nodes* of L (and of p); p is called *context-free* if it has no context nodes, and *contextual* otherwise.

We use a special form of standard double-pushout graph transformation [10] for applying productions: Let p be a production as above, and consider some graph G. An injective morphism $m: L \to G$ is called a *matching* for p in G. If such a matching exists, we say that p is *applicable* to the nonterminal $m(x) \in \bar{G}$. The *replacement* of $m(x)$ by R (via m) is then given as the graph H obtained from the disjoint union of $G - m(x)$ and R by identifying every node $v \in \dot{L}$ with $m(v)$. We write this as $G \Rightarrow_{m,p} H$, but omit m if it is irrelevant, and write $G \Rightarrow_{\mathcal{P}} H$ if $G \Rightarrow_p H$ for some p taken from a set $\mathcal{P}$ of productions.

This leads to the notion of a CHR grammar [3,4].

Definition 1 (CHR grammar). A *contextual hyperedge replacement grammar* $\Gamma = \langle \Sigma, \mathcal{N}, \mathcal{P}, S \rangle$ (*CHR grammar*) consists of alphabets Σ and $\mathcal{N}$ as above, a finite set $\mathcal{P}$ of productions over Σ, and a *start symbol* $S \in \mathcal{N}$ such that $rank(S) = 0$. The *language* generated by Γ is given as $\mathcal{L}(\Gamma) = \{G \in \mathcal{G}_{\Sigma \setminus \mathcal{N}} \mid S^\bullet \Rightarrow_{\mathcal{P}}^* G\}$. Γ is a (context-free) *hyperedge replacement grammar* (*HR grammar* [11]) if all productions in $\mathcal{P}$ are context-free.

CHR grammars can generate languages that cannot be generated by HR grammars. In particular, this includes languages of unbounded treewidth, like the language $\mathcal{G}_\Sigma$ of all graphs and our running example introduced next.

Example 1 (CHR grammar for dags). Figure 1 shows our running example, and introduces our conventions for drawing graphs and productions. Nodes are circles, nonterminal edges are rectangular boxes containing the corresponding labels, and terminal edges are shapes like ▷. (In this example, all nodes are labeled with the "invisible" label ␣, i.e., they are effectively unlabeled.) Edges

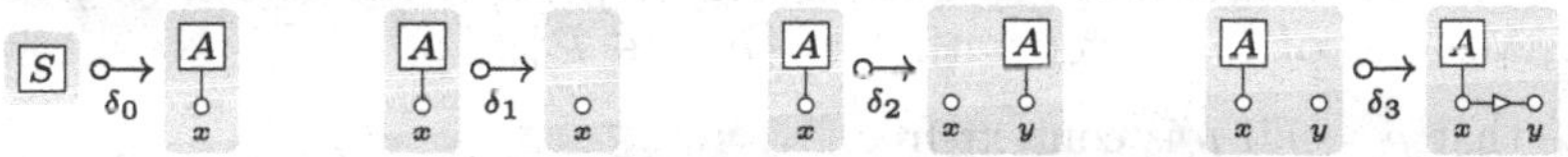

Fig. 1. Productions for generating dags (Example 1)

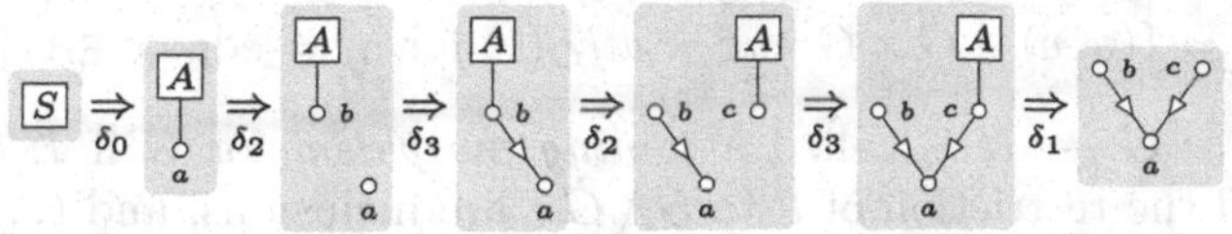

Fig. 2. A derivation with Δ

are connected to their attached nodes by lines ordered counter-clockwise around the edge, starting at noon. For productions (L, R), we draw L and R, and specify the inclusion of $\dot{L}$ in $\dot{R}$ by ascribing the same identifier to them, like x and y in our example.

Figure 1 defines the productions δ_0 to δ_3 of the CHR grammar Δ. S and A are nonterminal labels of rank 0 and 1, respectively, and $\triangleright$ is a binary terminal label. A derivation with this grammar is shown in Fig. 2.

It is easy to see that Δ derives only non-empty unlabeled acyclic graphs (*dags*, for short): In every derivation, the A-edge is attached to a node with indegree 0 so that no cycles may be introduced by production δ_3. Vice versa, every non-empty dag D can be generated with Δ: The nodes of D can be sorted topologically, e.g., as $v_1, \ldots, v_n$. Then every v_i can be generated with production δ_2, and its outgoing edges can be generated with production δ_3 since the targets of these edges must be nodes v_j with $j < i$. So $\mathcal{L}(\Delta)$ is indeed the set of all non-empty dags.

In previous work, we have devised efficient parsing algorithms for HR grammars [5,8]. These algorithms apply productions in canonical order (analogous to leftmost and rightmost derivations in string grammars). When extending these algorithms to CHR grammars, a production may only be applied when its context nodes have been created in previous steps. This may be in conflict with the canonical application orders. In [9], we have shown that there is a close relationship between a CHR grammar Γ and its so-called borrowing (HR) grammar $\hat{\Gamma}$: every graph $H \in \mathcal{L}(\Gamma)$ is a "contraction" of a graph $G \in \mathcal{L}(\hat{\Gamma})$. Moreover, the converse is also true as long as Γ is acyclic, a notion to be recalled later.

In the following, we assume that $\Sigma \setminus \mathcal{N}$ contains two auxiliary edge labels that are not used elsewhere in Γ: edges carrying the unary label $\circledcirc$ will mark borrowed nodes, and binary edges labeled $\neq$ will connect borrowed nodes with other nodes in the same right-hand side, to signify that they must not be identified with each other by contraction later on.

Definition 2 (Borrowing grammar). Let $\Gamma = \langle \Sigma, \mathcal{N}, \mathcal{P}, S \rangle$ be a CHR grammar. For $p = (L, R) \in \mathcal{P}$, its *borrowing production* $\hat{p} = (\hat{L}, \hat{R})$ is obtained by (1) removing every context node from $\hat{L}$ and (2) constructing $\hat{R}$ from R as follows: for every context node v of p, attach a new $\circledcirc$-edge to v, and add $\neq$-edges from v to every other node with the label $lab_L(v)$. The *borrowing grammar* $\hat{\Gamma} = \langle \Sigma, \mathcal{N}, \hat{\mathcal{P}}, S \rangle$ of Γ is given with $\hat{\mathcal{P}} = \{\hat{p} \mid p \in \mathcal{P}\}$.

Note that $\hat{p} = p$ if p is context-free.

Definition 3 (Contraction). For a graph G let

$$\dot{G}_{\circledcirc} = \{v \in \dot{G} \mid v = att_G(e) \text{ for a } \circledcirc\text{-edge } e \in \bar{G}\} \text{ and}$$
$$\neq_G = \{(u, v) \in \dot{G} \times \dot{G} \mid uv = att_G(e) \text{ for a } \neq\text{-edge } e \in \bar{G}\}.$$

A morphism $\mu \colon G \to H$ is called a *joining morphism* for G if $\dot{H} = \dot{G} \setminus \dot{G}_{\circledcirc}$, $\bar{H} = \bar{G}$, $\bar{\mu}$ and the restriction of $\dot{\mu}$ to $\dot{G} \setminus \dot{G}_{\circledcirc}$ are inclusions, and $(v, \dot{\mu}(v)) \notin \neq_G$ for every $v \in \dot{G}_{\circledcirc}$. The graph *core*$(H)$ obtained from H by removing all edges with labels $\circledcirc$ and $\neq$ is called the *μ-contraction* of G or just *a contraction* of G.

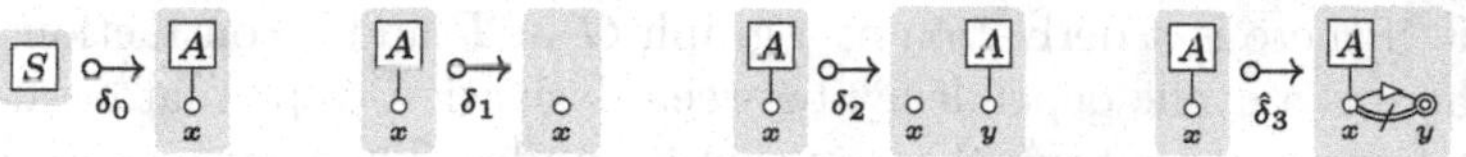

Fig. 3. Borrowing productions for generating dags

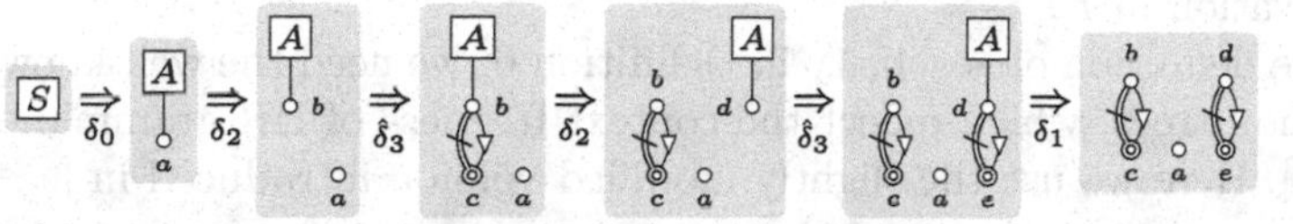

Fig. 4. A derivation with $\hat{\Delta}$

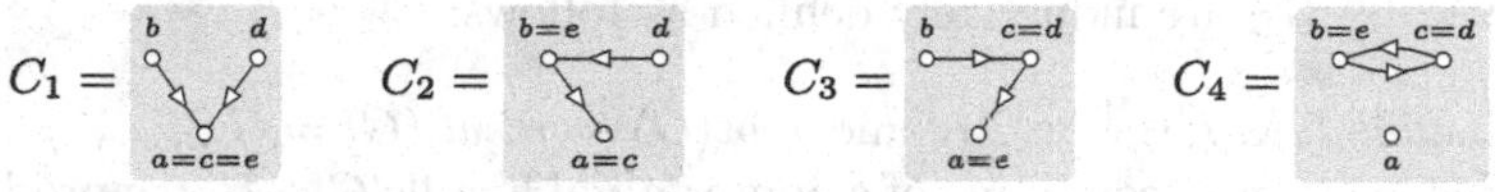

Fig. 5. The four contractions of the graph derived in Fig. 4

Example 2 (Borrowing grammar). Figure 3 shows the borrowing productions for the productions of the CHR grammar of dags in Fig. 1, where the contextual production δ_3 is replaced by the borrowing production $\hat{\delta}_3$.

A derivation with the borrowing grammar $\hat{\Delta}$ is shown in Fig. 4; the resulting terminal graph can be contracted in four possible ways, to the graphs C_1 to C_4 shown in Fig. 5. The contraction C_4 yields a cyclic graph; it is the only one of these four that cannot be generated with the productions of the CHR grammar Δ in Example 1.

Definition 4 (Borrowing version of a derivation). Let $\Gamma = \langle \Sigma, \mathcal{N}, \mathcal{P}, S \rangle$ be a CHR grammar and $\hat{\Gamma}$ its borrowing grammar. A derivation

$$S^\bullet \Rightarrow^{\hat{m}_1}_{\hat{p}_1} H_1 \Rightarrow^{\hat{m}_2}_{\hat{p}_2} H_2 \Rightarrow^{\hat{m}_3}_{\hat{p}_3} \cdots \Rightarrow^{\hat{m}_n}_{\hat{p}_n} H_n$$

in $\hat{\Gamma}$ is a *borrowing version* of a derivation

$$S^\bullet \Rightarrow^{m_1}_{p_1} G_1 \Rightarrow^{m_2}_{p_2} G_2 \Rightarrow^{m_3}_{p_3} \cdots \Rightarrow^{m_n}_{p_n} G_n$$

in Γ if the following hold, for $i = 1, 2, \ldots, n$ and $p_i = (L, R)$:

1. $\hat{p}_i$ is the borrowing production of p_i,
2. if $\bar{L} = \{e\}$ then $\hat{m}_i(e) = m_i(e)$, and
3. for every $x \in \bar{R} \cup (\dot{R} \setminus \dot{L})$, the images of x in G_i and H_i are the same.

By a straightforward induction, it follows that every derivation in Γ has a borrowing version in $\hat{\Gamma}$, and G_i is the μ_i-contraction of H_i for $i \in [n]$, where the joining morphism μ_i is uniquely determined by $\bar{\mu}_i(e) = e$ for all $e \in \bar{H}_i$.

Theorem 1 will show that the converse is also true, i.e., that every contraction of a graph in $\mathcal{L}(\hat{\Gamma})$ can also be derived in Γ, provided that Γ is acyclic. Informally,

Γ is cyclic if there is a derivation of a graph G in $\hat{\Gamma}$ and a contraction H of G so that there is a cyclic dependency between derivation steps that create nodes and derivation steps that use them as context nodes. These cyclic dependencies then result in derivations of graphs in $\hat{\Gamma}$ having a contraction that cannot be derived in Γ because there is no reordering of the derivation steps that yields a valid derivation in Γ.

For the definition of acyclicity in Definition 6, we need the well-known notion of derivation trees, which reflect the context-freeness of HR grammars [2, Definition 3.3]. Here we use the slightly modified version introduced in [9].

Definition 5 (Derivation tree). Let $\Gamma = \langle \Sigma, \mathcal{N}, \mathcal{P}, S \rangle$ be a HR grammar. The set $\mathbb{T}_\Gamma$ of *derivation trees* over Γ and the mappings $root\colon \mathbb{T}_\Gamma \to \mathcal{H}_\Sigma$ as well as $result\colon \mathbb{T}_\Gamma \to \mathcal{G}_\Sigma$ are inductively defined as follows:

- Each handle $G \in \mathcal{H}_\Sigma$ is in $\mathbb{T}_\Gamma$, and $root(G) = result(G) = G$.
- A triple $t = \langle G, p, c \rangle$ consisting of a nonterminal handle $G \in \mathcal{H}_\mathcal{N}$, a production $p \in \mathcal{P}$, and a sequence $c = t_1 t_2 \cdots t_n \in \mathbb{T}_\Gamma^*$ is in $\mathbb{T}_\Gamma$ if the union graphs $G' = G^\circ \cup \bigcup_{i=1}^n root(t_i)$ and $G'' = G^\circ \cup \bigcup_{i=1}^n result(t_i)$ exist, $G \Rightarrow_p G'$, and $nodes(result(t_i)) \cap nodes(result(t_j)) = nodes(root(t_i)) \cap nodes(root(t_j))$ for all distinct $i, j \in [n]$, where $nodes(H)$ denotes the node set of a graph H. Furthermore, we let $root(t) = G$ and $result(t) = G''$.

We assume the ordering of the subtrees in $c = t_1 t_2 \cdots t_n$ within a derivation tree $t = \langle G, p, c \rangle$ to be chosen arbitrarily, but kept fixed.

Let t, t' be any derivation trees. We call t a parent tree of t', written $t \succ t'$, if $t = \langle G, p, t_1 t_2 \cdots t_n \rangle$ and $t' = t_i$ for some i, and we call t' a subtree of t if $t' = t$ or $t = \langle G, p, t_1 t_2 \cdots t_n \rangle$ and t' is a subtree of t_i for some i. A derivation tree t *introduces* a node u (at its root) if $t = \langle G, p, t_1 t_2 \cdots t_n \rangle$ and $u \in nodes(root(t_i)) \setminus \dot{G}$ for some i. The set of all these nodes is denoted by $intro(t)$.

The following theorem is equivalent to Theorem 3.4 in [2]:

Lemma 1 (See [9, Theorem 1]). *Let $\Gamma = \langle \Sigma, \mathcal{N}, \mathcal{P}, S \rangle$ be a HR grammar, $H \in \mathcal{H}_\Sigma$ a handle and $G \in \mathcal{G}_\Sigma$ a graph. There is a derivation tree $t \in \mathbb{T}_\Gamma$ with $root(t) = H$ and $result(t) = G$ iff $H \Rightarrow_\mathcal{P}^* G$.*

Note that derivation trees are defined only for HR grammars. In the contextual case, any properly labeled node can be used as a context node as long as it has been created earlier in a derivation. This fact produces dependencies between derivation steps which do not exist in HR derivations.

In order to describe these additional dependencies, let us define the relation $\sqsubset_\mu$ on subtrees of a derivation tree $t \in \mathbb{T}_{\hat{\Gamma}}$ described by a joining morphism μ for $result(t)$. For any two subtrees t', t'' of t, we let $t' \sqsubset_\mu t''$ iff there is a node $u \in intro(t'')$ so that $\dot{\mu}(u) \neq u$ and $\dot{\mu}(u) \in intro(t')$.

Informally, $t' \sqsubset_\mu t''$ means that t' describes a derivation step (the topmost one that transforms the root handle of t'), which creates a node used as a contextual node in the corresponding topmost contextual derivation step described by t''. This restricts the set of all borrowing versions of derivations characterized by t:

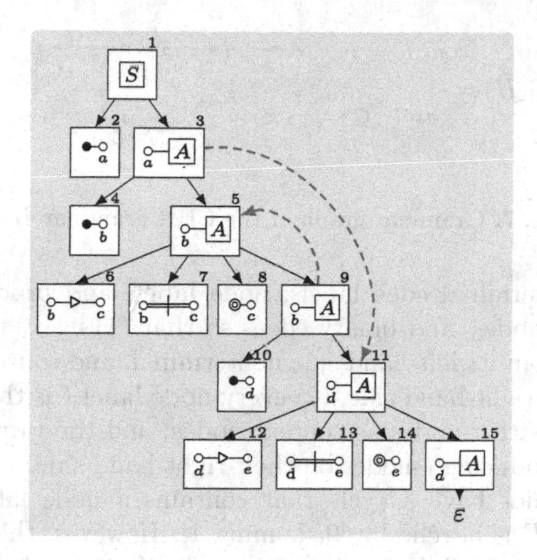

Fig. 6. Derivation tree of the derivation in Fig. 4

the derivation step described by t'' must occur after the one described by t'. However, the order of derivation steps must obey the parent tree relation $\succ$ as well. This motivates the following:

Definition 6 (Acyclic CHR grammar). A CHR grammar Γ is *acyclic* if $(\succ \cup \sqsubset_\mu)^+$ is irreflexive for all derivation trees $t \in \mathbb{T}_{\hat{\Gamma}}$ over $\hat{\Gamma}$ and all joining morphisms μ for $result(t)$. Otherwise, Γ is *cyclic*.

Example 3 (Derivation tree for dags). The derivation tree of the borrowing derivation in Fig. 4 is shown in Fig. 6 in black. Edges between derivation tree nodes represent the parent tree relation $\succ$. The thick dashed arrows drawn in red represent the relation $\sqsubset_\mu$ for the joining morphism μ defined by $\mu(e) = b$ and $\mu(c) = d$, yielding contraction C_4 in Fig. 5. This implies $t_3 \sqsubset_\mu t_{11}$ and $t_9 \sqsubset_\mu t_5$, respectively, when derivation subtrees are referred to by the numbers at their root nodes. Relations $\succ$ and $\sqsubset_\mu$ thus introduce a cycle affecting t_5 and t_9.

Theorem 1 (See [9, Theorem 2]). *Let Γ be a CHR grammar and $\hat{\Gamma}$ its borrowing grammar. For every graph $H \in \mathcal{L}(\Gamma)$, there is a graph $G \in \mathcal{L}(\hat{\Gamma})$ so that H is a contraction of G. Moreover, every contraction of a graph in $\mathcal{L}(\hat{\Gamma})$ is in $\mathcal{L}(\Gamma)$ if Γ is acyclic.*

3 Acyclicity of CHR Grammars is Decidable

Definition 6 does not provide effective means to check whether a CHR grammar is acyclic or not. In [9], we have devised a decidable sufficient criterion for acyclicity based on the so-called *grammar graph* $\mathsf{GG}(\Gamma)$ of a CHR grammar Γ (see [9, Definition 11]).

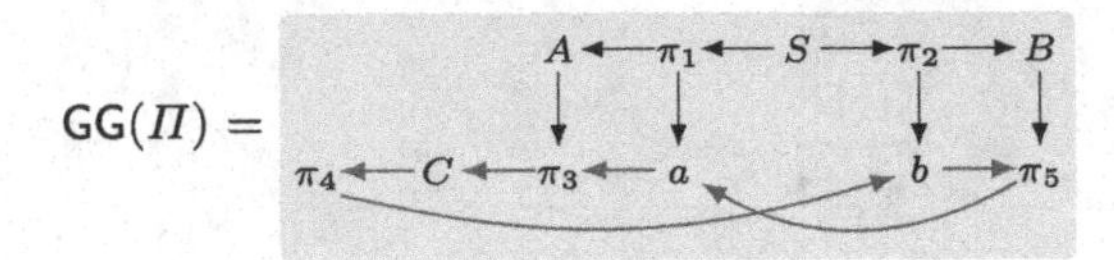

Fig. 7. Grammar graph of the CHR grammar Π

$\mathsf{GG}(\Gamma)$ has nonterminal edge labels, node labels and productions (or production names) as nodes, and binary edges so that (1) every production is the target of an edge from its left-hand side nonterminal, and source of edges to the nonterminals on its right-hand side, (2) every node label ℓ is the source of edges to all productions with ℓ-nodes as context nodes, and the target of edges from all productions introducing ℓ-nodes on their right-hand side.

If $\mathsf{GG}(\Gamma)$ does not have a cycle that contains a node label as node, one can conclude that Γ is acyclic by [9, Lemma 1]. However, this criterion is not necessary, i.e., one cannot be sure that Γ is cyclic if $GG(\Gamma)$ has such a cycle, as the following (pathological) example shows.

Example 4 (Acyclic CHR grammar with cyclic grammar graph). Let Π be the CHR grammar with the following productions over nullary nonterminal labels S (the start symbol), A, B, C, and node labels a, b:

$$\boxed{S} \underset{\pi_1}{\circ\!\!\to} \boxed{A}\,ⓐ \qquad \boxed{S} \underset{\pi_2}{\circ\!\!\to} \boxed{B}\,ⓑ \qquad \boxed{A}\,ⓐ \underset{\pi_3}{\circ\!\!\to} \boxed{C}\,ⓐ \qquad \boxed{C} \underset{\pi_4}{\circ\!\!\to} ⓑ \qquad \boxed{B}\,ⓑ \underset{\pi_5}{\circ\!\!\to} ⓐ\,ⓑ$$

Figure 7 shows the grammar graph $\mathsf{GG}(\Pi)$ that has clearly a cycle containing node labels a and b as nodes. Even so, Π is acyclic as one can see as follows: The borrowing grammar $\hat{\Pi}$ has only the following two derivations, each of them with one possible contraction:

$$\boxed{S} \underset{\pi_1}{\Rightarrow} \boxed{A}\,ⓐ \underset{\hat{\pi}_3}{\Rightarrow} \boxed{C}\,ⓐ\,◎ \underset{\pi_4}{\Rightarrow} ⓑ\,ⓐ\,◎ \to ⓑ\,ⓐ \qquad \boxed{S} \underset{\pi_2}{\Rightarrow} \boxed{B}\,ⓑ \underset{\hat{\pi}_5}{\Rightarrow} ⓐ\,ⓑ\,◎ \to ⓐ\,ⓑ$$

Both of them do not borrow any node before it has been created, i.e., $(\succ \cup \sqsubset_\mu)^+$ is irreflexive. Therefore, Π is acyclic.

We will now provide a decidable criterion for acyclicity of CHR grammars that is both sufficient *and* necessary, by turning a CHR grammar into a HR grammar whose language contains cyclic graphs iff the CHR grammar is cyclic. We start by motivating the construction of the HR grammar.

The dependency grammar Γ^{D} of a CHR grammar Γ is a HR grammar that has as its language graphs that contain derivation trees of the borrowing grammar $\hat{\Gamma}$. To be more precise, if there is a derivation of a graph H in $\hat{\Gamma}$, Γ^{D} derives a graph D that has a subgraph D' whose nodes correspond to the nonterminal derivation tree nodes of H, and its edges from parent to child nodes represent the parent tree relation $\succ$ on derivation trees. D contains additional nodes that represent nodes of H created by derivation steps, and additional edges. Whenever there is a possibility for a joining morphism μ to merge a node n with a borrowed node n' in H, D will contain a path between those nodes t and t' in D' that represent the creation of n and n', respectively. This path represents

$t \sqsubset_\mu t'$, and thus D contains a cycle iff $(\succ \cup \sqsubset_\mu)^+$ is reflexive, characterizing the cyclicity of Γ. This makes acyclicity of Γ decidable because it is decidable whether the language of an HR grammar contains cyclic graphs.

Let us first introduce some auxiliary concepts before we formally define dependency grammars: a production $p = (L, R)$ *borrows* a node label $\ell \in \dot{\Sigma}$ if L has a context node labeled ℓ; p *creates* ℓ if it creates a node labeled ℓ, i.e., $\ell = lab(v)$ for some $v \in \dot{R} \setminus \dot{L}$. We denote the sets of node labels borrowed and created by p by $B(p)$ and $C(p)$, respectively.

Definition 7 (Dependency grammar). Let $\Gamma = \langle \Sigma, \mathcal{N}, \mathcal{P}, S \rangle$ be a CHR grammar, where $\dot{\Sigma} = \{a_1, \ldots, a_m\}$. The *dependency grammar* of Γ is the HR grammar $\Gamma^{\mathsf{D}} = \langle \Sigma^{\mathsf{D}}, \mathcal{N}^{\mathsf{D}}, \mathcal{P}^{\mathsf{D}}, S \rangle$ consisting of the following components:

$$\mathcal{N}^{\mathsf{D}} = \{A^{\mathsf{D}} \mid A \in \mathcal{N}\} \cup \{S\} \qquad \bar{\Sigma}^{\mathsf{D}} = \mathcal{N}^{\mathsf{D}} \cup \{\mathtt{a}, \mathtt{b}, \mathtt{c}, \mathtt{d}, \mathtt{e}, \mathtt{f}, \mathtt{t}\}$$

$$\dot{\Sigma}^{\mathsf{D}} = \mathcal{N} \cup \{\uparrow a_1, \ldots, \uparrow a_m, \downarrow a_1, \ldots, \downarrow a_m\} \quad \mathcal{P}^{\mathsf{D}} = \{(S^\bullet, \mathsf{dep}(S))\} \cup \{p^{\mathsf{D}} \mid p \in \mathcal{P}\}$$

where $rank(A^{\mathsf{D}}) = 2m + 1$ for all $A \in \mathcal{N}$, S is nullary, and all other edge labels are binary.

To define the productions, let us denote by $\mathsf{dep}(A)$, for $A \in \mathcal{N}$, the A^{D}-handle consisting of an A^{D}-edge e which is attached to nodes $n_0, \ldots, n_{2m}$ labeled by $A, \uparrow a_1, \ldots, \uparrow a_m, \downarrow a_1, \ldots, \downarrow a_m$, respectively. For each of the labels $\ell \in \{A, \uparrow a_1, \ldots, \uparrow a_m, \downarrow a_1, \ldots, \downarrow a_m\}$, the unique node attached to e which is labeled ℓ will be denoted by $e.\ell$. We call $e.A$ the *main node* of the handle and the nodes $e.\uparrow a_i$ and $e.\downarrow a_i$ its *satellites*. (We also consider $e.A$ to be its own satellite.)

Now, for $p = (L, R) \in \mathcal{P}$ the *dependency production* $p^{\mathsf{D}} = (L^{\mathsf{D}}, R^{\mathsf{D}})$ is defined as follows: Suppose that the nonterminal edge of L has the label $A_0 \in \mathcal{N}$ and the nonterminal edges of R, ordered arbitrarily, have the labels $A_1, \ldots, A_k \in \mathcal{N}$. Then $L^{\mathsf{D}} = \mathsf{dep}(A_0)$ and R^{D} is the disjoint union of all handles $\mathsf{dep}(A_i)$ for $i \in [k]$, the set $\dot{L}^{\mathsf{D}}$ of all left-hand side nodes, and additional binary edges as specified below, where we denote the nonterminal edge in L^{D} by e_0 and that of $\mathsf{dep}(A_i)$ in R^{D} by e_i (for $i \in [k]$). Denoting a binary x-edge from $e_i.\ell$ to $e_j.\ell'$ by $e_i.\ell \rightarrow_x e_j.\ell'$, the terminal edges in R are:

$$e_0.A_0 \rightarrow_{\mathtt{t}} e_i.A_i \qquad \text{for all } i \in [k] \tag{1}$$

$$e_0.A_0 \rightarrow_{\mathtt{a}} e_0.\uparrow a \qquad \text{for all } a \in C(p) \tag{2}$$

$$e_i.\uparrow a \rightarrow_{\mathtt{b}} e_0.\uparrow a \qquad \text{for all } i \in [k] \tag{3}$$

$$e_i.\uparrow a \rightarrow_{\mathtt{c}} e_j.\downarrow a \qquad \text{for all } i, j \in [k]\ ,\ i \neq j \tag{4}$$

$$e_0.\downarrow a \rightarrow_{\mathtt{d}} e_i.\downarrow a \qquad \text{for all } i \in [k] \tag{5}$$

$$e_0.\downarrow a \rightarrow_{\mathtt{e}} e_0.A_0 \qquad \text{for all } a \in B(p) \tag{6}$$

$$e_i.\uparrow a \rightarrow_{\mathtt{f}} e_0.A_0 \qquad \text{for all } i \in [k] \text{ and } a \in B(p) \tag{7}$$

Note that each node in the right-hand side of a production is attached to one and only one nonterminal edge e, i.e., its denotation as $e.\ell$ is unique. Therefore, the derivation of a graph $D \in \mathcal{L}(\Gamma^{\mathsf{D}})$ induces a unique partition of $\dot{D}$ into subsets, each consisting of a main node and its satellites that have at some point during the derivation been attached to the same nonterminal.

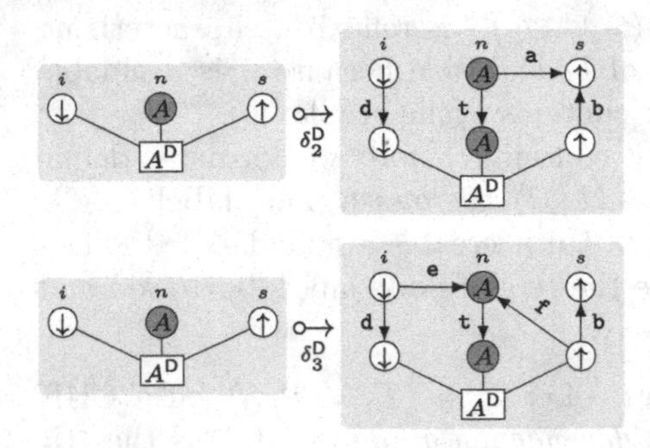

Fig. 8. Dependency productions for productions δ_2 and δ_3 of grammar Δ

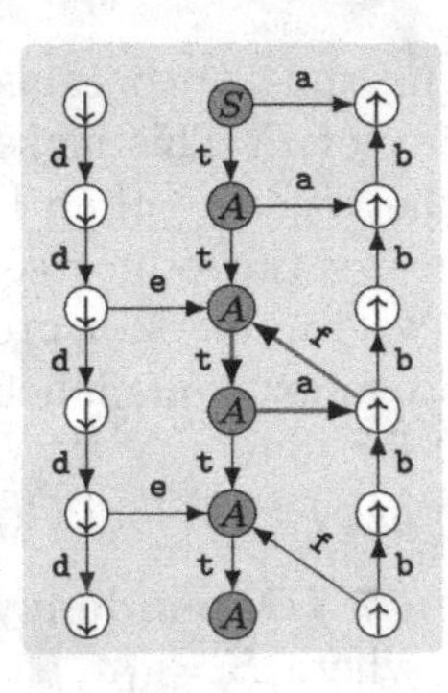

Fig. 9. Dependency graph of the derivation in Fig. 4

Derivations in borrowing grammars and dependency grammars are closely related: Consider a CHR grammar Γ, its borrowing grammar $\hat{\Gamma}$, and its dependency grammar Γ^{D}. The first step of a derivation in Γ^{D} yields $\mathsf{dep}(S)$. For $p \in \mathcal{P}$, the nonterminals in p^{D} (both in the left- and right-hand side) correspond bijectively to those in p, where each label A in p has been replaced by A^{D} in p^{D}. Thus, every derivation tree $t \in \mathbb{T}_{\hat{\Gamma}}$ corresponds to a unique derivation tree $t^{\mathsf{D}} \in \mathbb{T}_{\Gamma^{\mathsf{D}}}$, and vice versa. We call the graph $\mathit{result}(t^{\mathsf{D}})$ the *dependency graph* of t. By the above discussion, each graph derived by Γ^{D} is the dependency graph of some derivation tree $t \in \mathbb{T}_{\hat{\Gamma}}$.

Now, given such a dependency graph $D = \mathit{result}(t^{\mathsf{D}})$, consider the subgraph D' of D induced by its t-edges (defined by (1) in Definition 7). By (1), the nodes of D' are the main nodes of D, i.e., those carrying a label $A \in \mathcal{N}$. As explained above, in the derivation of D each such node was – together with its satellites – attached to a unique nonterminal edge, and this edge carried the corresponding label A^{D}. This means that D' is a tree which is isomorphic to the derivation tree t of which D is the dependency graph, provided that we disregard the leaves of t. The isomorphism relates each node n of D' to a unique subtree $T(n) = \langle H, p, c\rangle$ of t, and the node label of n coincides with the nonterminal label of H.

Example 5. We consider CHR gammar Δ for dags again (see Example 1). Production δ_2 does not borrow any node, but creates the "invisible" node label $\sqcup$, i.e., $C(\delta_2) = \{\sqcup\}$ and $B(\delta_2) = \emptyset$. Production δ_3 does not create any node, but borrows the node label $\sqcup$, i.e., $C(\delta_3) = \emptyset$ and $B(\delta_3) = \{\sqcup\}$. Figure 8 shows the corresponding dependency productions where $\uparrow$ and $\downarrow$ denote $\uparrow_\sqcup$ and $\downarrow_\sqcup$, respectively. The main nodes are drawn with a blueish background.

Figure 9 shows a dependency graph D created by Δ^{D}. It corresponds to the derivation shown in Fig. 4. The main nodes of D are drawn with a blueish background again; they are also the nodes of tree D' induced by the t-edges.

In the following, consider an arbitrary CHR grammar Γ, its dependency grammar Γ^{D}, and any dependency graph $D \in \mathcal{L}(\Gamma^{\mathsf{D}})$.

Lemma 2. *Every cycle in D contains an edge $\to_{\mathtt{e}}$ or $\to_{\mathtt{f}}$.*

Proof. We first show that every cycle in D contains a node labeled A for some $A \in \mathcal{N}$. If all the nodes in a cycle carried labels of the form $\uparrow a$ or $\downarrow a$, then the cycle can only contain edges labeled b, c, or d. But the definition of these edges in (3)–(5) prohibits such a cycle. Since the t-edges in D form a tree, the incoming edge of some A-node ($A \in \mathcal{N}$) in the cycle is not a t-edge, and must thus be labeled e or f. □

In the following, we write paths as an alternating sequence of nodes and edges like $n_0 \to_{a_1} n_1 \to_{a_2} \cdots \to_{a_l} n_l$. We may omit nodes between consecutive edges, and use the shorthand notation $n \to_u^* n'$ and $n \to_u^+ n'$ if there is a path from node n to n' consisting exclusively of $\to_u$-edges. $n \to_u^* n'$ also permits the empty path, i.e., $n = n'$, but $n \to_u^+ n'$ requires a path with at least one edge. We define further relations $\rightarrowtail$ and $\rightsquigarrow$ on nodes of D: $n \rightarrowtail n'$ iff there is a path $n \to_{\mathtt{t}}^+ \to_{\mathtt{a}} \to_{\mathtt{b}}^* \to_{\mathtt{f}} n'$, and $n \rightsquigarrow n'$ iff there is a path $n \to_{\mathtt{t}}^* \to_{\mathtt{a}} \to_{\mathtt{b}}^* \to_{\mathtt{c}} \to_{\mathtt{d}}^* \to_{\mathtt{e}} n'$.

Lemma 3. *D is cyclic iff $n \rightarrowtail n$ or $n \rightsquigarrow^k n$ for some $n \in \dot{D}$ and $k > 1$.*

Proof. D is obviously cyclic if it contains such a node n.

If D is cyclic, let D' be the tree induced by the t-edges of D. We distinguish between two cases: First assume that D contains an f-edge, say $n' \to_{\mathtt{f}} n$. Node n' must be labeled $\uparrow a$ for some $a \in \dot{\Sigma}$ and, by (2) and (3), there must be nodes n'', n''' such that $n'' \to_{\mathtt{a}} n''' \to_{\mathtt{b}}^* n' \to_{\mathtt{f}} n$. Note that n and n'' carry nonterminal labels in $\mathcal{N}$ and are thus main nodes of D. Now let m' and m''' be the main nodes of satellites n' and n''', respectively. Then n, m', n'', m''' are nodes of D'. By (2), (3), and (7), $n'' = m'''$, m''' is a (not necessarily proper) descendant of m', and m' is a child of n, i.e., n'' is a proper descendant of n, and therefore $n \to_{\mathtt{t}}^+ n''$. This shows that $n \rightarrowtail n$.

Now consider the case where D does not contain an edge $\to_{\mathtt{f}}$. We select any cycle of D, which must contain a node n with an incoming edge $\to_{\mathtt{e}}$ by Lemma 2. Removing all occurrences of $\to_{\mathtt{e}}$ in the cycle decomposes it into k paths. These paths have the form $\to_{\mathtt{t}}^* \to_{\mathtt{a}} \to_{\mathtt{b}}^* \to_{\mathtt{c}} \to_{\mathtt{d}}^*$, i.e., we have $n \rightsquigarrow^k n$ for some $k \geq 1$. But then we must have $k > 1$, which can be seen as follows. Consider a path $n_1 \to_{\mathtt{t}}^* n_2 \to_{\mathtt{a}} n_3 \to_{\mathtt{b}}^* n_4 \to_{\mathtt{c}} n_5 \to_{\mathtt{d}}^* n_6 \to_{\mathtt{e}} n_7$, and let m_i be the main node of satellite n_i, for $i \in [7]$. Then each m_i is a node of D'. In D', the path from m_1 to m_2 descends down the tree (by (1)), then stays at $m_2 = m_3$ (by (2)), and ascends to an ancestor m_4 of m_3 (by (3)). Now, by (4), m_5 is a proper sibling of m_4, and $m_6 = m_7$ is a descendant of m_5 (by (5) and (6)). Since D' is a tree, this implies that $n_1 \neq n_7$. Therefore, we cannot have $n \rightsquigarrow n$ for any node n. □

Lemma 4. *Let Γ be a CHR grammar and Γ^{D} its dependency grammar. Γ is cyclic iff the language of Γ^{D} contains a cyclic graph.*

Proof. Let $\Gamma = \langle \Sigma, \mathcal{N}, \mathcal{P}, S \rangle$ be a CHR grammar, $\hat{\Gamma}$ its borrowing grammar, and Γ^{D} its dependency grammar.

Let us first assume that the language of Γ^{D} contains a cyclic graph D, which is the dependency graph of a derivation tree $t \in \mathbb{T}_{\hat{\Gamma}}$ over $\hat{\Gamma}$. Mirroring Lemma 3, we distinguish two cases:

Case 1: D contains a node n such that $n \rightsquigarrow^k n$ for some $k > 1$, i.e., there are nodes $n_0, n_1, \ldots, n_k \in \dot{D}$, $n'_i, n''_i, n'''_i \in \dot{D}$, subtrees $\langle H_i, p_i, c_i \rangle$ and $\langle H'_i, p'_i, c'_i \rangle$ of t, and node labels $a_i, \ldots, a_k \in \dot{\Sigma}$ such that $n_0 = n = n_k$ and $lab_D(n''_i) = \uparrow a_i$, $lab_D(n'''_i) = \downarrow a_i$, $T(n_i) = \langle H_i, p_i, c_i \rangle$, $T(n'_i) = \langle H'_i, p'_i, c'_i \rangle$, and $n_{i-1} \rightarrow^*_{\mathtt{t}} n'_i \rightarrow_{\mathtt{a}} n''_i \rightarrow^*_{\mathtt{b}} \rightarrow_{\mathtt{c}} \rightarrow^*_{\mathtt{d}} n'''_i \rightarrow_{\mathtt{e}} n_i$ for $i \in [k]$.

By the condition in (6), $n'''_i \rightarrow_{\mathtt{e}} n_i$ implies that $a_i \in B(p_i)$, i.e., p_i has a context node labeled a_i. By the condition in (2), $n'_i \rightarrow_{\mathtt{a}} n''_i$ further implies that $a_i \in C(p'_i)$, i.e., p'_i creates a node labeled a_i. Therefore, by merging the corresponding nodes, there is a joining morphism μ_i such that $T(n'_i) \sqsubset_{\mu_i} T(n_i)$. Note that $n'_i \neq n_i$ follows from the fact that the path from n'_i to n_i contains an edge $\rightarrow_{\mathtt{c}}$, which implies that there is no $\neq$-edge between those a_i-nodes, and thus that these nodes are not prevented from becoming merged. Finally, $n_{i-1} \rightarrow^*_{\mathtt{t}} n'_i$ implies that $T(n'_i)$ is a subtree of $T(n_{i-1})$ in t, and therefore $T(n_{i-1}) \succ^* T(n'_i) \sqsubset_{\mu_i} T(n_i)$ for $i \in [k]$. It should be clear that there is a joining morphism that can act as μ_i for $i \in [k]$, and therefore, $T(n)(\succ \cup \sqsubset_\mu)^+ T(n)$, i.e., Γ is cyclic by Definition 6.

Case 2: D contains a node n such that $n \rightarrowtail n$. i.e., there is a path $n \rightarrow^+_{\mathtt{t}} \rightarrow_{\mathtt{a}} \rightarrow^*_{\mathtt{b}} \rightarrow_{\mathtt{f}} n$ in D. By arguments entirely analogous to Case 1, one can then conclude that there is a joining morphism μ such that $T(n)(\succ \cup \sqsubset_\mu)^+ T(n)$, i.e., Γ is cyclic.

Assume now that Γ is cyclic. By Definition 6, there is a derivation tree $t \in \mathbb{T}_{\hat{\Gamma}}$ and a joining morphism μ for $result(t)$ such that $(\succ \cup \sqsubset_\mu)^+$ is reflexive. Because $\succ^+$ is irreflexive, there must be subtrees $t_0, \ldots, t_k$ and $t'_1, \ldots, t'_k$ of t for some $k \geq 1$ such that $t_0 = t_k$ and $t_{i-1} \succ^* t'_i \sqsubset_\mu t_i$ for $i \in [k]$. Let us choose such subtrees $t_0, \ldots, t_k$ and $t'_1, \ldots, t'_k$ of t, with the additional condition that k is minimal (with respect to the given t), and let D be the dependency graph of t.

If $k = 1$, we have $t_0 \succ^* t'_1 \sqsubset_\mu t_0$, i.e., t'_1 is a (proper) subtree of t_0, and the topmost derivation step of t_0 uses a node with label $a \in \dot{\Sigma}$ as a context node that is created by the topmost derivation step of t'_1, indicated by μ merging the corresponding nodes. Let n and n' be the nodes of D such that $T(n) = t_0$ and $T(n') = t'_1$. Clearly, D contains the cycle $n \rightarrow^+_{\mathtt{t}} n' \rightarrow_{\mathtt{a}} n'' \rightarrow^*_{\mathtt{b}} n''' \rightarrow_{\mathtt{f}} n$ for some nodes n'', n''' with $lab_D(n'') = lab_D(n''') = \uparrow a$.

If $k > 1$, we can conclude that t'_i is not a subtree of t_i for any $i \in [k]$. Otherwise, we would have $t_i \succ^* t'_i \sqsubset_\mu t_i$, contradicting the selection of $k > 1$ being minimal for t.

Let $n_i \in \dot{D}$ such that $T(n_i) = t_i$ for $i = 0, \ldots, k$ and $n'_i \in \dot{D}$ such that $T(n'_i) = t'_i$ for $i \in [k]$ and assume that t_i uses a node with label $a_i \in \dot{\Sigma}$ as context node that is created by the topmost derivation step of t'_i for $i \in [k]$, indicated by μ merging the corresponding nodes. D thus contains paths $n_{i-1} \rightarrow^*_{\mathtt{t}} n'_i \rightarrow_{\mathtt{a}} n''_i \rightarrow^*_{\mathtt{b}} \rightarrow_{\mathtt{c}} \rightarrow^*_{\mathtt{d}} n'''_i \rightarrow_{\mathtt{e}} n_i$ where $n''_i, n'''_i \in \dot{D}$ such that $lab_D(n''_i) = \uparrow a_i$ and $lab_D(n'''_i) = \downarrow a_i$ for $i \in [k]$. These paths define a cycle in D because $n_0 = n_k$. □

Example 6. The dependency graph shown in Fig. 9 in Example 5 has a cycle drawn in red indicating that the CHR grammar Δ for dags is indeed cyclic.

Let us reconsider the "pathological" CHR grammar Π introduced in Example 4. The criterion devised in [9] does not help to decide whether Π is acyclic

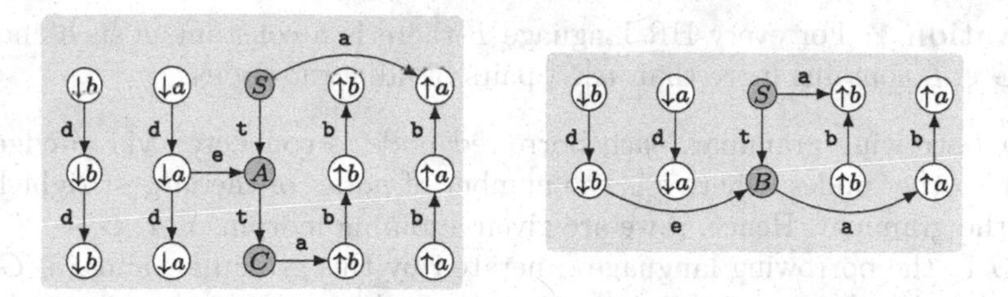

Fig. 10. Dependency graphs for the grammar in Example 4

or cyclic, but Lemma 4 does: Fig. 10 shows the only two dependency graphs of $\mathcal{L}(\Pi^{\mathsf{D}})$. Π is indeed acyclic because they do not contain any cycle.

Theorem 2. *Acyclicity of CHR grammars is decidable.*

Proof. Following Definition 7, the dependency grammar Γ^{D} of a CHR grammar Γ can effectively be constructed. Moreover, it is decidable whether the language of an HR grammar such as Γ^{D} contains cyclic graphs [12]. Therefore, acyclicity of CHR grammars is decidable by Lemma 4. □

The construction of the dependency grammar Γ^{D} of a CHR grammar Γ according to rules (1)–(6) can be made in polynomial time. However, a straightforward cyclicity check would inspect all combinations of the finite set of dependency productions $\mathcal{P}^{\mathsf{D}}$, taking exponential time. We are not aware of any more efficient cyclicity checks for HR grammars. However, this problem closely corresponds to the circularity problem of attribute grammars. Attribute grammars are a well-known formalism for adding semantic information to context-free string grammars, and a proper definition of their semantics requires acyclic dependencies between attributes [14]. Knuth's (corrected) algorithm for checking circularity of an attribute grammar has an exponential worst-case running time [15]; Jazayeri et al. [13] have further proved that any deterministic algorithm solving this problem requires exponential running time, so that an efficient algorithm for checking the cyclicity of CHR grammars is unlikely to exist.

4 Acyclicity Restricts Generative Power

In this section, we answer the second question left open in [9], namely whether the class of graph languages generated by acyclic CHR grammars is a proper subset of the class of all CHR languages. It was conjectured in [9] that the set of all dags (see Example 1) could not be generated by an acyclic CHR grammar, a proof of which would thus answer the question positively. We now show that this is indeed the case.

As usual, we call two nodes v, v' of a graph G adjacent if both v and v' occur in $att_G(e)$, for some edge $e \in \bar{G}$. It is well known that HR languages are graph languages of bounded treewidth. Since the number of (unordered) pairs of adjacent nodes in a graph G of bounded treewidth is linearly bounded by the number $|\dot{G}|$ of its nodes, the following observation is immediate.

Observation 1. For every HR language L there is a constant w such that no graph $G \in L$ contains more than $w|\dot{G}|$ pairs of adjacent nodes.

In a borrowing grammar, each borrowed node is connected via $\neq$-edges to at most s other nodes, where s is the number of nodes of the largest right-hand side of the grammar. Hence, if we are given a joining morphism $\mu\colon G \to H$ for a graph G in the borrowing language generated by this grammar, and $z \in \dot{G} \setminus \dot{H}$ is a borrowed node, then for all but at most s nodes $v \in \dot{H}$, we can instead map z to v to obtain another valid joining morphism.

To express this formally, let us first note that a joining morphism μ is uniquely determined by $\dot{\mu}$. More precisely, given a graph G, consider a function $f\colon \dot{G} \to \dot{G} \setminus \dot{G}_{\circledcirc}$ such that, for all $u, v \in \dot{G}$,

(J1) if $u \notin \dot{G}_{\circledcirc}$ then $f(u) = u$, and
(J2) $f(u) = f(v)$ only if u and v are not connected by a $\neq$-edge.

Then there is a unique joining morphism $\mu\colon G \to H$ with $\dot{\mu} = f$. In particular, H is uniquely determined by f. This also implies that, given a joining morphism μ and a borrowed node $z \in \dot{G}_{\circledcirc}$, we can modify μ so that it, instead of joining z with $\dot{\mu}(z)$, joins it with any other node $v \in \dot{G} \setminus \dot{G}_{\circledcirc}$, provided that condition (J2) is fulfilled. In the following, the resulting joining morphism is denoted by $\mu_{z\mapsto v}$, i.e., $\mu_{z\mapsto v}$ is the unique joining morphism such that, for all $u \in \dot{G}$,

$$\dot{\mu}_{z\mapsto v}(u) = \begin{cases} v & \text{if } u = z \\ \dot{\mu}(u) & \text{otherwise.} \end{cases}$$

Using this notation, the observations above can be stated formally as follows.

Observation 2. For every borrowing language L, there is a constant s such that the following holds. Let $G \in L$, and let $\mu\colon G \to H$ be a joining morphism for G. Then, for every node $z \in \dot{G}_{\circledcirc}$, there are at least $|\dot{H}| - s$ nodes $v \in \dot{H}$ such that $\dot{\mu}_{z\mapsto v}$ determines a valid joining morphism for G.

We can now show that the language $\mathcal{L}(\Delta)$ of Example 1 is a CHR language beyond the generative capacity of acyclic CHR grammars.

Lemma 5. *The language of all (unlabeled) dags can be generated by a CHR grammar but not by an acyclic CHR grammar.*

Proof. Example 1 shows that the language $\mathcal{D}$ of all directed acyclic graphs is indeed a CHR language. It remains to show that $\mathcal{D}$ cannot be generated by an acyclic CHR grammar. To prove this by contradiction, assume that there is an acyclic CHR grammar Γ auch that $L(\Gamma) = \mathcal{D}$. We can assume that $\hat{\Gamma}$ does not generate borrowed nodes that are only incident with $\circledcirc$- and $\neq$-edges, because it is well known that $\hat{\Gamma}$ can otherwise be modified to remove such nodes from its language, and by the definition of contraction, such nodes do not affect the result of a contraction.

For $m, n \in \mathbb{N}$, define H_{mn} to be the unlabeled graph with $\dot{H}_{mn} = \{u_1, \ldots, u_m\} \cup \{v_1, \ldots, v_n\} \cup \{u'_1, \ldots, u'_m\}$ with $u_m = v_1$ and $v_n = u'_1$, such that there are edges from u_i to u_{i+1} and from u'_i to u'_{i+1} for all $i \in [m-1]$, as well as from v_i to v_j for $1 \le i < j \le n$. Thus, H_{mn} consists of a complete graph on n nodes between two chains of m nodes from above and below (if edges point downwards). In the following, we consider pairs of graphs $G_{mn} \in L(\hat{\Gamma})$ and $H_{mn} \in L(\Gamma)$ such that G_{mn} contracts to H_{mn} via a joining morphism μ.

We first fix m in such a way that it is larger than the constant s of Observation 2 applied to $\hat{\Gamma}$. Then G_{mn} can have at most $2(m-1)+2(n-1)$ nodes u such that $\dot{\mu}(u) \in \{u_1, \dots, u_m, u'_1, \dots, u'_m\}$ (since each of these nodes has at least one incident edge while H_{mn} contains only $2(m-1)+2(n-1)$ edges incident with them in total). Since we have fixed m, this number is linear in n. However, the total number of edges of H_{mn} – and thus that of G_{mn} – grows quadratically in n, which by Observation 1 means that we can choose n sufficiently large to make sure that G_{mn} contains at least one node $z \in \dot{G}_{mn} \setminus \dot{H}_{mn}$ such that $\dot{\mu}(z) = v_i$ for some $i \in \{2, \dots, n-1\}$.

Now, consider such a node z. Since $m > s$, where s is the constant of Observation 2, by that same observation there is at least one $j \in [m]$ such that $\mu_{z \mapsto u_j}$ is also a valid joining morphism. The same holds for $u'_1, \dots, u'_m$ instead of $u_1, \dots, u_m$.

Let e be an edge incident with z that is neither a ◎- nor a $\neq$-edge. If z is the target of e, then its source is a node u such that $\dot{\mu}(u) = v_p$ for some $p < i$. Consider an appropriate $j \in [m]$ such that $\mu_{z \mapsto u_j}$ is a joining morphism (which, by the previous paragraph, exists). Thus, instead of being joined with v_i by μ, we join z with u_j by $\mu_{z \mapsto u_j}$. Since this leaves the edges originating from $u_j, \dots, u_m$ unaffected, the path from u_j to $u_m = v_1$ still exists in the contraction with respect to $\mu_{z \mapsto u_j}$, and so does the edge from v_1 to v_p. However, e now leads from v_p to u_j, which results in the cycle $u_j, \dots, u_m, v_p, u_j$.

The case where z is the source of e is symmetric, using $\mu_{z \mapsto u'_j}$ instead of $\mu_{z \mapsto u_j}$. Thus, both cases lead us to the conclusion that $L(\Gamma)$ contains a graph that has a cycle, contradicting the initial assumption that $L(\Gamma) = \mathcal{D}$. □

From Lemma 5, Example 1 and the fact (known from [9]) that acyclic CHR grammars can generate various graph languages that are not HR languages, we get the second main result of this paper as an immediate consequence.

Theorem 3. *The generative power of acyclic CHR grammars lies strictly between the generative powers of HR and CHR grammars.*

5 Conclusions

In this paper, we have established two main results: (1) acyclicity of CHR grammars is decidable and (2) the generative power of acyclic CHR grammars lies strictly between that of HR and CHR grammars. Since acyclicity is one condition for efficient parsing with the predictive top-down and predictive shift-reduce algorithms of [7,9], this is important for the practical use of CHR grammars.

Since this paper is on a very specific topic, related work is rare. We are only aware of Berglund's pumping lemma for CHR grammars [1], which shows their close relation to context-free hyperedge replacement. (It seems that this pumping lemma cannot be used to prove Lemma 5.)

In future work, we plan to compensate for the restricted generative power of acyclic CHR grammars by conditional contractions, which may require or forbid the existence of certain paths. Then, e.g., the language of all dags can be generated by a conditional acyclic CHR grammar that forbids that there is a path to a borrowed node from its contracted node. The specification of such paths could be based on the "navigational logic" proposed by Orejas et al. [16].

References

1. Berglund, M.: Analyzing and pumping hyperedge replacement formalisms in a common framework. In: Echahed, R., Plump, D. (eds.) Pre-Proceedings Tenth International Workshop on Graph Computation Models, GCM@STAF 2019, Eindhoven, The Netherlands, 17 July 2019, pp. 17–32 (2019)
2. Drewes, F., Habel, A., Kreowski, H.J.: Hyperedge replacement graph grammars. In: Rozenberg, G. (ed.) Handbook of Graph Grammars and Computing by Graph Transformation. Vol. I: Foundations, chap. 2, pp. 95–162. World Scientific, Singapore (1997)
3. Drewes, F., Hoffmann, B.: Contextual hyperedge replacement. Acta Informatica **52**(6), 497–524 (2015). https://doi.org/10.1007/s00236-015-0223-4
4. Drewes, F., Hoffmann, B., Minas, M.: Contextual hyperedge replacement. In: Schürr, A., Varró, D., Varró, G. (eds.) AGTIVE 2011. LNCS, vol. 7233, pp. 182–197. Springer, Heidelberg (2012). https://doi.org/10.1007/978-3-642-34176-2_16
5. Drewes, F., Hoffmann, B., Minas, M.: Predictive top-down parsing for hyperedge replacement grammars. In: Parisi-Presicce, F., Westfechtel, B. (eds.) ICGT 2015. LNCS, vol. 9151, pp. 19–34. Springer, Cham (2015). https://doi.org/10.1007/978-3-319-21145-9_2
6. Drewes, F., Hoffmann, B., Minas, M.: Predictive shift-reduce parsing for hyperedge replacement grammars. In: de Lara, J., Plump, D. (eds.) ICGT 2017. LNCS, vol. 10373, pp. 106–122. Springer, Cham (2017). https://doi.org/10.1007/978-3-319-61470-0_7
7. Drewes, F., Hoffmann, B., Minas, M.: Extending predictive shift-reduce parsing to contextual hyperedge replacement grammars. In: Guerra, E., Orejas, F. (eds.) ICGT 2019. LNCS, vol. 11629, pp. 55–72. Springer, Cham (2019). https://doi.org/10.1007/978-3-030-23611-3_4
8. Drewes, F., Hoffmann, B., Minas, M.: Formalization and correctness of predictive shift-reduce parsers for graph grammars based on hyperedge replacement. J. Log. Algebr. Methods Program. (JLAMP) **104**, 303–341 (2019). https://doi.org/10.1016/j.jlamp.2018.12.006
9. Drewes, F., Hoffmann, B., Minas, M.: Rule-based top-down parsing for acyclic contextual hyperedge replacement grammars. In: Gadducci, F., Kehrer, T. (eds.) ICGT 2021. LNCS, vol. 12741, pp. 164–184. Springer, Cham (2021). https://doi.org/10.1007/978-3-030-78946-6_9
10. Ehrig, H., Ehrig, K., Prange, U., Taentzer, G.: Fundamentals of Algebraic Graph Transformation. EATCS Monographs, Springer, Heidelberg (2006). https://doi.org/10.1007/3-540-31188-2
11. Habel, A.: Hyperedge Replacement: Grammars and Languages. LNCS, vol. 643. Springer, Heidelberg (1992). https://doi.org/10.1007/BFb0013875

12. Habel, A., Kreowski, H., Vogler, W.: Metatheorems for decision problems on hyperedge replacement graph languages. Acta Informatica **26**(7), 657–677 (1989). https://doi.org/10.1007/BF00288976
13. Jazayeri, M., Ogden, W.F., Rounds, W.C.: The intrinsically exponential complexity of the circularity problem for attribute grammars. Commun. ACM **18**(12), 697–706 (1975). https://doi.org/10.1145/361227.361231
14. Knuth, D.E.: Semantics of context-free languages. Math. Sys. Theory **2**(2), 127–145 (1968). https://doi.org/10.1007/BF01692511. Correction: [15]
15. Knuth, D.E.: Semantics of context-free languages: correction. Math. Sys. Theory **5**(2), 95–96 (1971). https://doi.org/10.1007/BF01702865
16. Navarro, M., Orejas, F., Pino, E., Lambers, L.: A navigational logic for reasoning about graph properties. J. Log. Algebraic Methods Program. **118**, 100616 (2021). https://doi.org/10.1016/j.jlamp.2020.100616

Graph Rewriting Components

Reiko Heckel[1], Andrea Corradini[2(✉)], and Fabio Gadducci[2]

[1] University of Leicester, Leicester, UK
rh122@le.ac.uk
[2] University of Pisa, Pisa, Italy
{andrea.corradini,fabio.gadducci}@unipi.it

Abstract. We introduce a component model for graph rewriting that allows to model a system as a network of components with interfaces representing shared views of internal states and transformations. Their composition assembles a global view whose behaviour is equivalent to the synchronised distributed execution of local components in the network. Formally, components are arrows in a category with interfaces as objects that, with suitable component connectors, forms a Frobenius algebra. This allows the use of string diagrams to model the architecture of basic components and connectors, such that their assembly is freely generated by the algebraic structure. The compositionality of the proposed model is reflected by Structural Operational Semantic rules.

Keywords: Graph transformation · Software components · String diagrams

1 Introduction

Software development relies on encapsulation, modularity, and reuse to manage complexity. At the level of software architecture, these principles are supported by components that provide the basic blocks from which larger systems are built. While languages, technologies, and architectural styles change over time and differ between domains, the main feature that separates components from lower-level (e.g., object-oriented) concepts is the use of interfaces describing not only the services provided by components but also their requirements towards their runtime context. This enables reuse of components across contexts that satisfy the stated requirements.

With the confluence of concepts from semantic web, graph databases, and model-based engineering, *knowledge graphs* [15] are emerging as key technology in enterprise and e-commerce applications, medical data management, cognitive digital twins, and social networks [16] to support data integration, sharing and mapping, graph-based analytics and machine learning [17]. In current applications, knowledge graphs lack the basic modularity, encapsulation and flexibility of deployment offered by most component models. But global centralised data

Research partly supported by MIUR PRIN project 2017FTXR7S “IT-MaTTerS”.

N. Behr and D. Strüber (Eds.): ICGT 2022, LNCS 13349, pp. 20–37, 2022.
https://doi.org/10.1007/978-3-031-09843-7_2

models make applications hard to evolve and maintain, hinder reuse, distributed development, analysis, and verification [8]. We need a discipline of *graph-based software engineering* using dedicated abstractions and language constructs to develop modular graph-based applications.

This paper addresses the theoretical foundations of components in *graph-based applications*, where graphs are central runtime artefacts to be shared, queried, mapped and synchronised, updated, transformed and analysed. Such operations can commonly be described by graph rewriting. The novel challenge for components of graph rewriting-based applications is that, while traditionally the internal state is fully encapsulated, graph data must be shared between organisations along with rights to query, change or analyse graphs locally and coordinate changes globally. Access to and operations on graphs should be offered as services ensuring data integrity. While maintaining local ownership, a virtual global graph should emerge as the central artefact for data integration and analytics [19].

We propose the architectural abstraction of *Graph Rewriting Components (GReCos)* as building blocks for graph-based systems, encapsulating graphs and their operations and offering these to other components and applications. This is realised by defining GReCos as graph transformation systems with interfaces for composition with other systems.

Formally, GReCos are cospans of morphisms between graph transformation systems with state, called *runtime systems*, where the central system represents the partially hidden implementation, the left interface describes the types, rules and graph provided and the right interface those required by the component. In particular, we are interested in *strict* components, where the interface graphs are projections of the internal state graph.

Morphisms between runtime systems that are strict in that sense reflect transformations, so interfaces provide a partial view of the behaviour of the implementation. We can compose components via pushouts of such morphisms, and if the given components are strict and satisfy a compatibility condition ensuring that their composition is strict, too, the resulting component represents a global view of the synchronised execution of its constituents. Vice versa, the global behaviour can be decomposed into matching local behaviours, allowing us to move freely between the two levels. This supports the need for a virtual global graph that can be used centrally without giving up localised representation.

To support flexible connections between components we establish the category of graph rewriting components as a symmetric monoidal category, specifically a Frobenius algebra [3], and use the associated syntax of string diagrams to represent the interconnection of components and interfaces. This view of the software architecture is analogous to component diagrams in UML. Given realisations of the basic components in terms of GReCos, architecture-level string diagrams are mapped freely to (basic and composite) GReCos, compiling the system from its architecture description and its basic components.

The approach thus represents a convergence of distributed graph transformation [18], service-oriented and modular graph transformation [7], and string diagrams [3]. We prove compositionality results relating local and composite behaviours. In particular, the behaviour of (disjoint) parallel compositions and (interface-based) functional compositions of components can be fully inferred from the behaviours of their constituents and basic architectural connectors. This supports the reuse of components in different contexts, guaranteeing that behavioural equivalence of components is maintained by composition.

2 Example

We model a simple architecture to motivate, illustrate and evaluate our concepts and results. The model consists of three components: a Client C, a Service S and a Database D. The component diagram below gives a high-level view of the architecture. The components are connected via three interfaces. The Service Interface SI describes the operations provided by S and used by C. Conversely, the Client Interface CI is implemented by C and used by S. The idea is that C sends a requests through SI to be executed by S which, in turn, replies via the callback interface CI. While executing the request, S calls on D to verify and update the data.

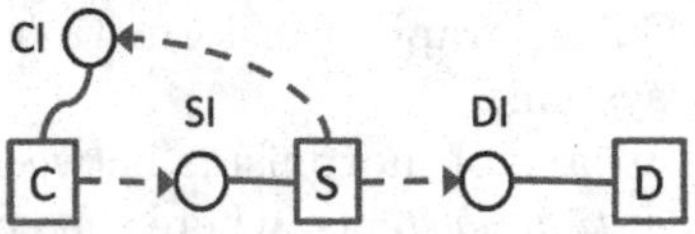

Components and interfaces are typed graph transformation systems with states related by morphisms. For our architecture. they are shown in Fig. 1. For each system we have the type graph in the left, followed by the rules, and the state graph made up of a single customer and its contract as a minimal test case.

The morphisms mapping type, state graphs and rules between interfaces and components are indicated by vertical arrows on the left. They describe how internal type and state graphs are partially visible through the interfaces. Rules in the interfaces are subrules of projections of the rules in components to the interface types. If the projection results in a rule without effect, this rule can be dropped, e.g. the process rule is in S but not SI, unless we want to use it to synchronise actions between components, e.g. between S and D via DI. Rules that are vertically aligned are related by morphisms. We use the integrated rule notation where left and right-hand sides are shown in the same graph, with deleted and created elements distinguished by colours blue and green and labelled $\{delete\}$ and $\{new\}$ respectively. In the bottom we show the global system view Sys obtained by composing components over their shared interfaces.

The model describes a claims process where C represents an insurance company's customer interface used to issue a request for payment. S is the service processing the request by checking the data D of the contract and, if successful, marking the customer as OK. Then a decision is made to either accept or reject the request, where acceptance requires a successful check and results in removing the link between customer and contract, indicating that after a payout the contract needs to be renewed. Either decision results in deleting the request's link to the customer to avoid making a decision repeatedly.

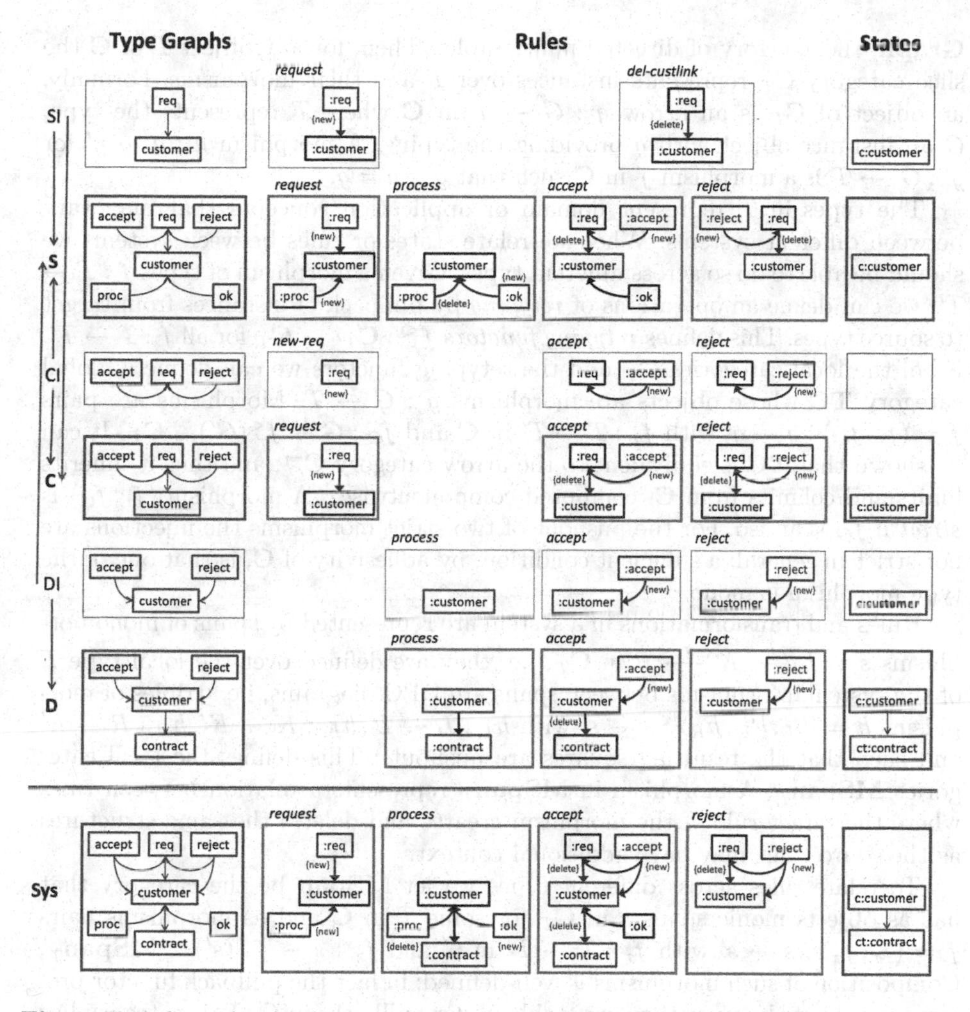

Fig. 1. Typed graph transformation systems with runtime states for components and interfaces.

Apart from the rules modelling operations that can be invoked through an interface, we distinguish change event rules, such as *new-req* and *del-custlink*, representing change events whose purpose is to notify a component that is sharing part of its state with another one that this other component has changed the shared state. This is conceptually different from an operation call, although it can be implemented as one, and is essential for keeping states synchronised between components.

3 Basic Notions

Assume an adhesive base category **C** with a strict initial object $\emptyset$ and arbitrary pushouts, where pushouts are stable under pullbacks; for example, let **C** be

Graph, the category of directed multigraphs. Then, for any object T of $\mathbf{C}$ the slice category $\mathbf{C}_T$ represents instances over T and their morphisms. Formally, an object of $\mathbf{C}_T$ is an arrow $g : G \to T$ in $\mathbf{C}$ where T represents the type, G an instance object with g providing the typing. A morphism $h : g \to g'$ for $g' : G' \to T$ is a morphism f in $\mathbf{C}$ such that $g' \circ h = g$.

The types in T represent domain or application concepts that may vary between different systems. When we relate states or rules between systems we should be able to do so across different types. Given a morphism of types $f : T \to T'$, we can define an operations of *retyping* by pullback of instances from target to source types. This defines *retyping functors* $f^< : \mathbf{C}_{T'} \to \mathbf{C}_T$ for all $f : T \to T'$. From the local categories $\mathbf{C}_T$ and the retyping functors we can define a global category $\mathbf{TC}$ whose objects are morphisms $g : G \to T$. Morphisms are pairs $f = \langle f_\tau, f_G \rangle : g \to g'$ with $f_\tau : T \to T'$ in $\mathbf{C}$ and $f_G : G \to f_\tau^<(G') \in \mathbf{C}_T$. It can be shown that $\mathbf{TC}$ is equivalent to the arrow category $\mathbf{C}^\to$, and thus it inherits limits and colimits from $\mathbf{C}$, computed componentwise.[1] A morphism $\langle f_\tau, f_G \rangle$ is *strict* if f_G is an iso. For the pushout of two strict morphisms the injections are not strict in general: a sufficient condition, by adhesivity of $\mathbf{C}$, is that one of the type morphism is mono.

Rules and transformations in a system are represented by spans of monomorphisms $s = L \xleftarrow{l} K \xrightarrow{r} R$ in $\mathbf{C}_T$, i.e. they are defined over the local type T of the system. Morphisms between spans are DPO diagrams, i.e., triples of morphisms $h = \langle h_L, h_K, h_R \rangle : s \to s'$ with $h_L : L \to L', h_K : K \to K', h_R : R \to R'$ and such that the resulting squares are pushouts. This defines the local categories $\mathbf{MSpan}_T$. A morphism in $\mathbf{MSpan}_T$ represents a relation between rules where the target rule of the morphism creates and deletes the same structures as the source, but may have additional context.

To relate rules across different types we let **MSpan** be the category that has as objects monic spans s in $\mathbf{C}_T$ for some T in $\mathbf{C}$ and as morphisms pairs $f = \langle f_\tau, f_\pi \rangle : s \to s'$ with $f_\tau : T \to T'$ in $\mathbf{C}$ and $f_\pi : s \to f_\tau^<(s') \in \mathbf{MSpan}_T$. Composition of such morphisms is well-defined: in fact the pullback functor preserves pushouts because they are stable under pullbacks in $\mathbf{C}$. This category has pullbacks and is finitely co-complete thanks to the properties of $\mathbf{C}$. In particular, the initial object is span $\langle \emptyset \leftarrow \emptyset \to \emptyset \rangle$, called the *empty rule*, typed over $\emptyset$.

Another interpretation of morphisms in $\mathbf{MSpan}_T$ is as DPO transformations, with the source representing the rule applied and the target the state transformation. Sometimes we want to relate such transformations, and for this purpose we introduce $\mathbf{DPO}_T$, the arrow category $\mathbf{MSpan}_T^\to$, which has local $\mathbf{MSpan}_T$ morphisms (i.e., DPO diagrams over T) $d : s_1 \to s_2$ as objects and pairs of such morphisms $\langle f_{top}, f_{bot} \rangle : d \to d'$ as arrows where $f_{top} : s_1 \to s_1'$ and $f_{bot} : s_2 \to s_2'$ such that the resulting square in $\mathbf{MSpan}_T$ commutes.

We relate DPO diagrams across different types in a global category **DPO** that has as objects DPO diagrams d in $\mathbf{C}_T$ for some T in $\mathbf{C}$ and as morphisms

[1] **TC** is obtained by applying the Grothendieck construction to the indexed category $\mathbf{C}^{op} \to \mathbf{Cat}$, mapping each object T to category $\mathbf{C}_T$ and each arrow to the corresponding retyping functor.

pairs $f = \langle f_\mathcal{T}, f_d \rangle : d \to d'$ with $f_\mathcal{T} : T \to T'$ in $\mathbf{C}$ and $f_d : d \to f_\mathcal{T}^{<}(d') \in \mathbf{DPO}_T$. That means, objects in **DPO** represent DPO transformations in different systems, and morphisms are mappings between them allowed to extend types and rules. Composition, limits and colimits are defined component-wise in **MSpan**.

Categories **TC**, **MSpan**, and **DPO** are equipped with a functor to $\mathbf{C}$ mapping objects and morphisms to their type objects and morphisms, respectively, which we denote by $\_\mathcal{T} : \mathbf{X} \to \mathbf{C}$ for $\mathbf{X} \in \{\mathbf{TC}, \mathbf{MSpan}, \mathbf{DPO}\}$.

4 Transformation and Runtime Systems

We want to use the empty rule to model steps at an interface due to unobservable steps in the body of a component, but also to model idle steps in the body itself. To this aim we introduce the rule name ϕ that maps to the empty rule. Apart from this feature, the following definition is standard.

Definition 1 (transformation systems). *A* transformation systems *is a triple* $R = \langle T, P, \pi \rangle$ *where*

- $T \in |\mathbf{C}|$ *is a type object;*
- P *is a set of rule names, including the special rule name* ϕ*;*
- $\pi : P \to |\mathbf{MSpan}_T|$ *assigns a monic span over* T *to each rule name, such that* $\pi(\phi) = \emptyset \leftarrow \emptyset \to \emptyset$.

Assuming a second system $R' = \langle T', P', \pi' \rangle$*, a* morphism of transformation systems *is a triple* $f = \langle f_\mathcal{T}, f_P, f_\pi \rangle : R \to R'$ *of*

- *a morphism of types* $f_\mathcal{T} : T \to T'$
- *a mapping from target to source rule names* $f_P : P' \to P$
- *a* P'*-indexed family of* **MSpan** *morphisms* $f_\pi(p') : \pi(f_P(p')) \to \pi'(p')$

such that $f_\mathcal{T} = (f_\pi(p'))_\mathcal{T}$ *for all* $p' \in P'$. *This defines the category* **Sys***.*

Morphisms are defined to reflect behaviour, as discussed later. Observe that each rule name $p' \in P'$ of the target system S' is mapped to a rule name $f_P(p') \in P$ of the source system S, and there is an **MSpan** morphism $f_\pi(p')$ from the latter rule to the first one. Spelling out the definiton of **MSpan** morphism, there is a DPO morphism from $\pi(f_P(p'))$ to the retyped rule $f_\mathcal{T}^{<}(\pi'(p'))$. In particular, this implies that if $f_P(p') = \phi$, then the retyped rule must be a span of isomorphisms, i.e. it has no effect when applied to any graph.

To model a system at runtime, we include its current state.

Definition 2 (runtime systems). *A* runtime system $S = \langle R, G \rangle$ *consists of a transformation system* $R = \langle T, P, \pi \rangle$ *and a state object* G *in* $\mathbf{C}_T$. *A morphism of runtime systems* $f = \langle f_R, f_G \rangle : S \to S'$ *with* $S' = \langle \langle T', P', \pi' \rangle, G' \rangle$ *is a morphism of transformation systems* $f_R = \langle f_\mathcal{T}, f_P, f_\pi \rangle$ *augmented by a* **TC** *morphism* $f_G = \langle f_\mathcal{T}, f'_G \rangle : G \to G'$. *Morphism* $f : S \to S'$ *is* strict *if so is* f_G*, i.e. if* $f'_G : G \to f_\mathcal{T}^{<}(G')$ *is an isomorphism. This defines the category* **RSys** *of runtime systems.*

Coming back to the example, in Fig. 1 we show seven runtime systems (three components, three interfaces and one global system), each with their type graph, rule names and associated rule spans (in integrated notation), and runtime state. The morphisms indicated in the left margin are all strict, representing inclusions of type graphs, state graphs, and sets of rule names, except for change event rules where *request* in S and C both map to *new-req* in CI, and *accept* and *reject* in S and C all map to *del-custlink* in SI. Implicitly, component rules that do not have a corresponding rule in an interface map to the empty rule ϕ, i.e., *process* in S maps to ϕ in both SI and CI and *request* in S maps to ϕ in DI. As observed above, this is allowed because after retyping these rules along the injections of type graphs, the resulting rules are spans of isomorphisms.

Given a transformation system R with type T, a transformation via p in R, denoted $G \overset{p,m}{\Longrightarrow}_R H$, is a DPO diagram seen as an $\mathbf{MSpan}_T$ morphism $t = \langle t_L, t_K, t_R \rangle : \pi(p) \to s$ that relates the rule span $\pi(p) = L \leftarrow K \to R$ and the bottom span $s = (G \leftarrow D \to H)$, with match $t_L = m$. We also write $p/t : G \Rightarrow_R H$ or just $\Rightarrow_R$ for the set of transformations.

A transformation sequence $s = G_0 \overset{p_1,m_1}{\Longrightarrow} \dots \overset{p_n,m_n}{\Longrightarrow} G_n$ in R is a sequence of transformations.[2] We write $\Rightarrow_R^*$ for the set of transformation sequences in R.

Transformations in R' are reflected by **Sys** morphisms. That means, if p'/t' is a transformation in R' then $f^<(p'/t') = f_P(p')/f_T^<(t') \circ f_\pi(p')$ is a transformation in R because $f_T^<$ preserves DPO diagrams and DPOs compose vertically (as **MSpan** morphisms). This yields a function $f^< : (\Rightarrow_{R'}) \to (\Rightarrow_R)$ extending to sequences as $f^< : (\Rightarrow_{R'})^* \to (\Rightarrow_R)^*$.

Transformation sequences in a runtime system $S = \langle R, G \rangle$ are sequences in R that start from state G. The projection of sequences against morphisms extends to runtime systems as $f^< : (\Rightarrow_{S'})^* \to (\Rightarrow_S)^*$, provided that $f : S \to S'$ is strict. Strict morphisms are preserved by transformations, that is, if $f = \langle f_R, f_G \rangle : \langle R, G \rangle \to \langle R', G' \rangle$ is strict, $p'/t' : G' \Rightarrow H'$ in R' and $f^<(p'/t') : G \Rightarrow H$ in R, then $\langle f_R, id_H \rangle : \langle R, H \rangle \to \langle R', H' \rangle$ is strict, as $H = f_T^<(H')$.

Sys is finitely co-complete because it has an initial object $R_\varnothing = \langle \emptyset, \{\phi\}, \pi \rangle$, where $\pi(\phi) = \emptyset \leftarrow \emptyset \to \emptyset$, and pushouts are built component-wise as pushouts on types, pullbacks on sets of rule names, and using amalgamation (pushouts in **MSpan**) on rule spans.

Definition 3 (pushouts of systems). *Given a span of transformation systems* $R_1 \xleftarrow{f_1} R_0 \xrightarrow{f_2} R_2$ *in* **Sys** *with* $R_i = \langle T_i, P_i, \pi_i \rangle$*, their pushout* $R_1 \xrightarrow{f_2^*} R \xleftarrow{f_1^*} R_2$ *with* $R = \langle T, P, \pi \rangle$*, is defined as follows.*

- $T_1 \xrightarrow{f_2{}^*_T} T \xleftarrow{f_1{}^*_T} T_2$ *is a pushout of* $T_1 \xleftarrow{f_{1T}} T_0 \xrightarrow{f_{2T}} T_2$ *in* $\mathbf{C}$
- $P_1 \xleftarrow{f_2{}^*_P} P \xrightarrow{f_1{}^*_P} P_2$ *is a pullback of* $P_1 \xrightarrow{f_{1P}} P_0 \xleftarrow{f_{2P}} P_2$ *in* $\mathbf{Set}^\bullet$
- *for* $p \in P$ *with* $f_2{}^*_P(p) = p_1$, $f_1{}^*_P(p) = p_2$*, and* $f_{1P}(p_1) = p_0 = f_{2P}(p_2)$*, let* $f_2{}^*_\pi(p), f_1{}^*_\pi(p)$ *and* $\pi(p)$ *be defined by the pushout* $\pi_1(p_1) \xrightarrow{f_2{}^*_\pi(p)} \pi(p) \xleftarrow{f_1{}^*_\pi(p)} \pi_2(p_2)$ *of* $\pi_1(p_1) \xleftarrow{f_{1\pi}(p_1)} \pi_0(p_0) \xrightarrow{f_{2\pi}(p_2)} \pi_2(p_2)$ *in* **MSpan**.

[2] We may drop the reference to the system if this is clear from context.

For a similar span in **RSys** *with local state graphs* $G_1 \xleftarrow{f_{1G}} G_0 \xrightarrow{f_{2G}} G_2$*, the pushout in* **Sys** *of the underlying transformation systems is lifted to* **RSys** *by the pushout* $G_1 \xrightarrow{f_{2G}^*} G \xleftarrow{f_{1G}^*} G_2$ *over the span of state graphs.*

Set$^\bullet$ is the category of *pointed sets*, i.e., sets with a distinguished element that is preserved by mappings. In our case these are the sets of rule names with the distinguished name ϕ bound to the empty rule. It is easy to see that pushouts injections thus constructed are indeed **Sys** or **RSys** morphisms because their components are pushouts in **C**, **Set**$^\bullet$ and **MSpan**. The universal property follows directly from the component-wise construction.

Applying this to our example in Fig. 1, the pushout of S and D via DI results in a union of their type and state graphs and an amalgamation of rules over shared interface rules in DI. This leads to a disjoint parallel composition of rules where this interface rule is empty.

A coproduct of two systems R_1 and R_2 is a pushout over the empty system $R_\varnothing$, which is initial in **Sys**. By contravariance of mapping types and rule names, this results in a coproduct of types and a product of rule names, such that each pair of rule names in the product is assigned a coproduct of the associated rules.

We can compose and decompose transformations over pushouts of systems if the morphisms relating them are strict.

Theorem 1 (compositionality of transformations). *Assume*

- *a pushout* $S_1 \xrightarrow{f_2^*} S \xleftarrow{f_1^*} S_2$ *of runtime systems* $S_1 \xleftarrow{f_1} S_0 \xrightarrow{f_2} S_2$*, where all morphisms are strict,* $S_i = \langle R_i, G_i \rangle$ *for* $i \in \{0,1,2\}$ *and* $S = \langle R, G \rangle$*;*
- *a triple of transformations* $p_i/t_i : G_i \Rightarrow H_i$ *in* R_i*, whose DPO diagrams* t_i *are related by* **DPO** *morphisms* $t_1 \xleftarrow{f_{1\pi}(p_1)} t_0 \xrightarrow{f_{2\pi}(p_2)} t_2$ *and where* $f_{1P}(p_1) = p_0 = f_{2P}(p_2)$*.*

Then, the transformations can be composed by a pushout in **DPO** *to yield a transformation* $p/t : G \Rightarrow H$ *in* S *with* $f_{1P}^*(p) = p_2$ *and* $f_{2P}^*(p) = p_1$*.*

Vice versa, a transformation $p/t : G \Rightarrow H$ *in* $\langle R, G \rangle$ *decomposes into a pushout of transformations over* $S_1 \xleftarrow{f_1} S_0 \xrightarrow{f_2} S_2$ *as* $p_1/t_1 = f_2^{*<}(p/t)$ *in* S_1*,* $p_2/t_2 = f_1^{*<}(p/t)$ *in* S_2 *and* $f_1^<(p_1/t_1) = p_0/t_0 = f_2^<(p_2/t_2)$ *in* S_0*.*

Proof (sketch). Both directions require that pushouts are stable under pullbacks, which is true in **C** by assumption.

Applying a sequence of *accept, process, accept* to the state graph in S, the result is a graph that looks like the right-hand side (preserved black and new green parts) of *accept*. In D the first step is an application of the empty rule ϕ, the second step has no effect on the graph but extends the match of *process* in S to check for a contract linked to the customer, and the third step deletes that link and adds the accept node and its edge, leaving a graph that looks like the right-hand side *accept* in D.

Since the state graph of S in Fig. 1 is a subgraph of that of D, when the pushout of runtime systems S and D via DI merges their state graphs, the resulting graph is isomorphic to that of D. The result above ensures that we can either transform this graph in the pushout system of S and D over DI, or do so in S and D with shared transformations in DI and then merge the resulting graphs, i.ė. synchronised local transformations exist if and only if there is a global transformation, and they have the same effect.

Theorem 1 ensures the compositionality of the operational semantics of components in Sect. 7, where transformations compose along composition of components and transformations in a composite component can be decomposed into synchronised transformations in its constituents.

5 Components

A component has runtime systems as *body*, *left* and *right interfaces*. Both interfaces are equipped with a morphism to the body. Formally, components are defined as abstract cospans in **RSys**. Components with matching left-right interfaces can be composed using pushouts in **RSys**. Note that, in our running example, the left and right interfaces are conceptually the *provided* and *required* interfaces of components. However, we stick to the typical left/right terminology of cospans, instead of using the provide/require terminology of software components, because in our operational semantic introduced later both interfaces behave identically, allowing to synchronize the transformations of the components they are connected to. The provided/required terminology suggests instead an invocation-based semantics, where a component can trigger through the required interface the execution of another component connected to the matching provided interface. This kind of semantics is topic of future work.

Components can also be composed in parallel using coproducts. The resulting structure is a symmetric monoidal category **Com** having the same objects of **RSys** and components as arrows (from the left to the right interface). This category is shown to be also a Frobenius algebra, implying that one can define arbitrary topologies of components.

Here we focus on the static interconnections of components, while in Sect. 7 we discuss their operational semantics based on transformations. We anticipate that the rich structure of the category of components cannot be fully exploited for the operational semantics, because only *strict* morphisms reflect transformations. We introduce *strict* components, where morphism to the body are strict, and discuss conditions ensuring that strictness is preserved by composition.

Cospans $c = (A \xrightarrow{a} C \xleftarrow{b} B)$ and $c' = (A \xrightarrow{a'} C' \xleftarrow{b'} B)$ are isomorphic if there is an isomorphism $i : C \to C'$ commuting the resulting triangles. We denote by $A[\xrightarrow{a} C \xleftarrow{b}]B$ the isomorphism class of c, called an *abstract cospan.*

Definition 4 (components). *A component is an abstract cospan* $c = \langle Li[\xrightarrow{li} Bd \xleftarrow{ri}]Ri\rangle$ *in* **RSys**. *Morphisms* $li : Li \to Bd$ *and* $ri : Ri \to Bd$ *map the left and right interfaces to the body. Component* c *is* strict *if both* li *and* ri *are strict.*

The category **Com** *of components has runtime systems as objects and components as arrows. The composition of components* $c_2 \circ c_1$, *for* $c_i = \langle Li_i [\xrightarrow{li_i} Bd_i \xleftarrow{ri_i}] Ri_i \rangle$ *and* $i = 1, 2$, *is defined if* $Li_2 = Ri_1$. *Then*

$$c_2 \circ c_1 = \langle Li_1 [\xrightarrow{li} Bd \xleftarrow{ri}] Ri_2 \rangle : Li_1 \to Ri_2$$

is the isomorphism class of the cospan obtained by a pushout $Bd_1 \xrightarrow{li_2^*} Bd \xleftarrow{ri_1^*} Bd_2$ *of* ri_1 *and* li_2 *with* $li = li_2^* \circ li_1$ *and* $ri = ri_1^* \circ ri_2$. *If* c_1 *and* c_2 *are both strict, then they are* compatible *if the pushout injections* li_2^* *and* ri_1^* *are strict. In this case also* $c_2 \circ c_1$ *is strict because strict morphisms compose. If* c_1 *and* c_2 *are strict and compatible we will denote their* strict composition *also by* $c_2 \circ_s c_1$.

For a runtime system S *in* $|\mathbf{RSys}|$ *its identity component is given by* $id_S = \langle S [\xrightarrow{id_S} S \xleftarrow{id_S}] S \rangle$, *and it is strict.*

Composition over shared interfaces allows to connect strict components by synchronising their transformations. In **Com** our example's components are represented as arrows $C : CI \to SI$, $S : SI \to CI + DI$ and $D : DI \to R_\varnothing$, their composition realised by the composition in **Com**, e.g., $S \circ C : CI \to CI + DI$ is the composition of C and D over SI. In order to link interface CI from the right of S to the left of C (as required for its use as a callback interface) we need the additional structure of parallel composition and component connectors.

Definition 5 (parallel composition in Com). *The* parallel composition $c_1 + c_2$ *of components* $c_i = \langle Li_i [\xrightarrow{li_i} Bd_i \xleftarrow{ri_i}] Ri_i \rangle$ *for* $i = 1, 2$ *is defined as the isomorphism class of the cospans obtained by a coproduct of the interface and body systems in* **RSys**

$$c_1 + c_2 = \langle Li_1 + Li_2 [\xrightarrow{li_1 + li_2} Bd_1 + Bd_2 \xleftarrow{ri_1 + ri_2}] Ri_1 + Ri_2 \rangle.$$

This defines a monoidal functor $+ : \mathbf{Com} \times \mathbf{Com} \to \mathbf{Com}$. *Furthermore, for each* S, S', *let* $\sigma_{S,S'} : \langle S + S' [\xrightarrow{[inr_S, inl_{S'}]} S' + S \xleftarrow{id_{S'+S}}] S' + S \rangle$ *be their* symmetry *component.*

The parallel composition of two strict components can be shown to be strict, and so are the symmetries. We can represent a component c, the composition $c_2 \circ c_1$ and the parallel composition $c_1 + c_2$ in an intuitive graphical way as:

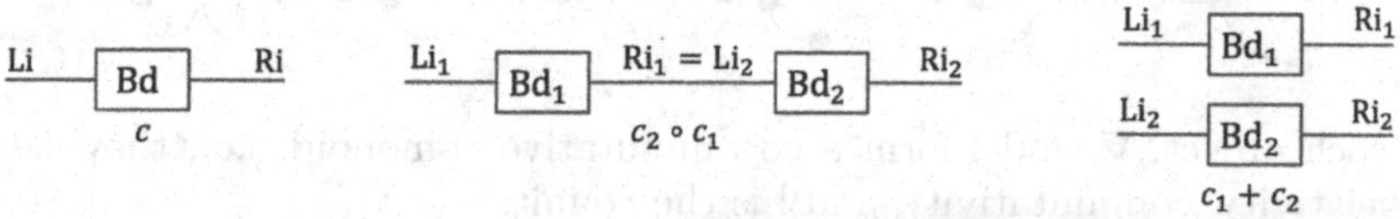

Li Bd Ri
c

Li_1 Bd_1 $Ri_1 = Li_2$ Bd_2 Ri_2
$c_2 \circ c_1$

Li_1 Bd_1 Ri_1
Li_2 Bd_2 Ri_2
$c_1 + c_2$

Identity and symmetry components are seen as *connectors* passing actions from one interface to the other. Other such connectors can be defined, for every systems S and S', by exploiting suitable morphisms in **RSys**.

- The *duplicator* is component $\nabla_S = \langle S[\xrightarrow{id_S} S \xleftarrow{[id_S, id_S]}]S + S\rangle$;
- The *co-duplicator* is component $\Delta_S = \langle S + S[\xrightarrow{[id_S, id_S]} S \xleftarrow{id_S}]S\rangle$;
- The *discharger* is component $!_S = \langle S[\xrightarrow{id_S} S \xleftarrow{\emptyset}]R_\emptyset\rangle$;
- The *co-discharger* is component $?_S = \langle R_\emptyset[\xrightarrow{\emptyset} S \xleftarrow{id_S}]S\rangle$.

Graphically, we show such connector components, which are all strict, as:

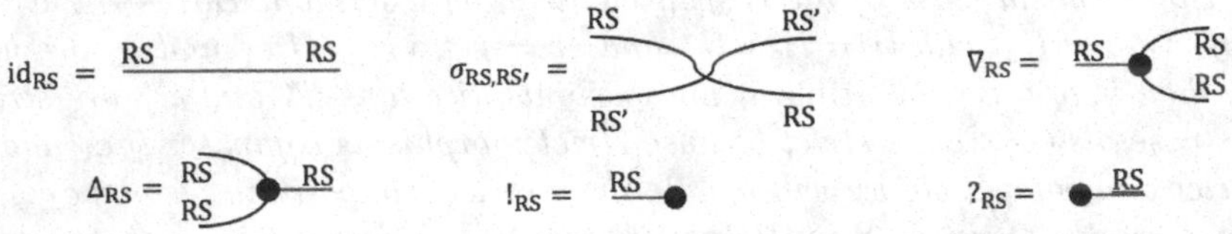

Theorem 2 (Com as Frobenius algebra). *Category* **Com** *with the monoidal functor* $+$ *and the family* σ *of symmetries of Definition 5 is a symmetric monoidal category. Furthermore, equipped with the families of connector components* ∇, Δ, $!$ *and* $?$ *as defined above* **Com** *is a Frobenius algebra.*

Proof. The category of abstract cospans built from a category with coproducts inherits a monoidal structure, induced by coproducts, which satisfies the laws for Frobenius algebras: see e.g. [3, Section 2.2].

6 Architectural Models

Due to Theorem 2 we can depict networks of components as *string diagrams* [3], a graphical syntax for structures whose basic elements take multiple inputs and outputs. The axioms of Frobenius algebras are sound and complete for string diagrams, in the sense that the diagrams representing two terms of the algebra can be topologically deformed into each other without cutting or joining wires if and only if the two terms are provably equal by the axioms.

Thanks to the axioms of symmetric monoidal categories (which we omit for brevity) the axioms of Frobenius algebras can be depicted as follows.

- for each object, Δ and $?$ form a commutative monoid, i.e., they satisfy associativity, commutativity, and $?$ is the unit:

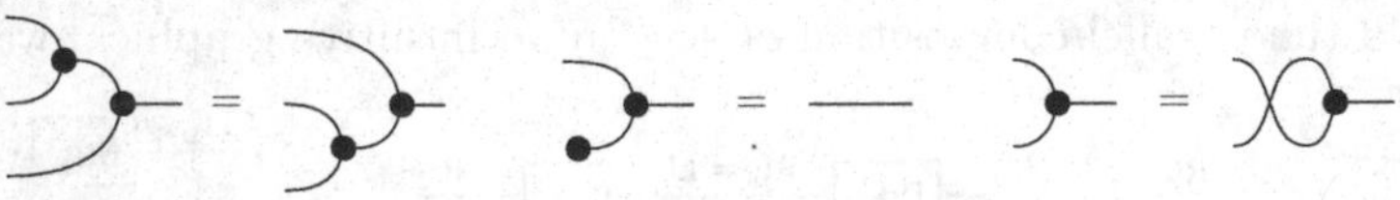

- for each object, ∇ and $!$ form a cocommutative comonoid, i.e., they satisfy associativity, commutativity, and $!$ is the counit:

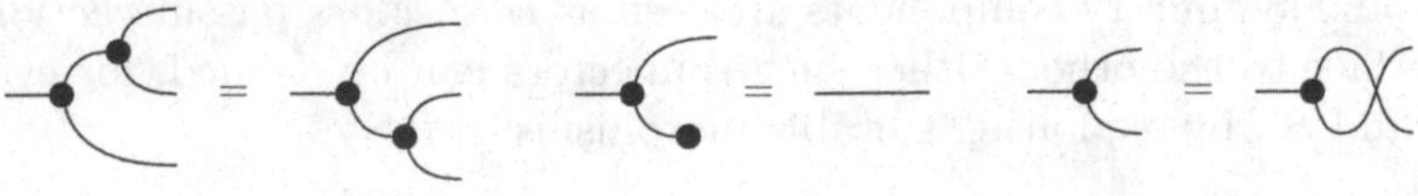

- the monoid and comonoid structures satisfy the Frobenius and special laws:

Since **Com** satisfies the axioms of Frobenius algebras, we can specify a complex component in **Com** by connecting the interfaces of its basic components. Any such drawings representing the same connections between interfaces are equivalent, such as two the string diagrams on the left below, both representing the component diagram of Sect. 2 with basic components C, S, and D.

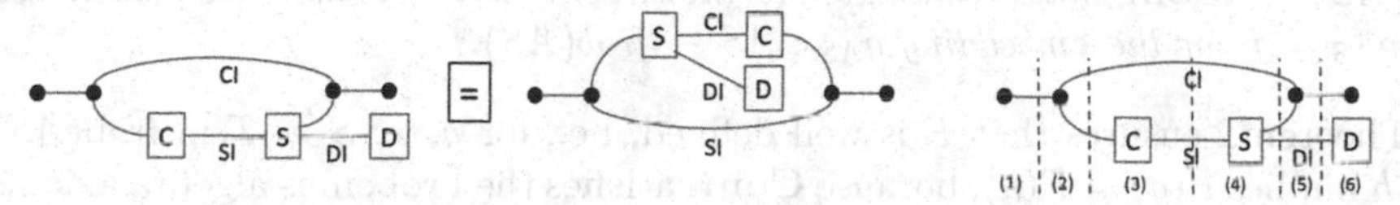

On the right we show how the left string diagram arises as sequential composition $?_{CI}$; ∇_{CI}; $id_{CI}+C$; $id_{CI}+S$; $\Delta_{CI}+id_{DI}$; $!_{CI}+D$ of expressions in the algebra of components and connectors, with vertical dashed lines in the figure representing ";". Based on the interpretation of basic components and connectors in **Com**, the constituent expressions represent the following cospans.

1. $?_{CI} = \langle R_\varnothing [\xrightarrow{\emptyset} CI \xleftarrow{id_{CI}}] CI\rangle$;
2. $\nabla_{CI} = \langle CI [\xrightarrow{id_{CI}} CI \xleftarrow{[id_{CI}, id_{CI}]}] CI + CI\rangle$;
3. $id_{CI} + C = \langle CI + CI [\xrightarrow{id_{CI}+cic} CI + C \xleftarrow{id_{CI}+sic}] CI + SI\rangle$;
4. $id_{CI} + S = \langle CI + SI [\xrightarrow{id_{CI}+sis} CI + S \xleftarrow{id_{CI}+[cis, dis]}] CI + CI + DI\rangle$;
5. $\Delta_{CI} + id_{DI} = \langle CI + CI + DI [\xrightarrow{[id_{CI}, id_{CI}]+id_{DI}} CI + DI \xleftarrow{id_{CI}+id_{DI}}] CI + DI\rangle$;
6. $!_{CI} + D = \langle CI + DI [\xrightarrow{id_{CI}+di_D} CI + D \xleftarrow{[\emptyset, \emptyset]}] R_\varnothing\rangle$.

Thus string diagrams serve as a bridge between the network-level description of an architecture in a component diagram and its "implementation" in graph rewriting components. The result of composing cospans 1–6 is the global system Sys in Fig. 1 with global rules emerging as amalgamation of component over interfaces rules and global state as pushout of component over interface states.

String diagrams providing a syntax for component networks are generated freely from an *architectural signature* of basic components and interfaces, just as term syntax for algebras is generated freely from an algebraic signature.

Definition 6 (architectural signature). *An* architectural signature $AS = \langle I, C, dom, cod\rangle$ *consists of sets of interface names I and component names C with functions $dom, cod : C \to I^*$ assigning each component name their sequences of names of left and right interfaces.*

The free Frobenius algebra frob(AS) over AS is a category that has sequences I^* as objects. Morphisms are directed hypergraphs with sorted interface nodes, called *network graphs*. They play the role of terms in algebraic signatures. Named components are represented by hyberedges distinguishing attachments of left and right interfaces. An architectural model assigns interpretations to interface and component names.

Definition 7 (interpretation, model). *An* interpretation *for a signature* AS *is a hypergraph morphism* $f = \langle f_I, f_C \rangle : AS \to \mathbf{Com}$, *i.e. a pair of mappings compatible with the domains and codomains of component names in* AS *and components in* $\mathbf{Com}$. *That means, each* $c : Li_1 \dots Li_n \to Ri_1 \dots Ri_m$ *in* AS *is mapped to an arrow* $f_C(c) : f_I(Li_1) + \dots + f_I(Li_n) \to f_I(Ri_1) + \dots + f_I(Ri_m)$.

The architectural model *for interpretation* f *is given by the functor* $F : frob(AS) \to \mathbf{Com}$ *that freely extends the given interpretation, i.e., such that* $F \circ \eta_{AS} = f$ *for the embedding* $\eta_{AS} : AS \to frob(AS)$.

Theorem 2 ensures that F is well defined, i.e., for $g, h : S \to T$ in frob(AS), $g = h$ implies $F(g) = F(h)$, because **Com** satisfies the Frobenius algebra axioms.

Category frob(AS) and model functor F represent the space of all component networks over a given collection of basic components with their interpretations. Since frob(AS) is free over AS, the extension F is unique and can be represented finitely by the hypergraph morphism $f : AS \to \mathbf{Com}$. If we consider the states of components only, this is similar to distributed graphs where a network graph forms the shape of a diagram in a category of local graphs, except that in our case graphs with interfaces are (part of) the arrows rather than the objects of the categories involved. However, in addition to states, we distribute entire runtime systems with interfaces along a network graph given by a morphism g in frob(AS). For a model F, a *configuration* consists of g and its interpretation $F(g)$ mapping the components named in g to their implementation in **Com**.

In our example, interface names are $I = \{si, ci, di\}$ and component names are $C = \{c, s, d\}$ with *dom* and *cod* given by $c : ci \to si, s : si \to ci\, di$ and $d : di \to \epsilon$ (the empty sequence). Interpretation f is defined by replacing lower with upper case characters, e.g., $f(s : si \to ci\, di) = S : SI \to CI + DI$.

7 Structural Operational Semantics

We exploit the compositionality of runtime system transformations for defining a structural operational semantics that derives the behaviour of complex components from that of basic ones and Frobenius algebra connectors. Since only morphims that are strict reflect transformations between runtime systems, we will focus on strict components only.

When presenting an architecture model, a basic component with n left interfaces and m right interfaces is shown as a diagram in **RSys** of shape $D = \langle Li_i \xrightarrow{li_i} Bd \xleftarrow{ri_j} Ri_j \rangle$ with $1 \leq i \leq n$ and $1 \leq j \leq m$. In **Com** this basic component is an abstract cospan constructed by the coproducts of their left and right interfaces as

$$cospan(D) = \langle Li_1 + \dots + Li_n \xrightarrow{li} Bd \xleftarrow{ri} Ri_1 + \dots + Ri_m \rangle$$

where $li = [li_1, \ldots, li_n]$ and $ri = [ri_1, \ldots, ri_m]$ are the co-pairings of the interface morphisms, induced by the universality of the respective coproducts. It is sufficient to require that all the interface morphisms are strict: the strictness of the co-pairing morphisms can be shown easily.

Since strict components are based on strict **RSys** morphisms, they have an internal state in the body projected to corresponding states of the interfaces. Let $c = \langle Li[\xrightarrow{li} Bd \xleftarrow{ri}]Ri\rangle$ be a strict component. When the state changes through an internal transformation $s : G_{Bd} \Rightarrow_c G_{Bd'}$ of the body, s is only partly hidden because it is reflected by the strict morphisms li and ri to interface transformation $a = li^{<}(s)$ in Li and $b = ri^{<}(s)$ in Ri, that we call *(left and right) observations*. This defines a strict component transformation that we denote as

$$s : c \underset{b}{\overset{a}{\Longrightarrow}} c'$$

where c' is the resulting component that shares types and rules with c, and may only differ for the states. Note that this notation only makes sense if c is strict, thus its use establishes an assumption or a proof obligation, depending on the context. The strictness of c' follows by that of c and because strict morphisms are preserved by transformations.

If strict components $c_i = \langle Li_i[\xrightarrow{li_i} Bd_i \xleftarrow{ri_i}]Ri_i\rangle$ for $i = 1, 2$ are connected through $Li_2 = Ri_1$, internal transformations of c_1 and c_2 need to synchronize by projecting the same observation to the shared interface, that is if

$$c_1 \underset{b_1}{\overset{a_1}{\Longrightarrow}} c'_1 \quad \text{and} \quad c_2 \underset{b_2}{\overset{a_2}{\Longrightarrow}} c'_2$$

then we must have $b_1 = a_2$. If c_1 and c_2 are compatible (and thus $c_2 \circ c_1$ is strict, see Definition 4) then also c'_1 and c'_2 can be shown to be compatible, and this results in a composed transformation of $c_2 \circ_s c_1$, projecting to interfaces Li_1 and Ri_2 the same observation projected by the transformations of c_1 and c_2, respectively.

For the parallel composition of strict components, which is strict, recall that the set of rule names of a coproduct of systems $R_1 + R_2$ is a product $P_1 \times P_2$. Therefore, a transformation $a = p/t$ in $R_1 + R_2$ is an application of a rule pair $p = \langle p_1, p_2\rangle \in P_1 \times P_2$ with $p_i \in P_i$. The rule span $\pi(\langle p_1, p_2\rangle) = \pi_1(p_1) + \pi_2(p_2)$ is a coproduct in **MSpan** and the DPO diagram $t = t_1 + t_2$ a coproduct in **DPO**. Hence, a represents the disjoint parallel occurrence of transformations $a_i = p_i/t_i$ in R_i for $i = 1, 2$, which we write using juxtaposition as $a_1\, a_2$.

Summarizing, for strict and parallel composition we have the rules

$$\frac{c_1 \underset{b}{\overset{a}{\Longrightarrow}} c'_1 \;,\; c_2 \underset{c}{\overset{b}{\Longrightarrow}} c'_2 \;,\; c_1 \text{ and } c_2 \text{ compatible}}{c_2 \circ_s c_1 \underset{c}{\overset{a}{\Longrightarrow}} c'_2 \circ_s c'_1} \qquad \frac{c_1 \underset{b}{\overset{a}{\Longrightarrow}} c'_1 \;,\; c_2 \underset{d}{\overset{c}{\Longrightarrow}} c'_2}{c_1 + c_2 \underset{b\,d}{\overset{a\,c}{\Longrightarrow}} c'_1 + c'_2} \;.$$

For all connector components, we can easily infer from the definitions that their transformations are triggered by a transformation in the left or right interface. For interface transformations $a : S \Rightarrow S', a_i : S_i \Rightarrow S'_i$, we have the following connector component transformations:

- $id_S \overset{a}{\underset{a}{\Longrightarrow}} id_{S'}$ synchronises transformations between the two interfaces;
- $\sigma_{S_1,S_2} \overset{a_1 a_2}{\underset{a_2 a_1}{\Longrightarrow}} \sigma_{S_2',S_1'}$ crosses the wires between $S_1 + S_2$ and $S_2 + S_1$;
- $\nabla_S \overset{a}{\underset{aa}{\Longrightarrow}} \nabla_{S'}$ synchronises transformations of its left and two right interfaces;
- $\Delta_S \overset{aa}{\underset{a}{\Longrightarrow}} \Delta_{S'}$ synchronises transformations of its right and two left interfaces;
- $!_S \overset{a}{\underset{\emptyset}{\Longrightarrow}} !_{S'}$ allows arbitrary transformations on its left interface;
- $?_S \overset{\emptyset}{\underset{a}{\Longrightarrow}} ?_{S'}$ allows arbitrary transformations on its right interface.

Component transformations can be composed and decomposed along both strict and parallel composition of components.

Theorem 3 (composition and decomposition of transformations). *Assume strict components* $c_i = \langle Li_i[\xrightarrow{li_i} Bd_i \xleftarrow{ri_i}]Ri_i\rangle$ *for* $i = 1, 2$. *Then,*

$$s_1 : c_1 \overset{a_1}{\underset{b_1}{\Longrightarrow}} c_1' \text{ and } s_2 : c_2 \overset{a_2}{\underset{b_2}{\Longrightarrow}} c_2' \quad \text{if and only if} \quad s : c_1 + c_2 \overset{a_1\, a_2}{\underset{b_1\, b_2}{\Longrightarrow}} c_1' + c_2'.$$

For c_1, c_2 *as above such that* $Li_2 = Ri_1$ *and* c_1, c_2 *compatible,*

$$s_1 : c_1 \overset{u}{\underset{v}{\Longrightarrow}} c_1' \text{ and } s_2 : c_2 \overset{v}{\underset{w}{\Longrightarrow}} c_2' \quad \text{if and only if} \quad s : c_2 \circ_s c_1 \overset{u}{\underset{w}{\Longrightarrow}} c_2' \circ_s c_1'.$$

Proof. The parallel composition $c_1 + c_2$ is based on a component-wise coproduct of the body and interface runtime systems of c_1 and c_2. Viewing the coproduct as a pushout over the initial system $R_\varnothing$, we can use Theorem 1 to derive s as composition of s_1 and s_2, and s_1, s_2 as decomposition of s.

This means that rule and DPO diagram of s are coproducts of rules and DPO diagrams of s_1 and s_2, respectively. Since a_1, a_2, b_1, b_2 are defined by projections via pullbacks which, in an adhesive category with strict initial object, preserve coproducts, the same relation holds for the rules and transformations of interfaces. Hence a_1, a_2 and b_1, b_2 composes into $a_1\, a_2$ and $b_1\, b_2$ respectively, and vice versa. For strict composition we can apply Theorem 1 directly: The body of $c_2 \circ_s c_2$ is a pushout of those of c_1 and c_2 over the shared interface, and interface states and transformations are projections of those in the bodies, so s is the composition of s_1 and s_2 and vice versa.

With *bisimilarity* $\equiv$ of strict components as the largest relation satisfying $f \equiv g$ iff for all a, b $f \overset{a}{\underset{b}{\Longrightarrow}} h$ iff $g \overset{a}{\underset{b}{\Longrightarrow}} k$ and $h \equiv k$, we have the following result.

Theorem 4 (bisimilarity as congruence). *Bisimilarity* $\equiv$ *on strict components is a congruence for parallel composition* $+$ *and strict composition* $\circ_s$.

Proof. Assume $f \equiv f', g \equiv g'$. If $g \circ_s f \overset{a}{\underset{b}{\Longrightarrow}} k \circ_s h$ then $f \overset{a}{\underset{c}{\Longrightarrow}} h$ and $g \overset{c}{\underset{b}{\Longrightarrow}} k$ by Theorem 3 (decomposition). This implies $f' \overset{a}{\underset{c}{\Longrightarrow}} h'$ and $g' \overset{c}{\underset{b}{\Longrightarrow}} k'$ since $f \equiv f'$ and $g \equiv g'$, and then $g' \circ_s f' \overset{a}{\underset{b}{\Longrightarrow}} k' \circ_s h'$ by Theorem 3 (composition). Reversing the roles of f, g and f', g' we can show the inverse implication. Then, by coinduction, $h \circ_s k \equiv h' \circ_s k'$ implies $g \circ_s f \equiv g' \circ_s f'$. The proof for $+$ is analogous.

Concretely this means that, if a component works in a given context, e.g. C in the context of S and D as defined in our example architecture, and we replace that context by a behaviourally equivalent one, e.g., adding a second instance of D for redundancy, the resulting system will have an equivalent overall behaviour.

8 Conclusion and Related Work

We introduced a component model for graph rewriting systems that allows to represent a global system as a network of components with interfaces representing shared views of internal states and transformations, and such that their composition reconstructs the global system.

Formally and conceptually our model represents the convergence of three main ingredients: Distributed graph transformations [18] formalise synchronised transformations of distributed graphs. Various notions of morphisms between graph transformation systems, discussed in [7] with their semantic properties, support the modularisation of types and rules. Algebraic representations of (network) graphs as arrows in a symmetric monoidal category and their visualisation by string diagrams [3] provide a syntax for component architectures.

Early steps towards modularity of formal specifications have been made in algebraic specifications [5] where the body of a module is related by morphisms with its import and export interfaces defining, respectively, required and provided services. In graph rewriting, work on modularity was inspired by algebraic specifications, programming and software engineering concepts [5], resulting in a number of proposals surveyed in [13]. More recently, [11] also proposes a compositional approach to graph transformations where local graphs with shared interfaces are composed via colimits into a global graph, and rules acting on local graphs are composed into a global rule acting on the global graph. Differently from our approch, however, compositionality is addressed at the instance level, not at the type level. Several other contributions address compositionality in graph transformation at instance level, including among others synchronized hyperedge replacement [9,14], rule amalgamation [2], distributed graph transformation [18] and borrowed contexts [1,6]. An interesting topic for future work is to compare the expressive power of compositionality modeled at instance or at type level.

Modules of typed graph transformation systems [12] follow the structure of algebraic specification modules while [7] combines modularity and service-oriented concepts. None of the above include a notion of state, i.e. they structure the specification but not the runtime of a system. We consider this the main difference between modules and components. Conversely, distributed graph transformations capture the distribution of graphs, rules and transformations in a category of diagrams over graphs [18] but without modularity at specification level.

We provide for the first time a component model integrating these two features. In this more general setting we achieve compositionality like in distributed graph transformations, relating global and synchronised local transformations, and describe the network architecture using Frobenius algebras to provide a constructive "compilation" assembling complex components from basic constituents.

In the future we would like to make explicit the invocation-based intuition of components, using a type system and refined operational semantics to distinguish provided and required interfaces and caller/callee roles in the synchronised applications of rules. We will exploit and extend the Frobenius structure to (1) support architectural equations defining, e.g., derived components as expressions over basic ones or behavioural equalities between configurations; (2) allow architectural reconfiguration as string diagram rewriting; and (3) consider a bigraph-like network level with hierarchical components.

Our notion of bisimilarity over doubly-labelled transformations as a congruence is analogous to functoriality of tile bisimilarity, and we can indeed phrase our operational semantics as an instance of the tile model [10]. In [4] tile bisimilarity is extended to remain compositional under dynamic reconfiguration.

References

1. Baldan, P., Ehrig, H., König, B.: Composition and decomposition of DPO transformations with borrowed context. In: Corradini, A., Ehrig, H., Montanari, U., Ribeiro, L., Rozenberg, G. (eds.) ICGT 2006. LNCS, vol. 4178, pp. 153–167. Springer, Heidelberg (2006). https://doi.org/10.1007/11841883_12
2. Boehm, P., Fonio, H., Habel, A.: Amalgamation of graph transformations: a synchronization mechanism. J. Comput. Syst. Sci. **34**(2/3), 377–408 (1987). https://doi.org/10.1016/0022-0000(87)90030-4
3. Bonchi, F., Gadducci, F., Kissinger, A., Sobocinski, P., Zanasi, F.: String diagram rewrite theory I: rewriting with Frobenius structure. J. ACM **69**(2), 14:1–14:58 (2022)
4. Bruni, R., Montanari, U., Sassone, V.: Observational congruences for dynamically reconfigurable tile systems. Theoret. Comput. Sci. **335**(2–3), 331–372 (2005). https://eprints.soton.ac.uk/261844/
5. Ehrig, H., Mahr, B.: Fundamentals of Algebraic Specification 2: Module Specifications and Constraints. EATCS Monographs on Theoretical Computer Science, vol. 21. Springer Verlag, Berlin (1990). https://doi.org/10.1007/978-3-642-61284-8
6. Ehrig, H., König, B.: Deriving bisimulation congruences in the DPO approach to graph rewriting with borrowed contexts. Math. Struct. Comput. Sci. **16**(6), 1133–1163 (2006). https://doi.org/10.1017/S096012950600569X
7. Engels, G., Heckel, R., Cherchago, A.: Flexible interconnection of graph transformation modules. In: Kreowski, H.-J., Montanari, U., Orejas, F., Rozenberg, G., Taentzer, G. (eds.) Formal Methods in Software and Systems Modeling. LNCS, vol. 3393, pp. 38–63. Springer, Heidelberg (2005). https://doi.org/10.1007/978-3-540-31847-7_3
8. Evans, E.: Domain-Driven Design: Tackling Complexity in the Heart of Software. Addison-Wesley, Boston (2004)
9. Ferrari, G.L., Hirsch, D., Lanese, I., Montanari, U., Tuosto, E.: Synchronised hyperedge replacement as a model for service oriented computing. In: de Boer, F.S., Bonsangue, M.M., Graf, S., de Roever, W.-P. (eds.) FMCO 2005. LNCS, vol. 4111, pp. 22–43. Springer, Heidelberg (2006). https://doi.org/10.1007/11804192_2

10. Gadducci, F., Montanari, U.: The tile model. In: Plotkin, G.D., Stirling, C., Tofte, M. (eds.) Proof, Language, and Interaction, Essays in Honour of Robin Milner, pp. 133–166. The MIT Press, Cambridge (2000)
11. Ghamarian, A.H., Rensink, A.: Generalised compositionality in graph transformation. In: Ehrig, H., Engels, G., Kreowski, H.-J., Rozenberg, G. (eds.) ICGT 2012. LNCS, vol. 7562, pp. 234–248. Springer, Heidelberg (2012). https://doi.org/10.1007/978-3-642-33654-6_16
12. Groe-Rhode, M., Presicce, F.P., Simeoni, M.: Refinements and modules for typed graph transformation systems. In: Fiadeiro, J.L. (ed.) WADT 1998. LNCS, vol. 1589, pp. 138–151. Springer, Heidelberg (1999). https://doi.org/10.1007/3-540-48483-3_10
13. Heckel, R., Engels, G., Ehrig, H., Taentzer, G.: Classification and comparison of modularity concepts for graph transformation systems. In: Engels, G., Kreowski, H.J., Rozenberg, G. (eds.) Handbook of Graph Grammars and Computing by Graph Transformation, vol. 2, pp. 669–690. World Scientific (1999)
14. Dan, H., Ugo, M.: Synchronized hyperedge replacement with name mobility. In: Larsen, K.G., Nielsen, M. (eds.) CONCUR 2001. LNCS, vol. 2154, pp. 121–136. Springer, Heidelberg (2001). https://doi.org/10.1007/3-540-44685-0_9
15. Hogan, A., et al.: Knowledge Graphs. No. 22 in Synthesis Lectures on Data, Semantics, and Knowledge, Morgan & Claypool (2021). https://kgbook.org/
16. Lassila, O.: Graph abstractions matter, December 2021. https://2021.connected-data.world
17. Schad, J.: Graph powered machine learning: Part 1. ML Conference Berlin, October 2021. https://mlconference.ai/ml-summit/
18. Taentzer, G.: Distributed graphs and graph transformation. Appl. Categorical Struct. **7**(4), 431–462 (1999)
19. Xiao, G., Ding, L., Cogrel, B., Calvanese, D.: Virtual knowledge graphs: an overview of systems and use cases. Data Intell. **1**(3), 201–223 (2019). https://doi.org/10.1162/dint_a_00011

Decidability of Resilience for Well-Structured Graph Transformation Systems

Okan Özkan(✉)

University of Oldenburg, Oldenburg, Germany
o.oezkan@informatik.uni-oldenburg.de

Abstract. Resilience is a concept of rising interest in computer science and software engineering. For systems in which correctness w.r.t. a safety condition is unachievable, fast recovery is demanded. We ask whether we can reach a safe state in a bounded number of steps whenever we reach a bad state. In a well-structured framework, we investigate problems of this kind where the bad and safety conditions are given as upward/downward-closed sets. We obtain decidability results for graph transformation systems by applying our results for subclasses of well-structured transition systems. Moreover, we identify sufficient criteria of graph transformation systems for the applicability of our results.

Keywords: Resilience · Graph transformation systems · Decidability · Well-structured transition systems

1 Introduction

Resilience is a broadly used concept in computer science and software engineering (e.g., [11]). In general engineering systems, *fast recovery* from a degraded state is often termed as resilience, see, e.g., [17]. In view of the latter interpretation of resilience, we investigate on the question whether a SAFE state can be reached in a bounded number of steps from any BAD state (where BAD is not necessarily the complement of SAFE). This concept is meaningful for systems in which violation of SAFE cannot be avoided. Our notion of resilience generalizes correctness (e.g., [2,10,15]) w.r.t. a safety condition.

For modeling systems, we use *graph transformation systems (GTSs)* in the *single pushout approach (SPO)*, as considered, e.g., in [7], which provide visual interpretability but yet also a precise formalism. In this perception, system states are captured by graphs and state changes by graph transformations. Our goal is to obtain decidability results for GTSs by considering their induced *transition systems*. A transition system consists of a set of states of any kind (not necessarily graphs) and a transition relation on the state set.

Usually, the state set (set of graphs) is infinite. To handle infinite state sets, we employ the concept of well-structuredness studied, e.g., in [1,9]. A *well-structured transition system (WSTS)* is, informally, a transition system equipped

N. Behr and D. Strüber (Eds.): ICGT 2022, LNCS 13349, pp. 38–57, 2022.
https://doi.org/10.1007/978-3-031-09843-7_3

with a *well-quasi-order* satisfying that larger states simulate smaller states (also called *compatibility* condition) and that certain predecessor sets can be effectively computed. In this well-structured setting, *ideal-based* sets (*upward-* or *downward-closed sets*) play an important role. They enjoy a number of suitable properties for verification such as finite representation (of upward-closed sets) and closure properties. For WSTSs, the *ideal reachability (coverability) problem* is decidable [1,9], which is an integrant of our results.

Well-structuredness of GTSs is investigated, e.g., in [12] for several well-quasi-orders. The well-quasi-order we use is the subgraph order which permits strong compatibility but comes with the restriction of path-length-boundedness on the graph class.

We show decidability for subclasses of GTSs of bounded path length. Each subclass exhibits additional requirements, i.e., effectiveness or unreliability properties. Additionally, we identify sufficient criteria of GTSs for the applicability of the results.

More precisely, we consider the *explicit* resilience problem where the bound on the number of steps for recovery is given and the *bounded* resilience problem which asks whether there exists such a bound. These problems are formulated for marked GTSs each of which consists of a GTS together with a graph class closed under rule application and an INITial subset of graphs. We ask: Starting from any graph in INIT, whenever we reach a BAD graph, can we reach a SAFE graph in $\leq k$ (in a bounded number of) steps?

To illustrate the idea of our resilience concept, we give an example.

Example (circular process protocol). Consider a ring of three processes P_0, P_1, P_2 each of which has an unordered collection (multiset) containing commands. Each command belongs to a process and is labeled accordingly as c_0, c_1, or c_2. The protocol is described below. A formalization as GTS can be found in Sect. 4.

- The process P_0 *liberal*, i.e., it can *initiate* (generate) a command c_0 in the collection of the next process.
- Every process P_i can *forward* a command c_j, $i \neq j$, not belonging to itself.
- If a process P_i receives a command c_i, it is *enabled* and can
  1. *execute* its specific process action, or
  2. *clear* all commands in its collection and forward a command of the next process, or
  3. *leave* the process ring (if $i \neq 0$) and forward a command of the next process.

  Afterwards, the command c_i is deleted.
- Any command may get *lost* in any state.

The process action of P_0 is to forward two commands, c_1 and c_2. The process action of P_1 (P_2) is to forward a command c_2 (c_1). The topology of the process ring changes when a process leaves the ring. Processes P_1 and P_2 can leave the ring only if the other process has not left the ring before. In Fig. 1, the initial state where every process P_i has one command c_i in its collection and the three possible topologies are shown. A process P_i is represented by an edge labeled with P_i. The collections are represented by white nodes which may have loops labeled with c_i corresponding to the contained commands.

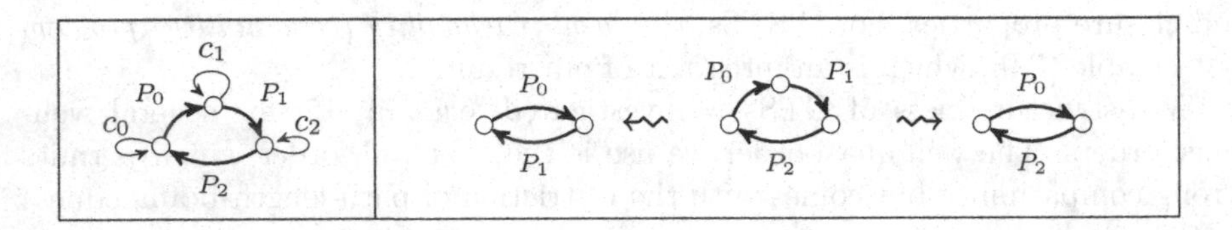

Fig. 1. Initial state and topologies of the circular process protocol.

Consider the following instances of the bounded resilience problem with the initial state as in Fig. 1:

BAD	¬AllEnabled	Command(c_2)	AllEnabled	NoCommand
SAFE	AllEnabled	Collection(c_0, c_1)	¬AllEnabled	No3Processes

For every instance of the bounded resilience problem, we are interested in a bound k for the number of steps needed for recovery. In the first instance, we ask whether we can reach a state where every process is enabled in $\leq k$ steps whenever we reach a state where this is not the case. In the second instance, we ask whether we can reach a state with a collection containing commands c_0, c_1 in $\leq k$ steps whenever we reach a state with a collection containing c_2. The third instance is the "dual" problem to the first one where the constraints for BAD and SAFE are exchanged. In the fourth instance, we ask whether we can reach a state containing no three processes in $\leq k$ steps whenever we reach a state without commands. One may ask:

- Does such a k exist? If so, what is the minimal k?
- Is there a generic method for problems of this kind?

We will answer these questions in Sect. 3 and 4.

This paper is organized as follows: In Sect. 2, we recall preliminary concepts of GTSs and (WS)TSs. We show decidability of the resilience problems for subclasses of marked WSTSs in Sect. 3. In Sect. 4, we apply our results to marked GTSs. In Sect. 5, we give sufficient, rule-specific criteria for the applicability of our results. We present related concepts in Sect. 6 and close with a conclusion in Sect. 7. The proofs in full length and a further example can be found in a technical report [13].

2 Preliminaries

We recall the concepts used in this paper, namely graph transformation systems [6,7] and (in particular well-structured) transition systems [9].

2.1 Graph Transformation Systems

In the following, we recall the definitions of graphs, graph constraints, rules, and graph transformation systems [6,7].

A directed, labeled graph consists of a finite set of nodes and a finite set of edges where each edge is equipped with a source and a target node and where each node and edge is equipped with a label. Note that this kind of graphs are a special case of the hypergraphs considered in [12].

Definition 1 (graphs & morphisms). A *(directed, labeled) graph* (over a finite label alphabet $\Lambda = \Lambda_V \cup \Lambda_E$) is a tuple $G = \langle V_G, E_G, src_G, tgt_G, lab_G^V, lab_G^E \rangle$ with finite sets V_G and E_G of *nodes* (or *vertices*) and *edges*, functions $src_G, tgt_G : E_G \to V_G$ assigning *source* and *target* to each edge, and *labeling functions* $lab_G^V : V_G \to \Lambda_V$, $lab_G^E : E_G \to \Lambda_E$. A *(simple, undirected) path* in G of length ℓ is a sequence $\langle v_1, e_1, v_2 \ldots, v_\ell, e_\ell v_{\ell+1} \rangle$ of nodes and edges s.t. $src_G(e_i) = v_i$ and $tgt_G(e_i) = v_{i+1}$, or $tgt_G(e_i) = v_i$ and $src_G(e_i) = v_{i+1}$ for every $1 \leq i \leq \ell$, and all contained nodes and edges occur at most once. Given graphs G and H, a *(partial graph) morphism* $g : G \rightharpoonup H$ consists of partial functions $g_V : V_G \rightharpoonup V_H$ and $g_E : E_G \rightharpoonup E_H$ which preserve sources, targets, and labels, i.e., $g_V \circ src_G(e) = src_H \circ g_E(e)$, $g_V \circ tgt_G(e) = tgt_H \circ g_E(e)$, $lab_G^V(v) = lab_H^V \circ g_V(v)$, and $lab_G^E(e) = lab_H^E \circ g_E(e)$ on all nodes v and egdes e, for which $g_V(v), g_E(e)$ is defined. Furthermore, if a morphism is defined on an edge, it must be defined on both incident nodes. The morphism g is *total* (*injective*) if both g_V and g_E are total (injective). If g is total and injective, we also write $g : G \hookrightarrow H$. The composition of morphisms is defined componentwise. A pair $\langle G \to C, G' \to C \rangle$ of morphisms is *jointly surjective* if every item of C has a preimage in G or G'.

Convention. We draw graphs as usual. Labels are indicated by a symbol or a color. In (partial) morphisms, we equip the image of a node with the same index. Nodes on which the morphism is undefined have no index.

We consider a special case of graph constraints [10,16], which are non-nested and based on positive ($\exists G$)/negative ($\neg\exists G$) constraints. For simplicity, we call them also positive (negative) constraints.

Definition 2 (positive & negative constraints). The class of *positive (negative) (graph) constraints* is the smallest class of expressions which contains $\exists G$ (negative: $\neg\exists G$) for every graph G and is closed under $\vee$ and $\wedge$. A graph G *satisfies* $\exists G'$ if there exists a total, injective morphism $G' \hookrightarrow G$. The semantics of the logical operators are as usual. We write $G \models c$ if G satisfies the positive/negative constraint c. For a positive/negative constraint c, we denote by $[\![c]\!]$ the set of all graphs G of the considered graph class with $G \models c$.

Using jointly surjective morphisms, every positive constraint can algorithmically be converted into an equivalent "$\vee$-normal form".

Fact 1 (from $\wedge$ to $\vee$). For every positive contraint c, we can effectively construct a positive constraint c' of the form $\bigvee_{1 \leq i \leq n} \exists G_i$ s.t. $[\![c]\!] = [\![c']\!]$ and there exists no total, injective morphism $G_i \hookrightarrow G_j$ for $i \neq j$.

We use the single pushout (SPO) approach [7] with injective matches for modeling graph transformations.

Definition 3 (graph transformation). A *(graph transformation) rule* $r = \langle L \rightharpoonup R \rangle$ is a partial morphism from a graph L to a graph R. A *graph transformation system (GTS)* is a finite set of rules. A *transformation* $G \Rightarrow H$ from a graph G to a graph H applying a rule r at a total, injective *match morphism* $g : L \hookrightarrow G$ is given by a *pushout* as shown in Fig. 2 (1) (for existence and construction of pushouts, see, e.g., [7]). We write $G \Rightarrow_r H$ to indicate the applied rule, and $G \Rightarrow_{\mathcal{R}} H$ if $G \Rightarrow_r H$ for a rule r in the rule set $\mathcal{R}$.

In Fig. 2 (2), an example for a transformation is shown.

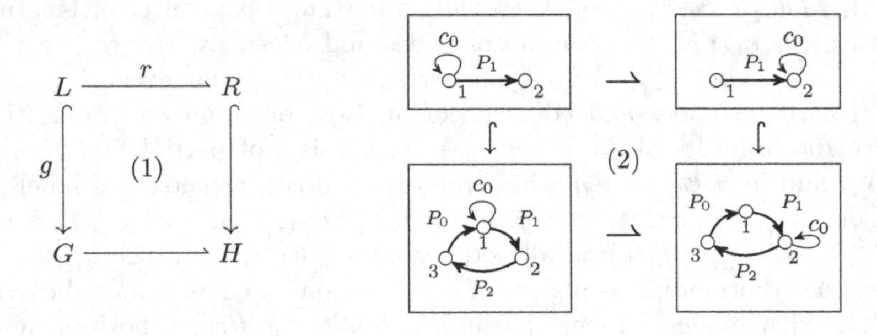

Fig. 2. Pushout scheme and example of a transformation.

2.2 Transition Systems and Well-structuredness

We recall the notion of transition systems.

Definition 4 (transition system). A *transition system (TS)* $\langle S, \rightarrow \rangle$ consists of a (possibly infinite) set S of *states* and a *transition relation* $\rightarrow \subseteq S \times S$. Let $\rightarrow^0 = \mathrm{Id}_S$ (identitiy on S), $\rightarrow^1 = \rightarrow$, and $\rightarrow^k = \rightarrow^{k-1} \circ \rightarrow$ for every $k \geq 2$. Let $\rightarrow^{\leq k} = \bigcup_{0 \leq j \leq k} \rightarrow^j$ for every $k \geq 0$. The *transitive closure* is given by $\rightarrow^* = \bigcup_{k \geq 0} \rightarrow^k$.

Often we are interested in the predecessors or successors of state set.

Definition 5 (pre- & postsets). Let $\langle S, \rightarrow \rangle$ be a transition system. For $S' \subseteq S$ and $k \geq 0$, we define $\mathrm{pre}^k(S') = \{s \in S \mid \exists s' \in S' : s \rightarrow^k s'\}$ and $\mathrm{post}^k(S') = \{s \in S \mid \exists s' \in S' : s' \rightarrow^k s\}$. Let $\mathrm{pre}^{\leq k}(S') = \bigcup_{j \leq k} \mathrm{pre}^j(S')$, $\mathrm{pre}^*(S') = \bigcup_{k \geq 0} \mathrm{pre}^k(S')$, $\mathrm{post}^{\leq k}(S') = \bigcup_{j \leq k} \mathrm{post}^j(S')$, and $\mathrm{post}^*(S') = \bigcup_{k \geq 0} \mathrm{post}^k(S')$. We abbreviate $\mathrm{post}^1(S')$ by $\mathrm{post}(S')$ and $\mathrm{pre}^1(S')$ by $\mathrm{pre}(S)$. A TS $\langle S, \rightarrow \rangle$ is *finite-branching* if $\mathrm{post}(s)$ is finite and computable for every given state s.

Several problems are undecidable for infinite-state TSs in general. However, interesting decidability results can be achieved if the system is well-structured [1,9,12]. A prerequisite for this concept is a well-quasi-order on the state set.

Definition 6 (well-quasi-order). A *quasi-order* is a reflexive, transitive relation. A *well-quasi-order (wqo)* over a set X is a quasi-order $\leq \subseteq X \times X$ s.t. every infinite sequence $\langle x_0, x_1, \ldots \rangle$ in X contains an increasing pair $x_i \leq x_j$ with $i < j$. A (well-)quasi-order is *decidable* if it can be decided whether $x \leq x'$ for all $x, x' \in X$.

In our setting, the subgraph order is of crucial importance.

Example 1 (subgraph order). The subgraph order $\leq$ is given by $G \leq H$ iff there is a total, injective morphism $G \hookrightarrow H$. Let S_ℓ be a graph class of bounded path length (with bound ℓ). The restriction of $\leq$ to S_ℓ is a wqo [4,12]. However, it is not a wqo on all graphs: The infinite sequence $\langle$, , , $\ldots\rangle$ of cyclic graphs of increasing length contains no increasing pair.

Assumption. From now on, we implicitly equip every set of graphs with the subgraph order. By "$\leq$" we mean either an abstract wqo or the subgraph order.

Upward- and downward-closed sets are of special interest.

Definition 7 (ideal & basis). Let X be a set and $\leq$ a quasi-order on X. For every subset A of X, we denote by $\uparrow A = \{x \in X \mid \exists a \in A : a \leq x\}$ the *upward-closure* and $\downarrow A = \{x \in X \mid \exists a \in A : x \leq a\}$ the *downward-closure* of A. An *ideal* $I \subseteq X$ is an upward-closed set, i.e., $\uparrow I = I$. An *anti-ideal* $J \subseteq X$ is a downward-closed set, i.e., $\downarrow J = J$. An (anti-)ideal is *decidable* if membership for every $x \in X$ is decidable. A *basis* of an ideal I is a subset $B \subseteq I$ s.t. (i) $\uparrow B = I$ and (ii) $b \neq b' \Rightarrow b \not\leq b'$ for all $b, b' \in B$.

Fact 2 (ideals of graphs). For every positive (negative) constraint c, $[\![c]\!]$ is an (anti-)ideal.

Ideals are, in general, infinite but can be represented by finite bases (a minimal generating set), similar to algebraic structures.

Fact 3 (finite basis [1, Lemma 3.3]). Every ideal has a basis and every basis is finite, provided that the superset is equipped with a wqo. Given a finite set A, a basis of $\uparrow A$ is computable, provided that the quasi-order is decidable.

Anti-ideals are the complements of ideals. Since an anti-ideal does not have an "upward-basis" in general, we will later demand that membership is decidable.

For well-structuredness, we demand that the wqo yields a simulation of smaller states by larger states. This condition is called compatibility.

Definition 8 (well-structured transition systems). Let $\langle S, \rightarrow \rangle$ be transition system and $\leq$ a decidable wqo on S. The tuple $\langle S, \leq, \rightarrow \rangle$ is a *well-structured transition system (WSTS)*, if:

(i) The wqo is *compatible* with the transition relation, i.e., for all $s_1, s_1', s_2 \in S$ with $s_1 \leq s_1'$ and $s_1 \rightarrow s_2$, there exists $s_2' \in S$ with $s_2 \leq s_2'$ and $s_1' \rightarrow^* s_2'$. If $s_1' \rightarrow^1 s_2'$, we say that it is *strongly compatible*. Both is illustrated in Fig. 3.
(ii) For every $s \in S$, a basis of $\uparrow \mathrm{pre}(\uparrow \{s\})$ is computable.

$$\forall\ \begin{array}{ccc} s_1 & \longrightarrow & s_2 \\ \mid\wedge & & \mid\wedge \\ s_1' & \overset{*}{\dashrightarrow} & s_2' \end{array}\ \exists \qquad \forall\ \begin{array}{ccc} s_1 & \longrightarrow & s_2 \\ \mid\wedge & & \mid\wedge \\ s_1' & \overset{1}{\dashrightarrow} & s_2' \end{array}\ \exists$$

(a) Compatibility (b) Strong compatibility

Fig. 3. Visualization of (strong) compatibility.

A *strongly WSTS (SWSTS)* is a WSTS with strong compatibility.

Remark. For GTSs, strong compatibility is achieved by applying the same rule to the bigger graph. In contrast to the double pushout (DPO) approach [6], SPO has the suitable property that every rule is applicable to the bigger graph.

Assumption. Let $\langle S, \leq, \rightarrow\rangle$ be a well-structured transition system.

The set of ideals of S is closed under preset, union, and intersection.

Fact 4 (stability of ideals [1, Lemma 3.2]). For ideals $I, I' \subseteq S$, the sets $\mathrm{pre}^*(I)$, $I \cup I'$, and $I \cap I'$ are ideals. For SWSTSs, the sets $\mathrm{pre}(I)$, $\mathrm{pre}^{\leq k}(I)$ for every $k \geq 0$ are ideals.

An important point in our argumentation is the observation that every infinite, ascending sequence of ideals w.r.t. a wqo eventually becomes stationary.

Lemma 1 (Noetherian state set [1, Lemma 3.4]). For every infinite, ascending sequence $\langle I_0 \subseteq I_1 \subseteq \ldots\rangle$ of ideals, $\exists k_0 \geq 0$ s.t. $I_k = I_{k_0}$ for all $k \geq k_0$.

Abdulla et al. [1] exploit Lemma 1 to show the decidability of ideal reachability (coverability) for SWSTSs. The idea is to iteratively construct the sequence of the ideals $I^k = \mathrm{pre}^{\leq k}(I)$ until it becomes stable. This construction is carried out by representing ideals by bases. This argumentation is similarly feasible for WSTSs, see, e.g., [9, proof of Thm. 3.6].

Lemma 2 (ideal reachability [1, Thm. 4.1]). Given a basis of an ideal $I \subseteq S$ and a state s of a SWSTS, we can decide whether we can reach a state $s_I \in I$ from s. In particular, $\mathrm{pre}^{\leq k}(I) = \mathrm{pre}^*(I) \iff \mathrm{pre}^{\leq k+1}(I) = \mathrm{pre}^{\leq k}(I)$, and a basis of $\mathrm{pre}^*(I)$ is computable.

3 Decidability

We show the decidability of resilience problems for subclasses of SWSTSs by extending the idea in [14] to a systematic investigation.

In our setting, ideal-based sets of states play an important role.

Definition 9 (ideal-based). A set is *ideal-based* if it is (i) an ideal with a given basis, or (ii) a decidable anti-ideal. We denote by

(i) $\mathcal{I}$ the set of ideals with given bases, (ii) $\mathcal{J}$ the set of decidable anti-ideals.

We formulate resilience problems for marked WSTSs, i.e., WSTSs with a specified set INIT of inital states starting from which we investigate resilience.

Definition 10 (marked WSTS). A *marked WSTS* is a tuple $\langle S, \leq, \rightarrow, \text{INIT}\rangle$ where $\langle S, \leq, \rightarrow\rangle$ is a WSTS and INIT $\subseteq S$. If INIT is finite, we call it *fin-marked.*

EXPLICIT RESILIENCE PROBLEM FOR WSTSs

Given: A marked WSTS $\langle S, \leq, \rightarrow, \text{INIT}\rangle$, ideal-based sets SAFE, BAD $\subseteq S$, a natural number $k \geq 0$.

Question: $\forall s \in \text{INIT} : \forall(s \rightarrow^* s' \in \text{BAD}) : \exists(s' \rightarrow^{\leq k} s'' \in \text{SAFE})$?

BOUNDED RESILIENCE PROBLEM FOR WSTSs

Given: A marked WSTS $\langle S, \leq, \rightarrow, \text{INIT}\rangle$, ideal-based sets SAFE, BAD $\subseteq S$.

Question: $\exists k \geq 0 : \forall s \in \text{INIT} : \forall(s \rightarrow^* s' \in \text{BAD}) : \exists(s' \rightarrow^{\leq k} s'' \in \text{SAFE})$?

For our further considerations, we regard requirements in order to obtain decidability, i.e., we consider the following subclasses of marked WSTSs.

Definition 11 (requirements). A marked WSTS $\langle S, \leq, \rightarrow, \text{INIT}\rangle$ is

(1) post*-effective if INIT is finite and a basis of $\uparrow\text{post}^*(\text{INIT})$ is computable,
(2) *lossy* if $\downarrow\text{post}^*(\text{INIT}) = \text{post}^*(\text{INIT})$,
(3) $\perp$*-bounded (bottom-bounded)* if there exists $\ell \geq 0$ s.t. $s_B \in \text{post}^{\leq \ell}(s)$ for every $s \in S$ and every element s_B of a basis of S with $s \geq s_B$.

The requirement of post*-effectiveness describes the computability of the smallest reachable states from the initital states. The notion of lossiness means that the set of reachable states from the initital states is downward-closed. This is an abstraction from the lossiness concept in [9, p. 83]. Usually, the term "lossy" describes the circumstance that (almost) any piece of information of a state may get lost. Another kind of unreliability is $\perp$-boundedness which means that from every state, every smaller basis element (the bottom underneath) is reachable in a bounded number of steps. Thereby (almost) all information of a state may get lost in a bounded of number of steps.

The following lemma is crucial for many following proofs.

Lemma 3 (ideal-inclusion [14, **Lemma 4**]**).** Let A be a set, I an ideal, and J an anti-ideal. Then, $A \cap J \subseteq I \iff \uparrow A \cap J \subseteq I$.

Applying this lemma to a basis B_I of an ideal I, we obtain that the inclusion $I \cap J \subseteq I'$ in an ideal I' can be checked by computing $B_I \cap J$ and then checking whether $B_I \cap J \subseteq I'$.

We give a characterization of post*-effectiveness via "anti-ideal reachability".

Proposition 1 (characterization of post*-effectiveness). For a class of finite-branching WSTSs, a basis of $\uparrow \mathrm{post}^*(s)$ is computable for every given state s iff the anti-ideal reachability problem is decidable, i.e., given a state s of a WSTS in the regarded class and a decidable anti-ideal J, it can be decided whether $\exists s' \in J : s \rightarrow^* s'$.

Proof (sketch) On one hand, we can decide the anti-ideal reachability problem by computing a basis of $\uparrow \mathrm{post}^*(s)$ and checking whether the intersection with the anti-ideal is empty (Lemma 3). One the other hand, we can compute a basis of $\uparrow \mathrm{post}^*(s)$ by computing the sequence of ideals $P_k = \uparrow \mathrm{post}^{\leq k}(s)$ until it becomes stationary (Lemma 1). The stop condition, i.e., the condition which guarantees that we can terminate the algorithm, is formalized as anti-ideal reachability. □

The characterization in Proposition 1 is used to show that Petri nets are post*-effective. It is well-known that Petri nets constitute SWSTSs [9, Thm. 6.1].

Example 2 (variations of Petri nets). (1) Petri nets (equipped with any finite set of initial states) are post*-effective by Proposition 1: Reachability for Petri nets is decidable and recursively equivalent to submarking reachability [8, p.6]. This corresponds to the anti-ideal reachability problem for Petri nets.
(2) *Lossy Petri nets* are Petri nets where in any state, one token may get lost at any place. Lossy Petri nets are lossy for every set of initial states.
(3) *Reset-lossy (mixed-lossy) Petri nets* are reset Petri nets [5] where in any state, all tokens (or one token) may get lost at any place. Reset-lossy (mixed-lossy) Petri nets are $\perp$-bounded (and lossy for every set of initial states).

For some results, we assume that a basis of the set of all states is given. This is only relevant if we use these basis elements for computations.

Notation. For a WSTS $\langle S, \leq, \rightarrow \rangle$ with a given basis of S, we write WSTS^B.

The next proposition shows how the requirements are related provided that a basis of the set of all states is given. The Venn diagram in Fig. 4 illustrates the relations of the subclasses corresponding to the requirements.

Proposition 2. Lossy ($\perp$-bounded) fin-marked WSTS^{B}s are post*-effective.

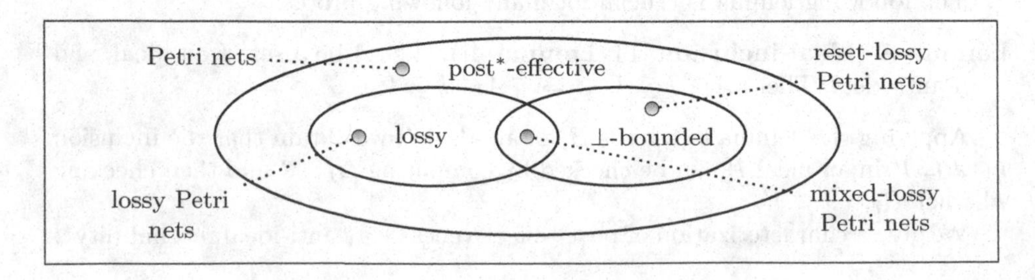

Fig. 4. Subclasses of fin-marked WSTS^{B}s.

Proof (sketch). Let $\langle S, \leq, \rightarrow, \text{INIT}\rangle$ be a lossy ($\perp$-bounded) fin-marked WSTS^B. To compute a basis of $\uparrow \text{post}^*(\text{INIT})$ for a finite set INIT, we look at the reachable elements of a basis of the set S of all states. Such a basis element is reachable iff its upward-closure is reachable. By Lemma 2, the latter is decidable. □

Our main result for fin-marked SWSTSs terms sufficient criteria under which the resilience problems are decidable.

Theorem 1 (decidability for fin-marked SWSTSs). Both resilience problems are decidable for fin-marked SWSTSs which are

(1) post*-effective if $\text{BAD} \in \mathcal{J}$, $\text{SAFE} \in \mathcal{I}$ (corresp. [14, Thm. 1]),
(2) lossy if $\text{BAD}, \text{SAFE} \in \mathcal{I}$.

The bounded resilience problem is decidable for fin-marked SWSTS^{B}s which are

(3) lossy and $\perp$-bounded if $\text{BAD} \in \mathcal{I}$, $\text{SAFE} \in \mathcal{J}$,
(4) $\perp$-bounded if $\text{BAD}, \text{SAFE} \in \mathcal{J}$.

Key Idea of the Proof. We compute a finite representation of $\text{post}^*(\text{INIT}) \cap \text{BAD}$ for checking inclusion in a decidable ideal I which is a predecessor set of SAFE.

The proof structure is shown in Fig. 5: Lemma 4 states that for post*-effective (lossy) fin-marked SWSTSs, a finite representation of $\text{post}^*(\text{INIT}) \cap \text{BAD}$ is computable, i.e., inclusion in a decidable ideal is decidable. In the case $\text{SAFE} \in \mathcal{I}$, the set $\text{pre}^{\leq k}(\text{SAFE})$ is a decidable ideal for every $k \geq 0$. (Lemma 5 shows the existence of bounds for the set of all predecessors of $\text{SAFE} \in \mathcal{J}$ provided that the SWSTS is $\perp$-bounded.) Proposition 3 shows that $\text{pre}^*(\text{SAFE})$ constitutes a decidable ideal in the case $\text{SAFE} \in \mathcal{J}$ if the SWSTS^B is $\perp$-bounded.

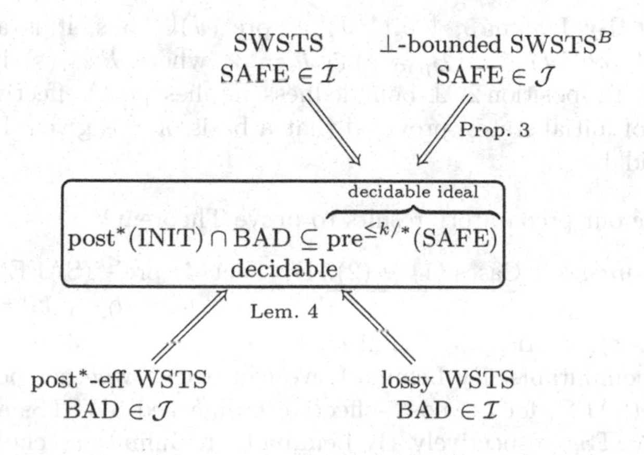

Fig. 5. Structure of the decidability proof for fin-marked SWSTSs.

The following lemma states that the inclusion of $\text{post}^*(\text{INIT}) \cap \text{BAD}$ in an decidable ideal is decidable if we consider post*-effective in the case $\text{BAD} \in \mathcal{J}$ or lossy fin-marked WSTSs in the case $\text{BAD} \in \mathcal{I}$.

Lemma 4 (checking inclusion). Let $\langle S, \leq, \rightarrow, \text{INIT}\rangle$ be a fin-marked WSTS, $\text{BAD} \subseteq S$, and $I \subseteq S$ be a decidable ideal. Then, it is decidable whether $\text{post}^*(\text{INIT}) \cap \text{BAD} \subseteq I$ provided that the fin-marked WSTS is (a) post*-effective and $\text{BAD} \in \mathcal{J}$, (b) lossy and $\text{BAD} \in \mathcal{I}$.

Proof (sketch). We compute a finite representation of $\text{post}^*(\text{INIT}) \cap \text{BAD}$ for checking inclusion in the decidable ideal I. To this aim, we use Lemma 3. In case (a), the finite representation is $B_{\text{post}} \cap \text{BAD}$ where B_{post} is a basis of $\uparrow \text{post}^*(\text{INIT})$. In case (b), the finite representation is $\downarrow \text{post}^*(\text{INIT}) \cap B_{\text{BAD}}$ where B_{BAD} is a basis of BAD. □

By the next lemma, $\perp$-boundedness implies that for any anti-ideal J, $\text{pre}^*(J)$ is an ideal and $\text{pre}^*(J) = \text{pre}^{\leq k}(J)$ for a $k \geq 0$.

Lemma 5 (existence of bounds). For every $\perp$-bounded SWSTS and every anti-ideal J, there exists a $k \geq 0$ s.t. $\text{pre}^*(J) = \uparrow \text{pre}^*(J) = \text{pre}^{\leq k}(J)$.

Proof (sketch). By Lemma 1, for every set A of states, there exists a $k_0 \geq 0$ s.t. $\uparrow \text{pre}^*(A) = \uparrow \text{pre}^{\leq k_0}(A)$. By strong compatibility and $\perp$-boundedness, there exists a constant $\ell \geq 0$ s.t. $\uparrow \text{pre}^{\leq k}(J) \subseteq \text{pre}^{\leq k+\ell}(J)$ for every anti-ideal J. Hence, $\text{pre}^*(J) \subseteq \uparrow \text{pre}^*(J) = \uparrow \text{pre}^{\leq k_0}(J) \subseteq \text{pre}^{\leq k_0+\ell}(J) \subseteq \text{pre}^*(J)$. □

The following proposition identifies sufficient prerequisites s.t. $\text{pre}^*(\text{SAFE})$ constitutes a decidable ideal in the case $\text{SAFE} \in \mathcal{J}$.

Proposition 3 (decidable ideals). For every $\perp$-bounded SWSTS^B and every decidable anti-ideal J, the set $\text{pre}^*(J)$ is a decidable ideal.

Proof (sketch). By Lemma 5, $\uparrow \text{pre}^*(J) = \text{pre}^*(J)$. Thus, it is an ideal. By Lemma 3, $s \notin \text{pre}^*(J) \iff B_{\text{post}}(s) \cap J = \varnothing$ where $B_{\text{post}}(s)$ is a basis of $\uparrow \text{post}^*(s)$. By Proposition 2, $\perp$-boundedness implies post*-effectiveness w.r.t. any finite set of initial states, provided that a basis of S is given. Hence, membership is decidable. □

We compile our preparatory results to prove Theorem 1.

Proof (of Theorem 1). Cases (1) & (2). By Fact 4, $\text{pre}^{\leq k}(\text{SAFE})$ is an ideal for every $k \geq 0$ since $\text{SAFE} \in \mathcal{I}$. For every $k \geq 0$, $\text{pre}^{\leq k+1}(\text{SAFE}) = \text{pre}(\text{pre}^{\leq k}(\text{SAFE})) \cup \text{SAFE}$. By Definition 8 and Fact 3, a basis of $\text{pre}^{\leq k}(\text{SAFE})$ is iteratively computable. By Lemma 4, we can decide whether $\text{post}^*(\text{INIT}) \cap \text{BAD} \subseteq \text{pre}^{\leq k}(\text{SAFE})$ for (1) post*-effective fin-marked SWSTSs and (2) lossy fin-marked SWSTSs, respectively. By Lemma 1, the infinite ascending sequence $\text{SAFE} \subseteq \text{pre}^{\leq 1}(\text{SAFE}) \subseteq \text{pre}^{\leq 2}(\text{SAFE}) \subseteq \ldots$ becomes stationary, i.e., there is a minimal $k_0 \geq 0$ s.t. $\text{pre}^{\leq k_0}(\text{SAFE}) = \text{pre}^*(\text{SAFE})$. By Lemma 2, we can also determine this k_0. Thus, we can determine the minimal number $k = k_{\min}$ s.t. $\text{post}^*(\text{INIT}) \cap \text{BAD} \subseteq \text{pre}^{\leq k}(\text{SAFE})$ (if it exists) and also whether it exists. Hence, we can decide the bounded resilience problem and given any k, we can check whether $k_{\min} \leq k$ to decide the explicit resilience problem.

Cases (3) & (4). By Lemma 5, for $\perp$-bounded SWSTSs, there exists a $k \geq 0$ s.t. $\text{pre}^*(\text{SAFE}) = \text{pre}^{\leq k}(\text{SAFE})$. Hence, checking bounded resilience is equivalent to testing inclusion in $\text{pre}^*(\text{SAFE})$. By Proposition 3, for $\perp$-bounded SWSTS^{B}s, $\text{pre}^*(\text{SAFE})$ is a decidable ideal since $\text{SAFE} \in \mathcal{J}$. By Lemma 4, we obtain that checking $\text{post}^*(\text{INIT}) \cap \text{BAD} \subseteq \text{pre}^*(\text{SAFE})$ is decidable for (3) lossy, $\perp$-bounded fin-marked SWSTS^{B}s and (4) post*-effective, $\perp$-bounded fin-marked SWSTS^{B}s, respectively. By Proposition 2, $\perp$-boundedness implies post*-effectiveness provided that a basis of the set of all states is given. □

4 Application to Graph Transformation Systems

We translate the results for WSTSs into the GTS setting.

The sets of positive and negative constraints are subsumed as ideal-based constraints.

Definition 12 (ideal-based constraints). We denote the set of positive (negative) constraints by $\mathcal{I}_c$ ($\mathcal{J}_c$). An *ideal-based constraint* is an element of $\mathcal{I}_c \cup \mathcal{J}_c$.

Recall that we consider the subgraph order as wqo. Path-length-boundedness on the graph class guarantees that the subgraph order yields a wqo.

Similarly to marked WSTSs, a marked GTS is a GTS together with a graph class closed under rule application and a subset INIT of graphs.

Definition 13 (marked GTS). A *marked GTS* is a tuple $\langle S, \mathcal{R}, \text{INIT}\rangle$ where S is a (possibly infinite) set of graphs, $\mathcal{R}$ is a GTS with $\Rightarrow_{\mathcal{R}} \subseteq S \times S$, and $\text{INIT} \subseteq S$. We speak of a marked GTS *of bounded path length*, shortly GTS^{bp}, if S is of bounded path length and there exist $I \in \mathcal{I}$, $J \in \mathcal{J}$ (in the class of all graphs) s.t. $S = I \cap J$.

Remark. Ususally, one considers S as a decidable anti-ideal, as, e.g., in [12]. Then, the basis of S is given by the empty graph. By allowing $S = I \cap J$, we can consider more arbitrary bases of graphs. This is relevant for lossiness and $\perp$-boundedness. A basis of S is given by $B_I \cap J$ where B_I is basis of I.

Example 3 (starry sky). The rules $\langle \varnothing \rightharpoonup (A) \rangle$ and $\langle 1(A) \rightharpoonup 1(A)\!\!\rightarrow\!\!\bullet \rangle$ together with the set of disjoint unions of unboundedly many star-shaped graphs (including isolated nodes) and any subset form a marked GTS^{bp}.

We formulate the resilience problems for marked GTSs.

EXPLICIT RESILIENCE PROBLEM FOR GTSS

Given: A marked GTS $\langle S, \mathcal{R}, \text{INIT}\rangle$, ideal-based constraints Safe, Bad, a natural number $k \geq 0$.

Question: $\forall G \in \text{INIT} : \forall (G \Rightarrow^* G' \models \text{Bad}) : \exists (G' \Rightarrow^{\leq k} G'' \models \text{Safe})$?

BOUNDED RESILIENCE PROBLEM FOR GTSs

Given: A marked GTS $\langle S, \mathcal{R}, \text{INIT}\rangle$, ideal-based constraints Safe, Bad.
Question: $\exists k \geq 0 : \forall G \in \text{INIT} : \forall (G \Rightarrow^* G' \models \text{Bad}) : \exists (G' \Rightarrow^{\leq k} G'' \models \text{Safe})$?

In the resilience problems for WSTSs, we considered ideal-based sets. We show that one can input ideal-based constraints instead.

Lemma 6 (ideal-based graph sets). Let $S = I \cap J$ be a graph class where $I \in \mathcal{I}$ and $J \in \mathcal{J}$. For every positive (negative) constraint c, $[\![c]\!] \in \mathcal{I}$ $(\mathcal{J})$.

Proof. By Fact 2, for every positive (negative) constraint c, the set $[\![c]\!]$ is an (anti-)ideal. Satisfaction ($\models$) of negative constraints is decidable. Let c be a positive constraint and $b = \bigvee_{G \in B} \exists G$ where B is a given basis of I. Then, $b \wedge c$ is a positive constraint. By Fact 1, we can compute a positive constraint c' s.t. $[\![c']\!] = [\![b \wedge c]\!]$ (in the class of all graphs) and c' is of the form $\bigvee_{1 \leq i \leq n} \exists G_i$ where $G_i \not\leq G_j$ for $i \neq j$. Since $J \in \mathcal{J}$, we can compute the set $\{G_i \in J : 1 \leq i \leq n\}$ which is a basis of $[\![c]\!]_S = \{G \in S : G \models c\}$. Hence, we can assume that a basis of $[\![c]\!]_S$ is given. □

Remark. More general constraints [10,16] do not constitute (anti-)ideals w.r.t. the subgraph order, in general. Consider, e.g., the "nested" constraint AllLoop = $\forall($ 1 ○, $\exists($ 1 ○⟲ $))$ expressing that every node has a loop. The graph consisting of one node and one loop satisfies the latter constraint. However, the bigger graph consisting of two nodes and one loop does not satisfy it. (The smaller graph consisting of a single node does not satisfy it either.) Thus, $[\![\text{AllLoop}]\!]$ is not an (anti-)ideal. Regarding the induced subgraph order [12], some "nested" constraints constitute ideals: The constraint $\exists (G, \bigwedge_{G^+ \in \text{Ext}(G)} \neg \exists (G \hookrightarrow G^+))$ expresses that the graph G is an induced subgraph of the considered graph. Here $\text{Ext}(G)$ is the set of all graphs G^+ obtained from G by adding one edge.

The following result of König & Stückrath terms a sufficient criterion for GTSs to be well-structured.

Lemma 7 (well-structured GTS [12, Prop. 7]). Every marked GTS^{bp} induces a marked SWSTS^B (equipped with the subgraph order).

In particular, they give an effective procedure for obtaining a basis of $\text{pre}(\uparrow \{G\})$ for every given graph G. Note that in [12], König & Stückrath consider labeled hypergraphs. However, the proof in our case is the same.

Convention. When speaking of a (fin-)marked GTS^{bp}, we consider the induced (fin-)marked SWSTS^B. We also adopt the terminology for "post*-effective", "lossy", and "$\perp$-bounded".

We apply our results from Sect. 3 to fin-marked GTS^{bp}.

Theorem 2 (decidability for fin-marked GTSs). Both resilience problems are decidable for fin-marked GTS^{bp}s which are

(1) post*-effective if Bad $\in \mathcal{J}_c$, Safe $\in \mathcal{I}_c$,
(2) lossy if Bad, Safe $\in \mathcal{I}_c$.

The bounded resilience problem is decidable for fin-marked GTS^{bp}s which are

(3) lossy and $\perp$-bounded if Bad $\in \mathcal{I}_c$, Safe $\in \mathcal{J}_c$,
(4) $\perp$-bounded if Bad, Safe $\in \mathcal{J}_c$.

Proof. By Lemma 7 [12], every fin-marked GTS^{bp} induces a fin-marked SWSTS^B. Thus, the statements of Theorem 1 apply to GTS^{bp}s with the respective requirements. By Lemma 6, one can input ideal-based constraints instead of ideal-based sets. □

We illustrate our decidability results by an example.

Example: Circular Process Protocol

In Fig. 6, the formalization as GTS of the circular process protocol in Sect. 1 is shown. Note that each `Clear`-rule is undefined on the node which has a c_i-labeled loop. In a rule application, this node will be deleted and recreated. Note also that each `Leave`-rule identifies two nodes.

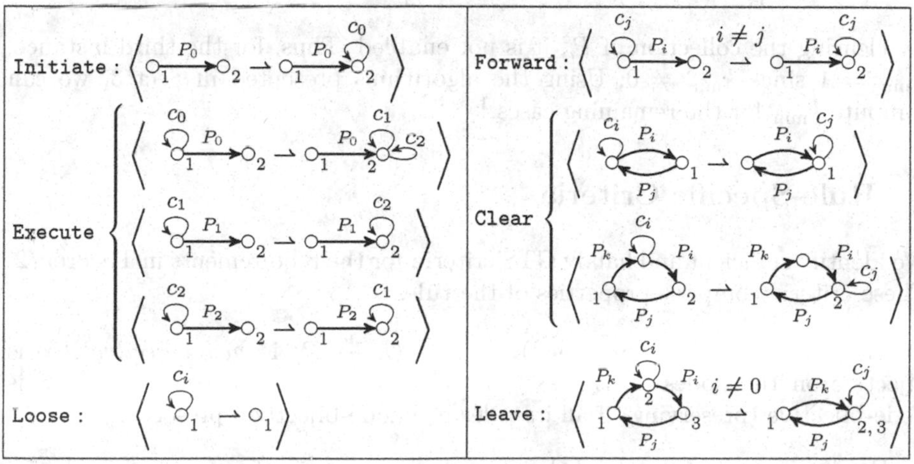

Fig. 6. Rules of the circular process protocol.

We consider all graphs with arbitrarily many commands of any kind (labeled with c_0, c_1, or c_2) in any collection, fitting to one of the topologies shown in Fig. 1. This graph class is of bounded path length. A basis of this graph class is given by the topologies without any commands as in Fig. 1. The marked GTS^{bp} is lossy since the rules for loosing a command c_i may be applied to any graph containing a command c_i. It is $\perp$-bounded since we can reach a graph with the same topology but containing no commands by (i) initiating a command c_0, (ii) forwarding it to the collection of P_0, (iii) clearing all collections one after another, and (iv) loosing the only remaining command. By Proposition 2, it is post*-effective.

The example constraints for BAD and SAFE in Sect. 1 can be expressed as positive/negative constraints:

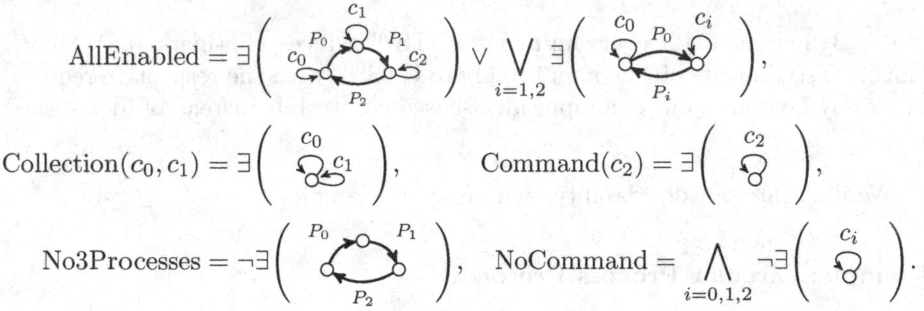

$$\text{AllEnabled} = \exists\left(c_1, P_0, P_1, c_0, c_2, P_2 \right) \vee \bigvee_{i=1,2} \exists\left(c_0, P_0, c_i, P_i \right),$$

$$\text{Collection}(c_0, c_1) = \exists\left(c_0, c_1 \right), \qquad \text{Command}(c_2) = \exists\left(c_2 \right),$$

$$\text{No3Processes} = \neg\exists\left(P_0, P_1, P_2 \right), \qquad \text{NoCommand} = \bigwedge_{i=0,1,2} \neg\exists\left(c_i \right).$$

It can be verified that the given k's in the following table are minimal.

BAD	¬AllEnabled	Command(c_2)	AllEnabled	NoCommand
SAFE	AllEnabled	Collection(c_0, c_1)	¬AllEnabled	No3Processes
k	6	4	1	5

By clearing the collection of P_i, it is not enabled. Thus, for the third instance, $k_{\min} = 1$ since $k_{\min} \neq 0$. Using the algorithms presented in Sect. 3, we can compute $k_{\min}$ for the remaining cases.[1]

5 Rule-Specific Criteria

We identify sufficient and handy GTS criteria for the requirements in Theorem 2. These criteria comprise properties of the rules.

Definition 14 (rule properties). A rule $\langle L \overset{p}{\rightharpoonup} R\rangle$ is *node-bijecitve* if p is bijective on the nodes. It is *preserving* if p is total and injective. A GTS is node-bijective (preserving) if all its rules are node-bijective (preserving).

For lossiness and $\perp$-boundedness, we consider sets of rules contained in a GTS in order to reach smaller graphs (but not smaller than basis elements).

Assumption. Let S be a graph class over Λ and B a basis of S.

[1] If SAFE $\in \mathcal{J} \setminus \mathcal{I}$, our method provides only the answer whether there is a bound k.

Each lossy rule deletes one item (node or edge/loop) outside of a basis element.[2] Therefore they are constructed s.t. in each rule, a basis element is present.

Construction 1 (lossy rules). The set $\mathcal{R}_{\text{loss}}(S)$ of *lossy rules* w.r.t. S are constructed as follows.

(1) For graphs $G \in B$, $H \in \mathcal{H}_\Lambda$, the set $\mathcal{C}(G, H)$ is defined as all graphs C s.t. $\exists\langle G \hookrightarrow C, H \hookrightarrow C\rangle$ jointly surjective, and

$$\mathcal{H}_\Lambda = \left\{ (x),\ (x) \xrightarrow{a} (y),\ (x)^{\circlearrowleft a} \;\middle|\; x, y \in \Lambda_V, a \in \Lambda_E \right\}.$$

(2) For every graph $C \in \mathcal{C}(G, H) \cap S$, every rule $\langle C \xrightarrow{p} p(C)\rangle$ where p is undefined on exactly one item (node or edge) which is not in (the image of) G and the identity otherwise, is a lossy rule.

A similar idea works for $\perp$-boundedness. Each bottom rule either deletes a node outside of a basis element, or deletes and recreates a node (with its incident edges) of a basis element.

Construction 2 (bottom rules). The set $\mathcal{R}_\perp(S)$ of *bottom rules* w.r.t. S are constructed as follows. For every basis element $G \in B$ and

(1) for every label $x \in \Lambda_V$ s.t. $G + (x) \in S$, the rule $\langle G + (x) \xrightarrow{p} G\rangle$ where p is undefined on the node (x) and the identity otherwise, is a bottom rule,[3]
(2) for every node $v \in V_G$, the rule $\langle G \xrightarrow{p} G\rangle$ where p is undefined on v and its incident edges, and the identity otherwise, is a bottom rule.

For $\perp$-boundedness, we additionally restrict the graph class. A graph class is *node-bounded* if the number of nodes in any graph of the class is bounded.

The following result shows that the rule-specific criteria are sufficient.

Theorem 3 (criteria). A marked GTS^{bp} $\langle S, \mathcal{R}, \text{INIT}\rangle$ is

(1) post*-effective if (INIT is finite and) $\mathcal{R}$ is node-bijective or preserving,
(2) lossy if $\mathcal{R}_{\text{loss}}(S) \subseteq \mathcal{R}$,
(3) $\perp$-bounded if S is node-bounded and $\mathcal{R}_\perp(S) \subseteq \mathcal{R}$.

Proof (sketch). (1) If $\mathcal{R}$ is preserving, the statement follows by Fact 3 since $\uparrow \text{post}^*(\text{INIT}) = \uparrow \text{INIT}$. If $\mathcal{R}$ is node-bijective, the statement follows by the reduction in the proof of [3, Prop. 10]. For any graph G, a Petri net with initial marking is constructed s.t. reachability and the wqo correspond to $G \Rightarrow^*_{\mathcal{R}}$ and the subgraph order, respectively. Petri nets are post*-effective, see Example 2.

[2] In [12], "lossy rules" w.r.t. the minor order, i.e., edge contraction rules, are considered in order to obtain well-structuredness for GTSs.
[3] The symbol "+" denotes the disjoint union of graphs.

(2) Using the lossy rules, we can delete any item in $G \setminus i(G_B)$ for every $G \in S$ and $G \geq G_B \in B$ where B is a basis of S and $i : G_B \hookrightarrow G$. Hence, a sequence of node and edge deletions from G to any smaller graph $G' \in S$ is also feasible via the lossy rules.

(3) Since S is node-bounded, we can delete all nodes in $G \setminus i(G_B)$ in a bounded number of steps for every $G \in S$ and $G \geq G_B \in B$ where B is a basis of S and $i : G_B \hookrightarrow G$. Then, by applying the rules for deleting and recreating, we can reach any smaller basis element in $\leq \max_{G_B \in B} |V_{G_B}|$ steps. □

Remark. A lossy/bottom rule intended for node deletion will delete dangling edges outside of (the image of) a basis element. A bottom rule of the "second type" is intended to delete dangling edges and restore items of the basis element.

Example 4 (criteria). (1) The GTS in Fig. 6 (circular process protocol) without the `Clear`- and `Leave`-rules is node-bijective. (2) The `Loose`-rules in Fig. 6 can be adapted (see Fig. 7) s.t. they fit in our definition of lossy rules, taking into account the three basis elements in Fig. 1.

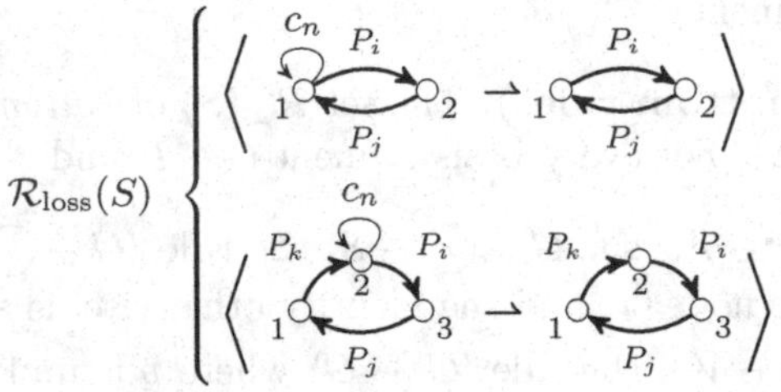

$$\mathcal{R}_{\text{loss}}(S) \left\{ \begin{array}{l} \langle \, G_1 \rightarrow G_1' \, \rangle \\ \langle \, G_2 \rightarrow G_2' \, \rangle \end{array} \right.$$

Fig. 7. The lossy rules of the circular process protocol.

(3) We adapt Example 3 (starry sky) s.t. the criterion for $\perp$-boundedness is fulfilled. Let D_n be the disjoint union of n A-labeled nodes and D_1^{loop} an A-labeled node with a single loop. We restrict the graph class to all graphs with exactly n A-labeled nodes and unboundedly many loops. The single basis element is the graph D_n. Consider the rule $\langle D_1 \hookrightarrow D_1^{\text{loop}} \rangle$ and the bottom rule $\langle D_n \xrightarrow{p} D_n \rangle$, i.e., deleting and recreating one node in D_n.

6 Related Concepts

The concept of resilience [11,17] is broadly used with varying definitions.

For modeling systems, we use SPO graph transformation as in [7].

Abdulla et al. [1] show the decidability of ideal reachability (coverability), eventuality properties and simulation in (labeled) SWSTSs. We use the presented algorithm as an essential integrant of our decidability proof.

Finkel and Schnoebelen [9] show that the concept of well-structuredness is ubiquitous in computer science by providing a large class of example models. They give several decidability results for well-structured systems with varying notions of compatibility, also generalizing the algorithm of [1] to WSTSs.

König and Stückrath [12] extensively study the well-structuredness of GTSs regarding three types of wqos (minor, subgraph, induced subgraph). All GTSs are strongly well-structured on graphs of bounded path length w.r.t. the subgraph order. This result enables us to apply our abstract results to GTSs. They regard Q-restricted WSTSs whose state sets have not to be a wqo but rather a subset Q of the states is a wqo. König & Stückrath develop a backwards algorithm based on [9] for Q-restricted WSTSs (GTSs).

Bertrand et al. [3] study the decidability of reachability and coverability for GTSs using, in parts, well-structuredness. A variety of rule-specific restrictions is investigated, e.g., containedness of node/edge-deletion rules. We use one of their results to obtain a sufficient criterion for post*-effectiveness. In contrast to [3], we stay in the framework of well-structured GTSs.

In Fig. 8, the main results of this paper (bold boxes) are placed in the context of known results. The arrows ($\longrightarrow$) mean "used for", the hooked arrows ($\hookrightarrow$) mean "instance of" or "generalized to". Our result for SWSTSs uses the well-known coverability algorithm [1,9] for (S)WSTSs which exploits the Noetherian property (a general concept for algebraic structures). For $\perp$-bounded SWSTSs, we also employ the Noetherian property. On the level of GTSs, we use the predecessor-basis procedure of [12]. To the best of the author's knowledge, the considered notion of resilience was first studied in [14]. We extended the latter to a systematic investigation. The result for SWSTSs in [14] (Thm. 1) corresponds to case (1) of our Theorem 1. The result for GTSs in [14] (Thm. 2) is slightly less general than case (1) of our Theorem 2. For case (1) of Theorem 3, we use a result in [3] and well-known results for Petri nets [8].

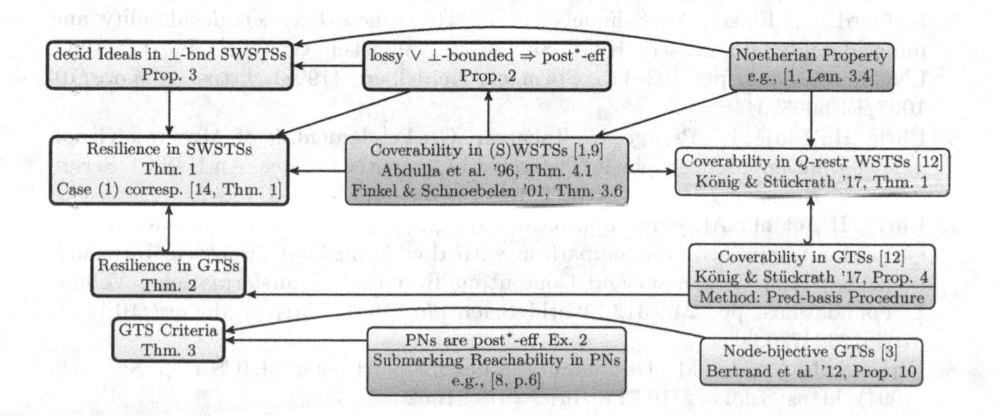

Fig. 8. Our results in the context of the theory of WSTSs and (ideal) reachability.

7 Conclusion

We provided a systematic investigation on resilience problems obtaining decidability results for subclasses of marked GTSs by using the concept of well-structuredness. The used well-quasi-order on graphs is the subgraph order, i.e., a prerequisite is the path-length-boundedness on the graph class. The requirements for decidability are post*-effectiveness or a kind of unreliability (lossy, $\perp$-bounded). We identified sufficient rule-specific criteria for these requirements.

For future work, we will consider (1) possibilities of a modified approach for typed graphs [6], (2) other proof methods to handle nested constraints [10], and (3) other well-quasi-orders on graphs, e.g., the induced subgraph order [12].

Acknowledgment. I am grateful to Annegret Habel, Nick Würdemann, and the anonymous reviewers for their helpful comments.

References

1. Abdulla, P.A., Cerans, K., Jonsson, B., Tsay, Y.K.: General decidability theorems for infinite-state systems. In: Proceedings of the LICS 1996, pp. 313–321. IEEE (1996). https://doi.org/10.1109/LICS.1996.561359
2. Apt, K.R., Olderog, E.: Verification of Sequential and Concurrent Programs. Texts and Monographs in Computer Science. Springer, Heidelberg (1991). https://doi.org/10.1007/978-1-4757-4376-0
3. Bertrand, N., Delzanno, G., König, B., Sangnier, A., Stückrath, J.: On the decidability status of reachability and coverability in graph transformation systems. In: 23rd International Conference on Rewriting Techniques and Applications (RTA 2012). LIPIcs, vol. 15, pp. 101–116 (2012). https://doi.org/10.4230/LIPIcs.RTA.2012.101
4. Ding, G.: Subgraphs and well-quasi-ordering. J. Graph Theory **16**(5), 489–502 (1992). https://doi.org/10.1002/jgt.3190160509
5. Dufourd, C., Finkel, A., Schnoebelen, P.: Reset nets between decidability and undecidability. In: Larsen, K.G., Skyum, S., Winskel, G. (eds.) ICALP 1998. LNCS, vol. 1443, pp. 103–115. Springer, Heidelberg (1998). https://doi.org/10.1007/BFb0055044
6. Ehrig, H., Ehrig, K., Prange, U., Taentzer, G.: Fundamentals of Algebraic Graph Transformation. Monographs in Theoretical Computer Science. An EATCS Series, Springer, Heidelberg (2006). https://doi.org/10.1007/3-540-31188-2
7. Ehrig, H., et al.: Algebraic approaches to graph transformation - part II: single pushout approach and comparison with double pushout approach. In: Handbook of Graph Grammars and Computing by Graph Transformations, Volume 1: Foundations, pp. 247–312. World Scientific (1997). https://doi.org/10.1142/9789812384720_0004
8. Esparza, J., Nielsen, M.: Decidability issues for petri nets. BRICS Rep. Ser. **1**(8) (1994). https://doi.org/10.7146/brics.v1i8.21662
9. Finkel, A., Schnoebelen, P.: Well-structured transition systems everywhere! Theor. Comput. Sci. **256**(1–2), 63–92 (2001). https://doi.org/10.1016/S0304-3975(00)00102-X

10. Habel, A., Pennemann, K.: Correctness of high-level transformation systems relative to nested conditions. Math. Struct. Comput. Sci. **19**(2), 245–296 (2009). https://doi.org/10.1017/S0960129508007202
11. Jackson, S., Ferris, T.L.J.: Resilience principles for engineered systems. Syst. Eng. **16**, 152–164 (2013). https://doi.org/10.1002/sys.21228
12. König, B., Stückrath, J.: Well-structured graph transformation systems. Inf. Comput. **252**, 71–94 (2017). https://doi.org/10.1016/j.ic.2016.03.005
13. Özkan, O.: Decidability of resilience for well-structured graph transformation systems. Technical report, Department of Computing Science, University of Oldenburg (2022). https://uol.de/fs/publikationen#c352844
14. Özkan, O., Würdemann, N.: Resilience of well-structured graph transformation systems. In: Proceedings of 12th International Workshop on Graph Computational Models. EPTCS, vol. 350, pp. 69–88 (2021). https://doi.org/10.4204/EPTCS.350.5
15. Poskitt, C.M., Plump, D.: Hoare-style verification of graph programs. Fundam. Informaticae **118**(1–2), 135–175 (2012). https://doi.org/10.3233/FI-2012-708
16. Rensink, A.: Representing first-order logic using graphs. In: Ehrig, H., Engels, G., Parisi-Presicce, F., Rozenberg, G. (eds.) ICGT 2004. LNCS, vol. 3256, pp. 319–335. Springer, Heidelberg (2004). https://doi.org/10.1007/978-3-540-30203-2_23
17. Trivedi, K.S., Kim, D.S., Ghosh, R.: Resilience in computer systems and networks. In: Proceedings of the ICCAD 2009, pp. 74–77. IEEE/ACM (2009). https://doi.org/10.1145/1687399.1687415

Probabilistic Metric Temporal Graph Logic

Sven Schneider[(✉)], Maria Maximova, and Holger Giese

Hasso Plattner Institute, University of Potsdam, Potsdam, Germany
{sven.schneider,maria.maximova,holger.giese}@hpi.de

Abstract. Cyber-physical systems often encompass complex concurrent behavior with timing constraints and probabilistic failures on demand. The analysis whether such systems with probabilistic timed behavior adhere to a given specification is essential. When the states of the system can be represented by graphs, the rule-based formalism of Probabilistic Timed Graph Transformation System (PTGTSs) can be used to suitably capture structure dynamics as well as probabilistic and timed behavior of the system. The model checking support for PTGTSs w.r.t. properties specified using Probabilistic Timed Computation Tree Logic (PTCTL) has been already presented. Moreover, for timed graph-based runtime monitoring, Metric Temporal Graph Logic (MTGL) has been developed for stating metric temporal properties on identified subgraphs and their structural changes over time.

In this paper, we (a) extend MTGL to the Probabilistic Metric Temporal Graph Logic (PMTGL) by allowing for the specification of probabilistic properties, (b) adapt our MTGL satisfaction checking approach to PTGTSs, and (c) combine the approaches for PTCTL model checking and MTGL satisfaction checking to obtain a Bounded Model Checking (BMC) approach for PMTGL. In our evaluation, we apply an implementation of our BMC approach in AutoGraph to a running example.

Keywords: cyber-physical systems · probabilistic timed systems · qualitative analysis · quantitative analysis · bounded model checking

1 Introduction

Cyber-physical systems often encompass complex concurrent behavior with timing constraints and probabilistic failures on demand [23,26]. Such behavior can then be captured in terms of probabilistic timed state sequences (or spaces) where time may elapse between successive states and where each step in such a sequence has a designated probability. The analysis whether such systems adhere to a given specification describing admissible or desired system behavior is essential in a model-driven development process.

Graph Transformation Systems (GTSs) [10] can be used for the modeling of systems when each system state can be represented by a graph and when

N. Behr and D. Strüber (Eds.): ICGT 2022, LNCS 13349, pp. 58–76, 2022.
https://doi.org/10.1007/978-3-031-09843-7_4

all changes of such states to be modeled can be described using the rule-based approach to graph transformation. Moreover, timing constraints based on clocks, guards, invariants, and clock resets have been combined with graph transformation in Timed Graph Transformation Systems (TGTSs) [4] and probabilistic aspects have been added to graph transformation in Probabilistic Graph Transformation Systems (PGTSs) [18]. Finally, the formalism of PTGTSs [21] combines timed and probabilistic aspects similar to Probabilistic Timed Automata (PTA) [20] and offers model checking support w.r.t. PTCTL [19,20] properties employing the PRISM model checker [19]. The usage of PTCTL allows for stating probabilistic real-time properties on the induced PTGT state space where each graph in the state space is labeled with a set of Atomic Propositions (APs) obtained by evaluating that graph w.r.t. e.g. some property specified using Graph Logic (GL) [12,26].[1]

However, structural changes over time in the state space cannot always be directly specified using APs that are *locally* evaluated for each graph.[2] To express such structural changes over time, MTGL [11,26] has been introduced based on GL. Using MTGL conditions, an unbounded number of subgraphs can be tracked over timed graph transformation steps in a considered state sequence once bindings have been established for them via graph matching. Moreover, MTGL conditions allow to identify graphs where certain elements have just been added to (removed from) the current graph. Similarly to MTGL, for runtime monitoring, Metric First-Order Temporal Logic (MFOTL) [3] (with limited support by the tool MONPOLY) and the non-metric timed logic EAGLE [1,14] (with full tool support) have been introduced operating on sets of relations and JAVA objects as state descriptions, respectively. In [7–9], sequences are monitored using model queries to identify complex event patterns of interest. In [15], the Quantified Temporal Logic (QTL) is introduced, which supports bindings and state representation similarly to MFOTL but supports only properties referring to the past and does not support metric bounds in its temporal operators. Besides these logic-based approaches, a multitude of further techniques have been developed based on e.g. automata for monitoring the system's behavior in the context of runtime monitoring (see [2] for a survey). Finally, note that runtime monitoring (as well as MTGL) focuses on the specification of single sequences whereas the analysis of probabilistic effects is meaningful only when considering a system with a branching behavior (due to non-determinism and/or probabilism).

Obviously, both logics PTCTL and MTGL have distinguishing key strengths but also lack bindings on the part of PTCTL and an operator for expressing

[1] Furthermore, UPPAAL [28] is an analysis tool for timed automata featuring model checking support for standard metric temporal properties and simulation-based support for cyber-physical systems extending timed automata but does not support probabilistic analysis. Lastly, the MODEST toolset [6,13] also provides analysis support for more complex cyber physical systems representable by e.g. stochastic hybrid automata w.r.t. probabilistic metric temporal requirements.

[2] For example, tracking (structural changes to) individual graph elements allows to express and analyze deadlines for each individual graph element whereas APs cannot distinguish between individual graph elements and hence cannot help in mapping each of them to their corresponding deadline.

probabilistic requirements on the part of MTGL.[3] Furthermore, specifications using both, PTCTL and MTGL conditions, are insufficient as they cannot capture phenomena based on probabilistic effects and the tracking of subgraphs at once. Hence, a more complex combination of both logics is required. Moreover, realistic systems often induce infinite or intractably large state spaces prohibiting the usage of standard model checking techniques. Bounded Model Checking (BMC) has been proposed in [16] for such cases implementing an on-the-fly analysis. Similarly, reachability analysis w.r.t. a bounded number of steps or a bounded duration have been discussed in [17].

To combine the strengths of PTCTL and MTGL, we introduce PMTGL by enriching MTGL with an operator for expressing probabilistic requirements as in PTCTL. Moreover, we present a BMC approach for PTGTSs w.r.t. PMTGL properties by combining the PTCTL model checking approach for PTGTSs from [21] (which is based on a translation of PTGTSs into PTA) with the satisfaction checking approach for MTGL from [11,26]. In our approach, we just support *bounded* model checking since the binding capabilities of PMTGL conditions require non-local satisfaction checks taking possibly the entire history of a (finite) path into account as for MTGL conditions. However, we obtain even *full* model checking support for two cases: (a) for the case of finite loop-free state spaces and (b) for the case where the given PMTGL condition does not need to be evaluated beyond a maximal time bound.

As a running example, we consider a system in which a sender decides to send messages at nondeterministically chosen time points, which have then to be transmitted to a receiver via a network of routers within a given time bound. In this system, transmission of messages is subject to a probabilistic failure on demand requiring a retransmission of a message that was lost at an earlier transmission attempt. For this scenario, we employ PMTGL to express the desired system property of timely message reception. Firstly, using the capabilities inherited from MTGL, we identify messages that have just been sent, track them over time, and check whether their individual deadlines are met. Secondly, using the probabilistic operator inherited from PTCTL, we specify lower and upper bounds for the probability with which such an identified message is transmitted to the receiver before the deadline expires. During analysis, we are interested in determining the *expected best-case and worst-case* probabilities for a successful multi-hop message transmission from sender to receiver. For our evaluation, we also consider further variants of the considered scenario where messages are dropped after n transmission failures.

This paper is structured as follows. In Sect. 2, we recall the formalism of PTA. In Sect. 3, we discuss further preliminaries including graph transformation, graph conditions, and the formalism of PTGTSs. In Sect. 4, we recall MTGL and present the extension of MTGL to PMTGL in terms of syntax and semantics. In Sect. 5, we present our BMC approach for PTGTSs w.r.t. PMTGL properties. In Sect. 6, we evaluate our BMC approach by applying its implementation in

[3] PTCTL model checkers such as Prism do not support the branching capabilities of PTCTL as of now due to the complexity of the corresponding algorithms.

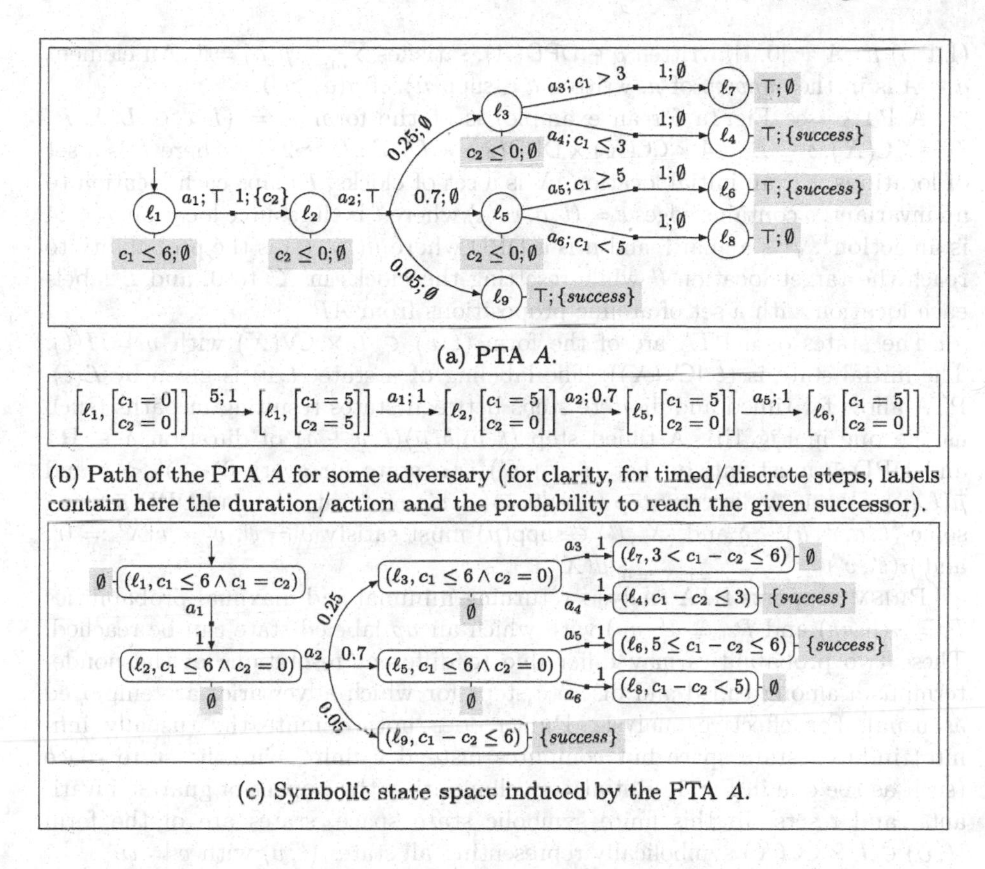

(a) PTA A.

(b) Path of the PTA A for some adversary (for clarity, for timed/discrete steps, labels contain here the duration/action and the probability to reach the given successor).

(c) Symbolic state space induced by the PTA A.

Fig. 1. PTA A, one of its paths, and its symbolic state space.

the tool AutoGraph [24] to our running example. Finally, in Sect. 7, we close the paper with a conclusion and an outlook on future work. Further details are given in a technical report [25].

2 Probabilistic Timed Automata

We briefly review PTA [20], which combine the use of clocks to capture real-time phenomena and probabilism to approximate/describe the likelihood of outcomes of certain steps, and PTA analysis as supported by Prism [19].

For a set of clocks X, clock constraints $\psi \in \mathsf{CC}(X)$ also called *zones* are finite conjunctions of clock comparisons $c_1 \sim n$ and $c_1 - c_2 \sim n$ where $c_1, c_2 \in X$, $\sim \in \{<, >, \leq, \geq\}$, and $n \in \mathbf{N} \cup \{\infty\}$. A clock valuation $(v : X \to \mathbf{R}_0^+) \in \mathsf{CV}(X)$ satisfies a zone ψ, written $v \models \psi$, as expected. The initial clock valuation $\mathsf{ICV}(X)$ maps all clocks to 0. For a clock valuation v and a set of clocks X', $v[X' := 0]$ is the clock valuation mapping the clocks from X' to 0 and all other clocks according to v. For a clock valuation v and a duration $\delta \in \mathbf{R}_0^+$, $v + \delta$ is the clock valuation mapping each clock x to $v(x) + \delta$. A Discrete Probability Distribution

(DPD) $\mu : A \rightarrow [0,1]$, written $\mu \in \mathsf{DPD}(A)$, satisfies $\sum_{a \in A} \mu(a) = 1$. An element $a \in A$ is in the *support* of μ, written $a \in \mathsf{supp}(\mu)$, if $\mu(a) > 0$.

A PTA (see Fig. 1a for an example) is of the form $A = (L, \bar{\ell} \in L, X, I : L \rightarrow \mathsf{CC}(X), \delta \subseteq L \times \mathcal{A} \times \mathsf{CC}(X) \times \mathsf{DPD}(2^X \times L), \mathcal{L} : L \rightarrow 2^{AP})$ where L is a set of locations, $\bar{\ell}$ is an initial location, X is a set of clocks, I maps each location to an invariant, δ contains edges $e = (\ell, a, \psi, \mu)$ where ℓ is the source location, $a \in \mathcal{A}$ is an action[4], ψ is a guard, and μ is a DPD where $\mu(X', \ell')$ is the probability to reach the target location ℓ' while resetting the clocks in X' to 0, and $\mathcal{L}$ labels each location with a set of atomic propositions from AP.

The states of a PTA are of the form $(\ell, v) \in L \times \mathsf{CV}(X)$ with $v \models I(\ell)$. The initial state is $(\bar{\ell}, \mathsf{ICV}(X))$. The labeling of a state (ℓ, v) is given by $\mathcal{L}(\ell)$. PTA allow for timed and discrete steps between states resulting in paths (such as the one in Fig. 1b). A timed step $(\ell, v)[\delta, \overline{\mu}\rangle(\ell, v + \delta)$ of duration $\delta \in \mathbf{R}^+$ and DPD $\overline{\mu}$ must satisfy that $(\ell, v + \delta')$ is a state for every $0 < \delta' < \delta$ and $\overline{\mu}(\ell, v + \delta) = 1$. A discrete step $(\ell, v)[0, \overline{\mu}\rangle(\ell', v')$ of duration 0 and DPD $\overline{\mu}$ using some $(\ell, a, \psi, \mu) \in \delta$ and $(X', \ell') \in \mathsf{supp}(\mu)$ must satisfy $v \models \psi$, $v' = v[X' := 0]$, and $\overline{\mu}(\ell', v') = \sum_{X', v' = v[X' := 0]} \mu(X', \ell')$.[5]

PRISM supports PTA analysis returning minimal and maximal probabilities $\mathcal{P}_{\mathsf{min}=?}(\mathsf{F}\ ap)$ and $\mathcal{P}_{\mathsf{max}=?}(\mathsf{F}\ ap)$ with which an *ap* labeled state can be reached. These two probabilities may differ due to different resolutions of the nondeterminism among timed and discrete steps for which adversaries are employed as usual. For effective analysis, PRISM does not compute the (usually infinite) induced state space but computes instead a finite symbolic state space (such as the one in Fig. 1c) intuitively eliminating the impact of guards, invariants, and resets. In this finite symbolic state space, states are of the form $(\ell, \psi) \in L \times \mathsf{CC}(X)$ symbolically representing all states (ℓ, v) with $v \models \psi$.

For example, the PTA A from Fig. 1a (for which adversaries only decide how much time to spend in location ℓ_1), $\mathcal{P}_{\mathsf{max}=?}(\mathsf{F}\ success) = 0.7 + 0.05$ using a probability maximizing adversary that lets $5 \leq \delta \leq 6$ time units elapse in ℓ_1 (0.25 is not added as ℓ_4 is not reachable using this adversary). Similarly, $\mathcal{P}_{\mathsf{min}=?}(\mathsf{F}\ success) = 0.05$ using a probability minimizing adversary that lets $3 < \delta < 5$ time units elapse in ℓ_1 (0.25 and 0.7 are not added as ℓ_4 and ℓ_6 are not reachable using this adversary).

---

[4] Actions in edges can be used to describe the purpose of the edge during modeling but also allow to define PTA based on a parallel composition of multiple PTA where these PTA synchronize on common actions. In [21], actions are used to store in PTA edges information about the PTGT steps from which they originate (as also stated in Sect. 3) while actions (being a standard part of PTA) do not play an important role in our BMC approach presented in Sect. 5.

[5] Here, $\overline{\mu}$ captures the unique successor state for timed steps and sums (possibly multiple non-zero) probabilities of domain elements of μ leading to a common successor state for discrete steps (in particular for the case that clocks to be reset have already the value 0).

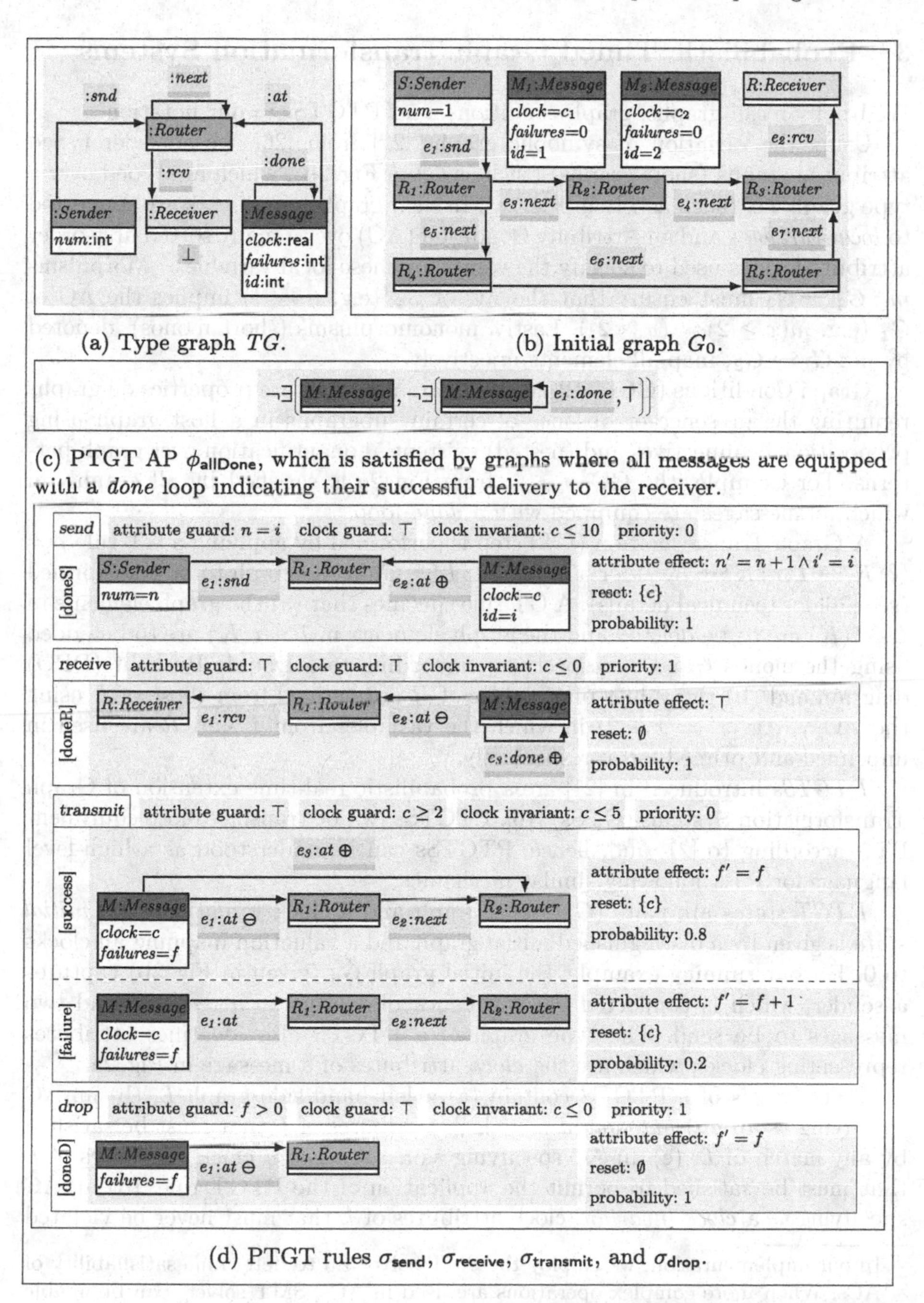

(a) Type graph TG.

(b) Initial graph G_0.

(c) PTGT AP ϕ_{allDone}, which is satisfied by graphs where all messages are equipped with a *done* loop indicating their successful delivery to the receiver.

(d) PTGT rules σ_{send}, $\sigma_{\mathsf{receive}}$, $\sigma_{\mathsf{transmit}}$, and σ_{drop}.

Fig. 2. Components of the PTGTS for the running example.

3 Probabilistic Timed Graph Transformation Systems

We briefly recall graphs, graph conditions, and PTGTSs in our notation.

Using the variation of symbolic graphs [22] from [26], we consider typed attributed graphs (short graphs) (such as G_0 in Fig. 2b), which are typed over a type graph TG (such as TG in Fig. 2a). In such graphs, attributes are connected to *local variables* and an Attribute Condition (AC) over a many sorted first-order attribute logic is used to specify the values for these local variables.[6] Morphisms $m : G_1 \rightarrow G_2$ must ensure that the AC of G_2 (e.g. $y = 4$) implies the AC of G_1 (e.g. $m(x \geq 2) = (y \geq 2)$). Lastly, monomorphisms (short monos), denoted by $m : G_1 \hookrightarrow G_2$, map all elements injectively.

Graph Conditions (GCs) [12,26] of GL are used to state properties on graphs requiring the presence or absence of certain subgraphs in a host graph using propositional connectives and (nested) existential quantification over graph patterns. For example, the GC ϕ_{allDone} from Fig. 2c is satisfied by all graphs, in which all messages are equipped with a *done* loop.

A Graph Transformation (GT) step is performed by applying a GT rule $\rho = (\ell : K \hookrightarrow L, r : K \hookrightarrow R, \gamma)$ for a match $m : L \hookrightarrow G$ on the graph to be transformed (see [26] for technical details). A GT rule specifies that (a) the graph elements in $L - \ell(K)$ are to be deleted and the graph elements in $R - r(K)$ are to be added using the monos ℓ and r, respectively, according to a Double Pushout (DPO) diagram and (b) the values of variables of R are derived from those of L using the AC γ (e.g. $x' = x + 2$) in which the variables from L and R are used in unprimed and primed form, respectively.[7]

PTGTSs introduced in [21] are a probabilistic real-time extension of Graph Transformation Systems (GTSs) [10]. PTGTSs can be translated into equivalent PTA according to [21] and, hence, PTGTSs can be understood as a high-level language for PTA following similar mechanics.

PTGT states are pairs (G, v) of a graph and a clock valuation. The *initial state* is given by a distinguished initial graph and a valuation mapping all clocks to 0. For our running example, the initial graph G_0 (given in Fig. 2b) captures a sender, which is connected via a network of routers to a receiver, and two messages to be send. The type graph of a PTGTS also identifies attributes representing clocks, which are the *clock* attributes of a message in Fig. 2a.

PTGT rules of a PTGTS contain (a) a left-hand side graph L, (b) an AC specifying as an *attribute guard* non-clock attributes of L that must be satisfied by any match of L, (c) an AC specifying as a *clock guard* clock attributes of L that must be satisfied to permit the application of the PTGT rule, (d) an AC specifying as a *clock invariant* clock attributes of L that must never be violated

[6] In our implementation, we employ the SMT solver Z3 to determine satisfiability of ACs. When more complex operations are used in ACs, SMT solvers can be unable to return definitive judgements in time, which does not happen for the running example. If this case would occur, the users would be inform accordingly.

[7] Nested application conditions given by GCs to further restrict rule applicability are straightforwardly supported by our approach but, to improve readability, not used in the running example and omitted subsequently.

for a match of L, (e) a natural number describing a *priority* preventing the application of the PTGT rule when a PTGT rule with higher priority can be applied, and (f) a nonempty set of tuples of the form $(\ell{:}K \hookrightarrow L, r{:}K \hookrightarrow R, \gamma, C, p)$ where (ℓ, r, γ) is an underlying GT rule, C is a set of clocks contained in R to be reset, and p is a real-valued probability from $[0, 1]$ where the probabilities of all such tuples must add up to 1.

For our running example, the PTGTS contains the four PTGT rules from Fig. 2d. The PTGT rules σ_{send}, $\sigma_{\mathsf{receive}}$, and σ_{drop} have each a unique underlying GT rule $\rho_{\mathsf{send,doneS}}$, $\rho_{\mathsf{receive,doneR}}$, and $\rho_{\mathsf{drop,doneD}}$, respectively, whereas the PTGT rule $\sigma_{\mathsf{transmit}}$ has two alternative underlying GT rules $\rho_{\mathsf{transmit,success}}$ and $\rho_{\mathsf{transmit,failure}}$. For each of these underlying GT rules, we depict the graphs L, K, and R in a single graph where graph elements to be removed and to be added are annotated with $\ominus$ and $\oplus$, respectively. Further information about the PTGT rule and its underlying GT rules are given in gray boxes. The PTGT rule σ_{send} is used to push the next message into the network by connecting it to the router that is adjacent to the sender. Thereby, the attribute *num* of the sender is used to push the messages in the order of their *id* attributes. The PTGT rule $\sigma_{\mathsf{receive}}$ has the higher priority 1 and is used to pull a message from the router that is adjacent to the receiver by marking the message with a *done* loop. The PTGT rule $\sigma_{\mathsf{transmit}}$ is used to transmit a message from one router to the next one. This transmission is successful with probability 0.8 and fails with probability 0.2. The clock guard and the clock invariant of $\sigma_{\mathsf{transmit}}$ (together with the fact that the clock of the message is reset to 0 whenever $\sigma_{\mathsf{transmit}}$ is applied or when the message was pushed into the network using σ_{send}) ensures that transmission attempts happen within 2–5 time units. Lastly, the PTGT rule σ_{drop} has priority 1 and is used to drop messages for which transmission has failed. In our evaluation in Sect. 6, we also consider the cases that messages are never dropped or not dropped before the second transmission failure by changing the attribute guard of σ_{drop} from $f > 0$ to $\bot$ and $f > 1$, respectively. PTGTS steps $(G, v)[\delta, \mu\rangle(G', v')$ are timed and discrete steps as for PTA.

PTGT APs are GCs ϕ and PTGT states (G, v) are labeled by ϕ when G satisfies ϕ. For our running example, the AP ϕ_{allDone} labels states where *each* message has been successfully delivered. Subsequently, we introduce PMTGL to identify relevant target states for analysis not relying on PTGT APs.

Besides translating a PTGTS into a PTA following [21], we can generate directly a symbolic state space (cf. Fig. 1c for the PTA case) using the tool AutoGraph [24] where each symbolic state (G, ψ) represents all states (G, v) with $v \models \psi$ and where ψ is encoded as a Difference Bound Matrix (DBM) [5].

4 Probabilistic Metric Temporal Graph Logic

Before introducing PMTGL, we recall MTGL [11,26] and adapt it to PTGTSs. To simplify our presentation, we focus on a restricted set of MTGL operators and conjecture that the presented adaptations of MTGL are compatible with full MTGL from [26] as well as with the orthogonal MTGL developments in [27].

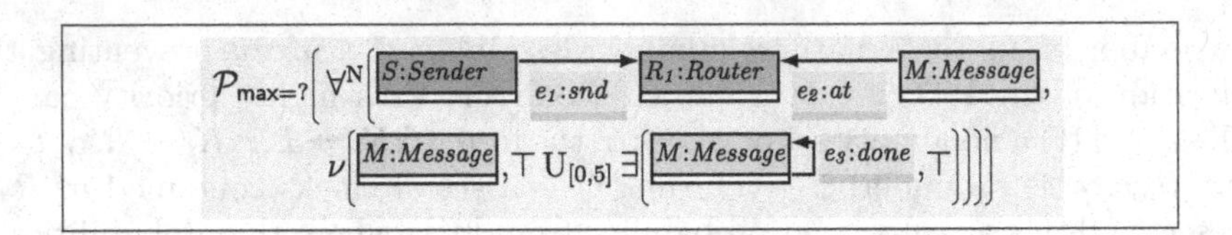

Fig. 3. PMTGC χ_{max} where the additional MTGL operator *forall-new* (written $\forall^{\text{N}}$) is derived from the operator *exists-new* by $\forall^{\text{N}}(f, \theta) = \neg\exists^{\text{N}}(f, \neg\theta)$.

The Metric Temporal Graph Conditions (MTGCs) of MTGL are specified using (a) the GC operators to express properties on a single graph in a path and (b) metric temporal operators to navigate through the path. For the latter, the operator $\exists^{\text{N}}$ (called *exists-new*) is used to extend a current match of a graph H to a supergraph H' in the future such that some additionally matched graph element could not have been matched earlier. Moreover, the operator U (called *until*) is used to check whether an MTGC θ_2 is eventually satisfied in the future within a given time interval while another MTGC θ_1 is satisfied until then.

Definition 1 (MTGCs). *For a graph H, $\theta_H \in \mathsf{MTGC}(H)$ is a* metric temporal graph condition (MTGC) *over H defined as follows:*

$$\theta_H ::= \top \mid \neg\theta_H \mid \theta_H \wedge \theta_H \mid \exists(f, \theta_{H'}) \mid \nu(g, \theta_{H''}) \mid \exists^{\text{N}}(f, \theta_{H'}) \mid \theta_H \, \text{U}_I \, \theta_H$$

where $f : H \hookrightarrow H'$ and $g : H'' \hookrightarrow H$ are monos and where I is an interval over $\mathbf{R}_0^+$.

For our running example, consider the MTGC given in Fig. 3 inside the operator $\mathcal{P}_{\text{max=?}}(\cdot)$. Intuitively, this MTGC states that (*forall-new*) whenever a message has just been sent from the sender to the first router, (*restrict*) when only tracking this message by match restriction (since at least the edge e_2 can be assumed to be removed in between), (*until*) eventually within 5 time units, (*exists*) this message is delivered to the receiver as indicated by the *done* loop.

In [11,26], MTGL was defined for timed graph sequences in which only discrete steps are allowed each having a duration $\delta > 0$. We now adapt MTGL to PTGTSs in which multiple graphs may occur at the same time point.

For tracking subgraphs in a path π over time using matches, we first identify the graph $\pi(\tau)$ in π at a position $\tau = (t, s) \in \mathbf{R}_0^+ \times \mathbf{N}$ where t is a total time point and s is a step index starting at 0 after every non-zero timed step.[8]

Definition 2 (Graph at Position). *A graph G is at position $\tau = (t, s)$ in a path π of a PTGTS S, written $\pi(\tau) = G$, if the auxiliary function* pos *defined below returns $\mathsf{pos}(\pi, i, t, s, \delta) = G$ for the* ith *step of π and delay δ (since the last change of the step index s).*

- *If $\pi_0 = ((G, v)[\delta, \mu\rangle(G', v'))$, then $\mathsf{pos}(\pi, 0, 0, 0, 0) = G$.*
- *If $\pi_i = ((G, v)[\delta, \mu\rangle(G', v'))$, $\mathsf{pos}(\pi, i, t, s, 0) = G$, and $\delta > 0$, then $\mathsf{pos}(\pi, i, t + \delta', 0, \delta') = G$ for each $\delta' \in (0, \delta)$ and $\mathsf{pos}(\pi, i + 1, t + \delta, 0, 0) = G'$*

[8] To compare positions, we define $(t, s) < (t', s')$ if either $t < t'$ or $t = t'$ and $s < s'$.

- *If* $\pi_i = ((G, v)[0, \mu\rangle(G', v'))$ *and* $\mathsf{pos}(\pi, i, t, s, \delta) = G$, *then* $\mathsf{pos}(\pi, i+1, t, s+1, 0) = G'$

A match $m : H \hookrightarrow \pi(\tau)$ into the graph at position τ can be propagated forwards/backwards over the steps in a path to the graph $\pi(\tau')$. Such a propagated match $m' : H \hookrightarrow \pi(\tau')$, written $m' \in \mathrm{PM}(\pi, m, \tau, \tau')$, can be obtained uniquely if *all* matched graph elements $m(H)$ are preserved by the considered steps, which is trivially the case for timed steps. When some graph element is not preserved, $\mathrm{PM}(\pi, m, \tau, \tau')$ is empty.

We now present the semantics of MTGL by providing a satisfaction relation, which is defined as for GL for the operators inherited from GL and as explained above for the operators *exists-new* and *until*.

Definition 3 (Satisfaction of MTGCs). *An MTGC* $\theta \in \mathsf{MTGC}(H)$ *over a graph* H *is* satisfied *by a path* π *of the PTGTS* S, *a position* $\tau \in \mathbf{R}_0^+ \times \mathbf{N}$, *and a mono* $m : H \hookrightarrow \pi(\tau)$, *written* $(\pi, \tau, m) \models \theta$, *if an item applies.*

- $\theta = \top$.
- $\theta = \neg\theta'$ *and* $(\pi, \tau, m) \not\models \theta'$.
- $\theta = \theta_1 \wedge \theta_2$, $(\pi, \tau, m) \models \theta_1$, *and* $(\pi, \tau, m) \models \theta_2$.
- $\theta = \exists(f : H \hookrightarrow H', \theta')$ *and* $\exists m' : H' \hookrightarrow \pi(\tau).\ m' \circ f = m \wedge (\pi, \tau, m') \models \theta$.
- $\theta = \nu(g : H'' \hookrightarrow H, \theta')$ *and* $(\pi, \tau, m \circ g) \models \theta'$.
- $\theta = \exists^{\mathrm{N}}(f : H \hookrightarrow H', \theta')$ *and there are* $\tau' \geq \tau$, $m' \in \mathrm{PM}(\pi, m, \tau, \tau')$, *and* $m'' : H' \hookrightarrow \pi(\tau')$ *s.t.* $m'' \circ f = m'$, $(\pi, \tau', m'') \models \theta$, *and for each* $\tau'' < \tau'$ *it holds that* $\mathrm{PM}(\pi, m'', \tau', \tau'') = \emptyset$.
- $\theta = \theta_1 \mathrm{U}_I\, \theta_2$, $\tau = (t, s)$, *and there are* $\delta \in I$ *and* $\tau' = (t + \delta, s')$ *s.t.*
  - $s' \geq s$ *if* $\delta = 0$,
  - *there is* $m' \in \mathrm{PM}(\pi, m, \tau, \tau')$ *s.t.* $(\pi, \tau', m') \models \theta_2$, *and*
  - *for every* $\tau \leq \tau'' < \tau'$ *there is* $m'' \in \mathrm{PM}(\pi, m, \tau, \tau'')$ *s.t.* $(\pi, \tau'', m'') \models \theta_1$.

Moreover, if $\theta \in \mathsf{MTGC}(\emptyset)$, $\tau = (0, 0)$, *and* $(\pi, \tau, \mathrm{i}(\pi(\tau))) \models \theta$, *then* $\pi \models \theta$.

We now introduce the Probabilistic Metric Temporal Graph Conditions (PMTGCs) of PMTGL, which are defined based on MTGCs.

Definition 4 (PMTGCs). *Each* probabilistic metric temporal graph condition (PMTGC) *is of the form* $\chi = \mathcal{P}_{\sim c}(\theta)$ *where* $\sim \in \{\leq, <, >, \geq\}$, $c \in [0, 1]$ *is a probability, and* $\theta \in \mathsf{MTGC}(\emptyset)$ *is an MTGC over the empty graph. Moreover, we also call expressions of the form* $\mathcal{P}_{\mathsf{min}=?}(\theta)$ *and* $\mathcal{P}_{\mathsf{max}=?}(\theta)$ *PMTGCs.*

The satisfaction relation for PMTGL defines when a PTGTS satisfies a PMTGC.

Definition 5 (Satisfaction of PMTGCs). *A PTGTS* S satisfies *the PMTGC* $\chi = \mathcal{P}_{\sim c}(\theta)$, *written* $S \models \chi$, *if, for any adversary Adv, the probability over all paths of Adv that satisfy* θ *is* $\sim c$. *Moreover,* $\mathcal{P}_{\mathsf{min}=?}(\theta)$ *and* $\mathcal{P}_{\mathsf{max}=?}(\theta)$ *denote the infimal and supremal expected probabilities over all adversaries to satisfy* θ.

For our running example, the evaluation of the PMTGC χ_{max} from Fig. 3 for the PTGTS from Fig. 2 results in the probability of $0.8^4 = 0.4096$ using a probability maximizing adversary *Adv* as follows. Whenever the first graph of the PMTGC can be matched, this is the result of an application of the PTGT rule σ_{send}. The adversary *Adv* ensures then that the matched message is transmitted as fast as possible to the destination router R_3 by (a) letting time pass only when this is unavoidable to satisfy the guard for the next transmission step and (b) never allowing to match the router R_4 by the PTGT rule $\sigma_{\mathsf{transmit}}$ as this leads to a transmission with 3 hops. For each message, the only transmission requiring at most 5 time units transmits the message via the router R_2 to router R_3 using 2 hops in at least $2+2$ time units. The urgently (i.e., without prior delay) applied PTGT rule $\sigma_{\mathsf{receive}}$ then attaches a *done* loop to the message as required by χ_{max}. Since the transmissions of the messages do not affect each other and messages are successfully transmitted only when both transmission attempts for each of the messages succeeded, the maximal probability to satisfy the inner MTGC is $(0.8 \times 0.8)^2 = 0.8^4$. Using $\mathcal{P}_{\mathsf{min}=?}(\cdot)$ results in a probability of 0 since there is e.g. the adversary *Adv'* that only allows a transmission with 3 hops via router R_4 exceeding the deadline.

5 Bounded Model Checking Approach

We now present our BMC approach in terms of an analysis algorithm for a fixed PTGTS S, PMTGC $\chi = \mathcal{P}_{\sim c}(\theta)$, and time bound $T \in \mathbf{R}_0^+ \cup \{\infty\}$. Using this algorithm, we analyze whether S satisfies χ when restricting the discrete behavior of S to the time interval $[0, T)$. In fact, we consider in this algorithm PMTGCs of the form $\mathcal{P}_{\mathsf{max}=?}(\theta)$ or $\mathcal{P}_{\mathsf{min}=?}(\theta)$ for computing expected probabilities since they are sufficient to analyze PMTGCs of the form $\mathcal{P}_{\sim c}(\theta)$.[9] In the subsequent presentation, we focus on the case of $\mathcal{P}_{\mathsf{max}=?}(\theta)$ and point out differences for the case of $\mathcal{P}_{\mathsf{min}=?}(\theta)$ where required.

Step 1: Encoding the Time Bound into the PTGTS
For the given PTGTS S and time bound T, we construct an adapted PTGTS S' into which the time bound T is encoded (for $T = \infty$, to be used when all paths derivable for the PTGTS are sufficiently short, we use $S' = S$). In S', we ensure that all discrete PTGT steps are disabled when time bound T is reached and that the PTGT invariants are then disabled. For this purpose, we (a) create a fresh local variable x_T of sort real and a fresh clock variable x_c (for which fresh types are added to the type graph to ensure non-ambiguous matching of variables during GT rule application), (b) add both variables and the attribute constraint $x_T = T$ to the initial graph of S, (c) add both variables to the graphs L, K, and R of each underlying GT rule $\rho = (\ell{:}K \hookrightarrow L, r{:}K \hookrightarrow R, \gamma)$ of each PTGT rule σ of S and add $x_c < x_T$ as an additional clock guard to each PTGT rule to prevent the application of PTGT rules beyond time bound T, and (d) add a PTGT rule σ_{BMC} with a clock guard $x_c \geq x_T$ and a clock invariant $x_c \leq x_T$, which (in its single underlying GT

[9] For example, $\mathcal{P}_{\mathsf{min}=?}(\theta) = c$ implies satisfaction of $\mathcal{P}_{\geq c'}(\theta)$ for any $c' \leq c$.

rule) deletes the variable x_T from the matched graph. The application of σ_{BMC} at time x_T ensures that no PTGT rule can be applied subsequently and that all PTGT invariants are disabled due to step (c).[10] For the resulting PTGTS S', we then solve the model checking problem for the given PMTGC χ.

Lemma 1 (Encoded BMC Bound). *If π is a path of the PTGTS S', then the time point of the last discrete step (if any exists) precedes T.*

Step 2: Construction of Symbolic State Space and Timing Specification Following the construction of a symbolic state space for a given PTA by the PRISM model checker (where states are given by pairs of locations and zones over the clocks of the PTA (cf. Sect. 2)), we may construct a symbolic state space for a given PTGTS where states are given by pairs of graphs and zones over the clocks contained in the graph. Paths $\hat{\pi}$ through such a symbolic state space are of the form $s_1[\mu_2\rangle s_2[\mu_3\rangle \dots s_n$ consisting of states and (nondeterministically selected) DPDs on successor states (i.e., $\mu_i(s_i) > 0$). Note again that each such path $\hat{\pi}$ is symbolic itself by not specifying the amount of time that elapses in each state. We call a path π of the form $s_1[\delta_2, \mu_2\rangle s_2[\delta_3, \mu_3\rangle \dots s_n$ a *timed realization* of $\hat{\pi}$ when the added delays $\delta_i \geq 0$ are a viable selection according to the zones contained in the states (e.g. for the symbolic state space in Fig. 1c, the zone $c_1 = c_2 \leq 6$ of the initial state allows any selection $\delta_1 \leq 6$).

As a deviation from the symbolic state space generation approach for PTA, we generate a *tree-shaped* symbolic state space M by not identifying isomorphic states. The absence of loops in M guaranteed by the tree-shaped form ensures that, as required by Step 3, every path of M is finite (on time diverging paths). Moreover, for each path $\hat{\pi}$ of M, guards, invariants, and clock resets have been encoded in the zones of the states also ensuring the existence of at least one timed realization π for each $\hat{\pi}$. For our analysis algorithm, ultimately deriving the resulting probabilities in Step 5, we now use the guards, invariants, and clock resets again to derive for each path[11] $\hat{\pi}$ of M a *timing specification* $\mathsf{TS}(\hat{\pi})$. This timing specification captures for a path $\hat{\pi}$ when each of its states has been reached (which may be impossible without the tree-shaped form of the symbolic state space) thereby characterizing all viable timed realizations π of $\hat{\pi}$. To define $\mathsf{TS}(\hat{\pi})$, we use time point clocks tpc_i for $1 \leq i \leq n$ where n is the maximal length of any path of M. For a path $\hat{\pi}$, tpc_i then represents in $\mathsf{TS}(\hat{\pi})$ the time point when state i has been just reached in $\hat{\pi}$. Hence, $\mathsf{TS}(\hat{\pi})$ ranges over tpc_i for $1 \leq i \leq m$ where m is the length of $\hat{\pi}$. In the following, we also use the notion of the *total time valuation* $\mathsf{ttv}(\pi)$ to be the AC equating the time point clock tpc_i and the time point $\sum_{1 \leq k < i} \delta_i$ of the ith step in π. Using this notion, we characterize that π is a timed realization of $\hat{\pi}$ (performing the same discrete steps) when $\mathsf{TS}(\hat{\pi}) \wedge \mathsf{ttv}(\pi)$ is satisfiable.

To define $\mathsf{TS}(\hat{\pi})$, we use a map $\mathsf{LastReset}(k, c) = k'$ returning for an index $1 \leq k \leq m$ and a clock c the largest index $k' \leq k$ where c was reset in $\hat{\pi}$ (which

[10] The additional PTGT rule σ_{BMC} is used since PTGT invariants cannot be disabled by changing them from γ to $\gamma \vee x_c \geq T$ due to the limited syntax of zones.

[11] We only consider paths starting in the initial state and ending in a leaf state.

can be easily computed by iterating once through $\hat{\pi}$). Recall that all clocks c are reset in the initial state, i.e., $\mathsf{LastReset}(1, c) = 1$. We include the ACs in $\mathsf{TS}(\hat{\pi})$ as follows for each state s_i. Firstly, when $i = 1$ (i.e., s_i is the initial state), we add $\mathsf{tpc}_1 = 0$ to $\mathsf{TS}(\hat{\pi})$. Secondly, when $i > 1$, we add $\mathsf{tpc}_{i-1} \leq \mathsf{tpc}_i$ to $\mathsf{TS}(\hat{\pi})$. Thirdly, when s_i was reached by respecting a guard ψ (implying $i > 1$), we add ψ to $\mathsf{TS}(\hat{\pi})$ after replacing each clock c contained in ψ by $\mathsf{tpc}_i - \mathsf{tpc}_{k'}$ where $k' = \mathsf{LastReset}(i-1, c)$.[12] Fourthly, when s_i was reached by respecting an invariant ψ', we add ψ' to $\mathsf{TS}(\hat{\pi})$ after replacing each clock c contained in ψ' by $\mathsf{tpc}_{i+1} - \mathsf{tpc}_{k'}$ where $k' = \mathsf{LastReset}(i, c)$.[13]

Lemma 2 (Sound Timing Specification). *If $\hat{\pi}$ is a path of the symbolic state space M constructed for the PTGTS S', then there is a one-to-one correspondence between valuations of the time point clocks tpc_i satisfying $\mathsf{TS}(\hat{\pi})$ and the time points at which states are reached in the timed realizations π of $\hat{\pi}$.*

For our running example (considering the restriction to a single message in the initial graph), for a path $\hat{\pi}_{\mathsf{ex}}$ where the message is sent to router R_1, transmitted to router R_2, transmitted to router R_3, and then received by receiver R, we derive (after simplification) $\mathsf{TS}(\hat{\pi}_{\mathsf{ex}})$ as the conjunction of $\mathsf{tpc}_1 = 0$, $0 \leq \mathsf{tpc}_2 \leq 10$, $\mathsf{tpc}_2 + 2 \leq \mathsf{tpc}_3 \leq \mathsf{tpc}_2 + 5$, and $\mathsf{tpc}_3 + 2 \leq \mathsf{tpc}_4 = \mathsf{tpc}_5 \leq \mathsf{tpc}_3 + 5$ essentially encoding the guards and invariants as expected.[14]

In the next two steps of our algorithm, we derive for the MTGC θ (contained in the given PMTGC $\mathcal{P}_{\mathsf{min}=?}(\theta)$ or $\mathcal{P}_{\mathsf{max}=?}(\theta)$) and a path $\hat{\pi}$ an AC describing timed realizations π of $\hat{\pi}$ satisfying θ. For our running example and the path $\hat{\pi}_{\mathsf{ex}}$ from above, this derived AC will be $\mathsf{tpc}_5 - \mathsf{tpc}_2 \leq 5$ expressing that the time elapsed between the sending of the message and its reception by the receiver is at most 5 time units as required by θ. Then, in Step 5 of the algorithm, we will identify (a) successful paths $\hat{\pi}$ to be those where $\mathsf{TS}(\hat{\pi})$ and the derived AC are satisfiable at once and (b) failing paths $\hat{\pi}$ to be those where $\mathsf{TS}(\hat{\pi})$ and the negated derived AC are satisfiable together.

Step 3: From MTGC Satisfaction to GC Satisfaction
Following the satisfaction checking approach for MTGL from [11,26], we translate the MTGC satisfaction problem into an equivalent, yet much easier to check, GC satisfaction problem using the operations fold and encode (presented below). The operation fold aggregates the information about the nature and timing of all GT steps of $\hat{\pi}$ into a single *Graph with History (GH)*. The operation encode translates the MTGC into a corresponding GC.[15] Technically, the MTGC θ is

[12] Intuitively, $\mathsf{tpc}_i - \mathsf{tpc}_{k'}$ is the duration between the last reset of c and the time point when the guard was checked upon state transition to s_i.

[13] Intuitively, $\mathsf{tpc}_{i+1} - \mathsf{tpc}_{k'}$ is the duration between the last reset of c and the time point at which the invariant was no longer checked due to the state transition to s_{i+1}.

[14] Note that $\mathsf{tpc}_4 = \mathsf{tpc}_5$ since the message reception by R takes no time.

[15] The operations fold and encode presented here are adaptations of the corresponding operations from [11,26] to the modified MTGL satisfaction relation for PTGTSs from Definition 3 allowing for successive discrete steps with zero-time delay in-between.

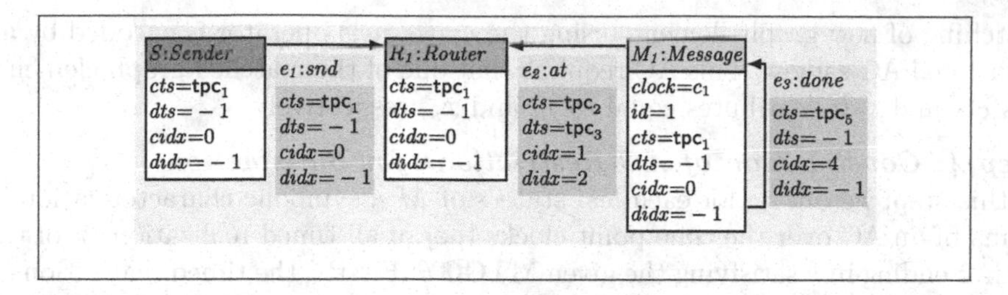

Fig. 4. A part of the GH G_H obtained using the operation fold for the path $\hat{\pi}_{\mathsf{ex}}$ of the running example.

satisfied by a timed realization π of a path $\hat{\pi}$ of M precisely when the encoded MTGC is satisfied by the folded GH G_H once the total time valuation AC $\mathsf{ttv}(\pi)$ is added to G_H (incorporating the precise timing of steps in π).

Theorem 1 (Soundness of fold and encode). *If θ is an MTGC over the empty graph,* $\mathsf{encode}(\theta) = \phi$*, $\hat{\pi}$ is a path through the symbolic state space constructed for the PTGTS S',* $\mathsf{fold}(\hat{\pi}) = G_H$*, π is a timed realization of $\hat{\pi}$ (i.e., a path through S'), and G'_H is obtained from G_H by adding the AC* $\mathsf{ttv}(\pi)$*, then $\pi \models \theta$ iff $G'_H \models \phi$.*

The operation fold generates for a path $\hat{\pi}$ the corresponding GH G_H by (a) constructing the union of all graphs of $\hat{\pi}$ where nodes/edges preserved in steps are identified and (b) recording for each node/edge in the resulting GH the position τ (cf. Definition 2) when it was created and deleted (if the node/edge is deleted at some point) in $\hat{\pi}$ using additional *creation/deletion time stamp attributes cts/dts* and *creation/deletion index attributes cidx/didx*. In particular, (i) nodes/edges contained in the initial state of $\hat{\pi}$ are equipped with attributes $cts = \mathsf{tpc}_1$ and $cidx = 0$, (ii) nodes/edges added in step i of $\hat{\pi}$ are equipped with attributes $cts = \mathsf{tpc}_i$ and $cidx = i$, (iii) nodes/edges deleted in step i of $\hat{\pi}$ are equipped with attributes $dts = \mathsf{tpc}_i$ and $didx = i$, and (iv) nodes/edges contained in the last state of $\hat{\pi}$ are equipped with attributes $dts = -1$ and $didx = -1$. For the path $\hat{\pi}_{\mathsf{ex}}$ from our running example, see Fig. 4 depicting the part of the GH G_H that is matched when checking the GC $\mathsf{encode}(\theta)$ against G_H.

The operation encode generates for the MTGC θ contained in the given PMTGC χ the corresponding GC ϕ (note that encode does not depend on a path and is therefore executed precisely once). Intuitively, it recursively encodes the requirements expressed using MTGL operators (see the items of Definition 3) on a timed realization π of a path $\hat{\pi}$ by using GL operators on the GH (obtained by folding $\hat{\pi}$) with additional integrated ACs. In particular, quantification over positions $\tau = (t, s)$ of global time t and step index s, as for the operators *exists-new* and *until*, is encoded by quantifying over additional variables x_t and x_s representing t and s, respectively. Also, matching of graphs, as for the operators *exists* and *exists-new*, is encoded by an additional AC alive. This AC requires that each matched node/edge in the GH has *cts*, *dts*, *cidx*, and *didx* attributes implying that this graph element exists for the position (x_t, x_s) in π. Lastly, matching of new graph elements using the *exists-new* operator is encoded by an additional AC earliest. This AC requires that one of the matched graph elements has *cts* and *cidx* attributes equal to x_t and x_s, respectively.

Step 4: Construction of AC-Restrictions for Satisfaction
In this step, we obtain for each leaf state s of M a symbolic characterization in terms of an AC over the time point clocks tpc_i of all timed realizations π of the path $\hat{\pi}$ ending in s satisfying the given MTGC θ. Firstly, the timed realizations π of the path $\hat{\pi}$ ending in s are characterized by the timing specification $\mathsf{TS}(\hat{\pi})$ as discussed in Step 2. Secondly, we refine the set of such timed realizations using an AC $\gamma_{\hat{\pi}}$ over the time point clocks tpc_i symbolically describing when such a timed realization satisfies the given MTGC θ. The AC $\gamma_{\hat{\pi}}$ is obtained by checking the GC $\mathsf{encode}(\theta) = \phi$ against the GH $\mathsf{fold}(\hat{\pi}) = G_H$. The conjunction of $\mathsf{TS}(\hat{\pi})$ and $\gamma_{\hat{\pi}}$ is then recorded in the set of *state conditions* $\mathsf{SC}(s)$ and is satisfied by precisely those valuations of the time point clocks tpc_i that correspond to timed realizations π ending in s satisfying the MTGC θ.[16] In Step 5, we also use the notion of *state probability* $\mathsf{SP}(s)$ assigning a probability of 1 to a state s when the AC in $\mathsf{SC}(s)$ is satisfiable and 0 otherwise.

Lemma 3 (Correct ACs). *If θ is an MTGC over the empty graph, $\hat{\pi}$ is a path of the symbolic state space constructed for the PTGTS S' ending in state s, and π is a timed realization of $\hat{\pi}$, then $\pi \models \theta$ iff $\mathsf{TS}(\hat{\pi}) \wedge \gamma_{\hat{\pi}} \wedge \mathsf{ttv}(\pi)$ is satisfiable.*

For our running example, when checking the encoded MTGC (cf. Fig. 3) for the GH partially given in Fig. 4, (a) the graph elements S, R_1, M_1, e_1, and e_2 are matched for the *forall-new* operator and (b) the graph elements M_1 and e_3 are matched for the *exists* operator. For (a), all matched graph elements are alive at the *symbolic position* $(\mathsf{tpc}_2, 1)$ characterizing all positions $(t, 1)$ where $\mathsf{tpc}_2 = t$. The ACs in the encoded MTGC then ensure that e.g. e_1 is alive since it was created not after tpc_2 ($cts = \mathsf{tpc}_1 \leq \mathsf{tpc}_2$ and $cidx = 0 \leq 1$) and it has never been deleted ($dts = -1$) whereas e.g. e_2 is alive since it was created at $(\mathsf{tpc}_2, 1)$ and it has been deleted strictly later ($dts = \mathsf{tpc}_2$ but $1 < didx$). Moreover, the matched graph elements are not alive earlier since e_2 was created at $(\mathsf{tpc}_2, 1)$. For (b), all matched graph elements are alive at $(\mathsf{tpc}_5, 4)$. Overall, we obtain (after simplification) the AC requiring that $\mathsf{tpc}_5 - \mathsf{tpc}_2 \leq 5$ as the encoded MTGC expresses the time bound ≤ 5 used in the *until* operator. For the last state of the path $\hat{\pi}_{\mathsf{ex}}$, we obtain the AC $\mathsf{TS}(\hat{\pi}_{\mathsf{ex}}) \wedge \mathsf{tpc}_5 - \mathsf{tpc}_2 \leq 5$, which is e.g. satisfied by the valuation $\{\mathsf{tpc}_1 = 0, \mathsf{tpc}_2 = 0, \mathsf{tpc}_3 = 2, \mathsf{tpc}_4 = 4, \mathsf{tpc}_5 = 4\}$ representing a timed realization π_{ex} of $\hat{\pi}_{\mathsf{ex}}$ where the message is transmitted as early as possible in both transmission steps.

Step 5: Computation of Resulting Probabilities
In this step, we compute the maximal/minimal probability for the satisfaction of the given MTGC θ, i.e., for reaching states s with clock valuation v satisfying the AC contained in the state conditions $\mathsf{SC}(s)$. However, this kind of specification of target states is not supported by PRISM, which requires a clock-independent

[16] For the case of $\mathcal{P}_{\mathsf{min}=?}(\theta)$, we define $\mathsf{SC}(s) = \{\mathsf{TS}(\hat{\pi}) \wedge \neg\gamma_{\hat{\pi}}\}$.

specification of target states. Therefore, we propose a custom analysis procedure to solve the analysis problem from above.

In the following, we first discuss, on an example, an analysis procedure for the case of a clock-independent labeling of states and then expand this procedure to the additional use of state conditions $\mathsf{SC}(s)$. For the symbolic state space in Fig. 1c, the maximal probability to reach a state labeled with *success* can be computed by propagating restrictions of valuations given by zones backwards. Initially, each state is equipped only with the zone given in the state space and the probability 1 when it is a target state. The zone/probability pairs $(c_1 - c_2 \leq 3, 1)$ and $(5 \leq c_1 - c_2 \leq 6, 1)$ of the ℓ_4-state and the ℓ_6-state are then propagated backwards without change to the ℓ_3-state and the ℓ_5-state, respectively. However, when steps have multiple target states, any subset of the target states is considered and the probabilities of pairs for the considered target states are summed up when the conjunction of their zones is satisfiable. For example, we obtain $(5 \leq c_1 - c_2 \leq 6, 0.75)$ for the ℓ_2-state since the conjunction of the zones obtained for the ℓ_5- and ℓ_9-states is satisfiable whereas the other subsets of target states result in unsatisfiable conjunctions or lower probabilities. When multiple zone/probability pairs with a common maximal probability are obtained, they are all retained for the source state of the step.

We now introduce our backward analysis procedure by adapting the procedure from above to the usage of the ACs contained in the state condition SC(s) instead of zones. Technically, our (fixed-point) backward analysis procedure updates the state conditions SC and state probability SP, which record the AC/probability pairs, until no further modifications can be performed according to the following definition.

Definition 6 (Backward Analysis Procedure). *The subsequent operation updating* SC *and* SP *is performed until a fixed-point is reached. When* SC, SP, *and I assign to each state s of M a set of ACs, a probability, and the depth of s in the tree-shaped state space M, respectively, (s, μ) is an edge of M, $S' \subseteq \mathsf{supp}(\mu)$ is a subset of the target states of μ, f selects for each target state $s' \in S'$ an AC from $\mathsf{SC}(s')$, $\gamma = \exists \mathsf{tpc}_{I(s)}. \bigwedge_{s' \in S'} f(s')$ is the AC derived for the state s based on the selections S' and f, γ is satisfiable, and $p = \sum_{s' \in S'} (\mu(s') \times \mathsf{SP}(s'))$ is the new probability for s based on the selections S' and f, then (a) $\mathsf{SC}(s)$ and $\mathsf{SP}(s)$ are changed to $\{\gamma\}$ and p when $p > \mathsf{SP}(s)$ recording the AC γ and the new maximal probability p derived for s and (b) $\mathsf{SC}(s)$ is changed to $\mathsf{SC}(s) \cup \{\gamma\}$ when $p = \mathsf{SP}(s)$ recording an additional AC γ and not changing the probability $\mathsf{SP}(s)$.*

Finally, using our BMC approach introduced in this section, we derive the expected maximal probability.[17]

Theorem 2 (Soundness of BMC Approach). *The presented BMC approach in terms of the presented 5-step analysis algorithm returns the correct probability for a given PTGTS S, PMTGC χ, and time bound T.*

[17] For $\mathcal{P}_{\mathsf{min}=?}(\theta)$, the procedure from Definition 6 returns $1 - \mathcal{P}_{\mathsf{min}=?}(\theta)$ maximizing the probability of failing paths by minimizing the probability for successful paths.

6 Evaluation

Our implementation of the presented BMC approach in AutoGraph [24] reports for all considered variations of our running example the expected best-case probability for timely message transmission of 0.8^{2n} (and the worst-case probability of 0) for n messages to be transmitted. For our experiments, we employed the time bound $T = 20$ corresponding to the maximum duration required for sending the message and transmitting it via the shortest connection. Note that an unbounded number of transmission retries for $T = \infty$ is unrealistic and would not allow for a finite state space M to be generated in Step 2. Also, any message transmission failure inevitably leads to a non-timely transmission of that message due to the time bound used in the PMTGC χ_{max}. However, the size of M is exponential in the number of messages to be transmitted as their transmission is independent from each other resulting in any resolution of their concurrent behavior to be contained in M. Hence, allowing for up to 10 transmission attempts via time bound $T = 20$ resulted in 31 states for $n = 1$ but exceeded our memory at 83000 states for $n = 2$. Using the drop rule to further limit the number of transmission retries allowed to analyze the variation of our running example in which two messages are transmitted but dropped after the second transmission failure resulting in 12334 states.

However, as of now, the bottle neck of our current implementation, which is faithful to our presentation from the previous section, is not the runtime but the memory consumption. To overcome this limitation, we plan to generate the tree-shaped state space M in a depth-first manner performing the subsequent steps of the analysis algorithm (Step 3–Step 5) on entirely generated subtrees of M (before continuing with the state space generation). This would allow to dispose paths from M that are no longer needed in subsequent steps of the algorithm. Also, when the memory consumption has been drastically reduced along this line, a multithreaded implementation would be highly beneficial due to the tree-shaped form of M and the independent analysis for its subtrees.

7 Conclusion and Future Work

We introduced PMTGL for the specification of cyber-physical systems with probabilistic timed behavior modeled as PTGTSs. PMTGL combines (a) MTGL with its binding capabilities for the specification of timed graph sequences and (b) the probabilistic operator from PTCTL to express best-case/worst-case probabilistic timed reachability properties. Moreover, we presented a novel BMC approach for PTGTSs w.r.t. PMTGL properties.

In the future, we plan to apply PMTGL and our BMC approach to the case study [21,23] of a cyber-physical system where, in accordance with real-time constraints, autonomous shuttles exhibiting probabilistic failures on demand navigate on a track topology. Moreover, we plan to extend our BMC approach by supporting the analysis of so-called optimistic violations introduced in [27].

References

1. Barringer, H., Goldberg, A., Havelund, K., Sen, K.: Rule-based runtime verification. In: Steffen, B., Levi, G. (eds.) VMCAI 2004. LNCS, vol. 2937, pp. 44–57. Springer, Heidelberg (2004). https://doi.org/10.1007/978-3-540-24622-0_5
2. Bartocci, E., et al.: Specification-based monitoring of cyber-physical systems: a survey on theory, tools and applications. In: Bartocci, E., Falcone, Y. (eds.) Lectures on Runtime Verification. LNCS, vol. 10457, pp. 135–175. Springer, Cham (2018). https://doi.org/10.1007/978-3-319-75632-5_5
3. Basin, D.A., Klaedtke, F., Müller, S., Zalinescu, E.: Monitoring metric first-order temporal properties. J. ACM **62**(2), 15:1-15:45 (2015). https://doi.org/10.1145/2699444
4. Becker, B., Giese, H.: On safe service-oriented real-time coordination for autonomous vehicles. In: 11th IEEE International Symposium on Object-Oriented Real-Time Distributed Computing (ISORC 2008), 5–7 May 2008, Orlando, Florida, USA, pp. 203–210. IEEE Computer Society (2008). ISBN 978-0-7695-3132-8. https://doi.org/10.1109/ISORC.2008.13. https://ieeexplore.ieee.org/xpl/mostRecentIssue.jsp?punumber=4519543
5. Bengtsson, J., Yi, W.: Timed automata: semantics, algorithms and tools. In: Desel, J., Reisig, W., Rozenberg, G. (eds.) ACPN 2003. LNCS, vol. 3098, pp. 87–124. Springer, Heidelberg (2004). https://doi.org/10.1007/978-3-540-27755-2_3
6. Bohnenkamp, H.C., D'Argenio, P.R., Hermanns, H., Katoen, J.: MODEST: a compositional modeling formalism for hard and softly timed systems. IEEE Trans. Software Eng. **32**(10), 812–830 (2006). https://doi.org/10.1109/TSE.2006.104
7. Búr, M., Szilágyi, G., Vörös, A., Varró, D.: Distributed graph queries for runtime monitoring of cyber-physical systems. In: Russo, A., Schürr, A. (eds.) FASE 2018. LNCS, vol. 10802, pp. 111–128. Springer, Cham (2018). https://doi.org/10.1007/978-3-319-89363-1_7
8. Dávid, I., Ráth, I., Varró, D.: Foundations for streaming model transformations by complex event processing. Softw. Syst. Model. **17**(1), 135–162 (2016). https://doi.org/10.1007/s10270-016-0533-1
9. Dávid, I., Ráth, I., Varró, D.: Streaming model transformations by complex event processing. In: Dingel, J., Schulte, W., Ramos, I., Abrahão, S., Insfran, E. (eds.) MODELS 2014. LNCS, vol. 8767, pp. 68–83. Springer, Cham (2014). https://doi.org/10.1007/978-3-319-11653-2_5
10. Ehrig, H., Ehrig, K., Prange, U., Taentzer, G.: Fundamentals of Algebraic Graph Transformation. Springer, Heidelberg (2006). https://doi.org/10.1007/3-540-31188-2
11. Giese, H., Maximova, M., Sakizloglou, L., Schneider, S.: Metric temporal graph logic over typed attributed graphs. In: Hähnle, R., van der Aalst, W. (eds.) FASE 2019. LNCS, vol. 11424, pp. 282–298. Springer, Cham (2019). https://doi.org/10.1007/978-3-030-16722-6_16
12. Habel, A., Pennemann, K.: Correctness of high-level transformation systems relative to nested conditions. Math. Struct. Comput. Sci. **19**(2), 245–296 (2009). https://doi.org/10.1017/S0960129508007202
13. Hahn, E.M., Hartmanns, A., Hermanns, H., Katoen, J.: A compositional modelling and analysis framework for stochastic hybrid systems. Formal Methods Syst. Des. **43**(2), 191–232 (2013). https://doi.org/10.1007/s10703-012-0167-z
14. Havelund, K.: Rule-based runtime verification revisited. Int. J. Softw. Tools Technol. Transf. **17**(2), 143–170 (2015). https://doi.org/10.1007/s10009-014-0309-2

15. Havelund, K., Peled, D.: Efficient runtime verification of first-order temporal properties. In: Gallardo, M.M., Merino, P. (eds.) SPIN 2018. LNCS, vol. 10869, pp. 26–47. Springer, Cham (2018). https://doi.org/10.1007/978-3-319-94111-0_2
16. Jansen, N., Dehnert, C., Kaminski, B.L., Katoen, J.-P., Westhofen, L.: Bounded model checking for probabilistic programs. In: Artho, C., Legay, A., Peled, D. (eds.) ATVA 2016. LNCS, vol. 9938, pp. 68–85. Springer, Cham (2016). https://doi.org/10.1007/978-3-319-46520-3_5
17. Katoen, J.: The probabilistic model checking landscape. In: Grohe, M., Koskinen, E., Shankar, N. (eds.) Proceedings of the 31st Annual ACM/IEEE Symposium on Logic in Computer Science, LICS 2016, 5–8 July 2016, pp. 31–45. ACM, New York (2016). https://doi.org/10.1145/2933575.2934574
18. Krause, C., Giese, H.: Probabilistic graph transformation systems. In: Ehrig, H., Engels, G., Kreowski, H.-J., Rozenberg, G. (eds.) ICGT 2012. LNCS, vol. 7562, pp. 311–325. Springer, Heidelberg (2012). https://doi.org/10.1007/978-3-642-33654-6_21
19. Kwiatkowska, M., Norman, G., Parker, D.: PRISM 4.0: verification of probabilistic real-time systems. In: Gopalakrishnan, G., Qadeer, S. (eds.) CAV 2011. LNCS, vol. 6806, pp. 585–591. Springer, Heidelberg (2011). https://doi.org/10.1007/978-3-642-22110-1_47
20. Kwiatkowska, M., Norman, G., Sproston, J., Wang, F.: Symbolic model checking for probabilistic timed automata. In: Lakhnech, Y., Yovine, S. (eds.) FORMATS/FTRTFT -2004. LNCS, vol. 3253, pp. 293–308. Springer, Heidelberg (2004). https://doi.org/10.1007/978-3-540-30206-3_21
21. Maximova, M., Giese, H., Krause, C.: Probabilistic timed graph transformation systems. J. Log. Algebr. Meth. Program. **101**, 110–131 (2018). https://doi.org/10.1016/j.jlamp.2018.09.003
22. Orejas, F.: Symbolic graphs for attributed graph constraints. J. Symb. Comput. **46**(3), 294–315 (2011). https://doi.org/10.1016/j.jsc.2010.09.009
23. RailCab Project. https://www.hni.uni-paderborn.de/cim/projekte/railcab
24. Schneider, S.: AutoGraph. https://github.com/schneider-sven/AutoGraph
25. Schneider, S., Maximova, M., Giese, H.: Probabilistic metric temporal graph logic. Technical report, 146. Hasso Plattner Institute, University of Potsdam (2022)
26. Schneider, S., Maximova, M., Sakizloglou, L., Giese, H.: Formal testing of timed graph transformation systems using metric temporal graph logic. Int. J. Softw. Tools Technol. Transfer **23**(3), 411–488 (2021). https://doi.org/10.1007/s10009-020-00585-w
27. Schneider, S., Sakizloglou, L., Maximova, M., Giese, H.: Optimistic and pessimistic on-the-fly analysis for metric temporal graph logic. In: Gadducci, F., Kehrer, T. (eds.) ICGT 2020. LNCS, vol. 12150, pp. 276–294. Springer, Cham (2020). https://doi.org/10.1007/978-3-030-51372-6_16
28. UPPAAL. Department of Information Technology at Uppsala University, Sweden and Department of Computer Science at Aalborg University, Denmark (2021). https://uppaal.org/

Categories of Differentiable Polynomial Circuits for Machine Learning

Paul Wilson[1,2](✉) and Fabio Zanasi[2]

[1] University of Southampton, Southampton, UK
paul@statusfailed.com
[2] University College London, London, UK
f.zanasi@ucl.ac.uk

Abstract. Reverse derivative categories (RDCs) have recently been shown to be a suitable semantic framework for studying machine learning algorithms. Whereas emphasis has been put on training methodologies, less attention has been devoted to particular *model classes*: the concrete categories whose morphisms represent machine learning models. In this paper we study presentations by generators and equations of classes of RDCs. In particular, we propose *polynomial circuits* as a suitable machine learning model. We give an axiomatisation for these circuits and prove a functional completeness result. Finally, we discuss the use of polynomial circuits over specific semirings to perform machine learning with discrete values.

1 Introduction

Reverse Derivative Categories [10] have recently been introduced as a formalism to study abstractly the concept of differentiable functions. As explored in [11], it turns out that this framework is suitable to give a categorical semantics for gradient-based learning. In this approach, models–as for instance neural networks–correspond to morphisms in some RDC. We think of the particular RDC as a 'model class'–the space of all possible definable models.

However, much less attention has been directed to actually defining the RDCs in which models are specified: existing approaches assume there is some chosen RDC and morphism, treating both essentially as a black box. In this paper, we focus on classes of RDCs which we call 'polynomial circuits', which may be thought of as a more expressive version of the boolean circuits of Lafont [17], with wires carrying values from an arbitrary semiring instead of $\mathbb{Z}_2$. Because we ensure polynomial circuits have RDC structure, they are suitable as machine learning models, as we discuss in the second part of the paper.

Our main contribution is to provide an algebraic description of polynomial circuits and their reverse derivative structure. More specifically, we build a presentation of these categories by operation and equations. Our approach will proceed in steps, by gradually enriching the algebraic structures considered, and culminate in showing that a certain presentation is *functionally complete* for the class of functions that these circuits are meant to represent.

N. Behr and D. Strüber (Eds.): ICGT 2022, LNCS 13349, pp. 77–93, 2022.
https://doi.org/10.1007/978-3-031-09843-7_5

An important feature of our categories of circuits is that morphisms are specified in the graphical formalism of *string diagrams*. This approach has the benefit of making the model specification reflect its combinatorial structure. Moreover, at a computational level, the use of string diagrams makes available the principled mathematical toolbox of *double-pushout rewriting*, via an interpretation of string diagrams as hypergraphs [6–8]. Finally, the string diagrammatic presentation suggests a way to encode polynomial circuits into datastructures: an important requirement for being able to incorporate these models into tools analogous to existing deep learning frameworks such as TensorFlow [1] and PyTorch [19].

Tool-building is not the only application of the model classes we define here. Recent neural networks literature [4,9] proposes to improve model performance (e.g. memory requirements, power consumption, and inference time) by 'quantizing' network parameters. One categorical approach in this area is [23], in which the authors define learning directly over boolean circuit models instead of training with real-valued parameters and then quantizing. The categories in our paper can be thought of as a generalisation of this approach to arbitrary semirings.

This generalisation further yields another benefit: while neural networks literature focuses on finding particular 'architectures' (i.e. specific morphisms) that work well for a given problem, our approach suggests a new avenue for model design: changing the underlying semiring (and thus the corresponding notion of arithmetic). To this end, we conclude our paper with some examples of finite semirings which may yield new approaches to model design.

Synopsis. We recall the notion of RDC in Sect. 2, and then study presentations of RDCs by operations and equations in Sect. 3. We define categories of polynomial circuits in Sect. 4, before showing how they can be made *functionally complete* in Sect. 5. Finally, we close by discussing some case studies of polynomial circuits in machine learning, in Sect. 6.

2 Reverse Derivative Categories

We recall the notion of reverse derivative category [10] in two steps. First we introduce the simpler structure of cartesian left-additive categories. We make use of the graphical formalism of *string diagrams* [20] to represent morphisms in our categories.

Definition 1. *A* ***Cartesian Left-Additive Category*** *([5, 10]) is a cartesian category in which each object A is equipped with a commutative monoid and zero map:*

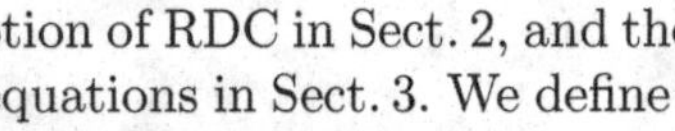

$$A, A \to A \qquad\qquad \bullet\!-\! A \tag{1}$$

so that

$$A \otimes B,\ A \otimes B \to A \otimes B \;=\; A, B, A, B \to A, B \qquad \to A \otimes B \;=\; \to A,\ \to B \tag{2}$$

Note that the category being cartesian means that: (I) it is symmetric monoidal, namely for each object A and B there are symmetries and identities A — A satisfying the laws of symmetric monoidal categories [20]; (II) each object A comes equipped with a *copy* and a *discard* map:

$$A \to A, A \qquad A \to \bullet \tag{3}$$

satisfying the axioms of commutative comonoids and natural with respect to the other morphisms in the category:

$$\tag{4}$$

Remark 1. Definition 1 is given differently than the standard definition of cartesian left-additive categories [10, Definition 1], which one may recover by letting addition of morphisms be $f + g :=$, and the zero morphism be $0 :=$. Equations of cartesian left-additive categories as given in [10, Definition 1]

$$x ⨟ (f+g) \;=\; (x ⨟ f) + (x ⨟ g) \qquad\qquad x ⨟ 0 \;=\; 0$$

are represented by string diagrams

and follow from Definition 1 thanks to the naturality of copy and discard, respectively. We refer to [5, Proposition 1.2.2 (iv)] for more details on the equivalence of the two definitions.

Now, Reverse Derivative Categories, originally defined in [10], are cartesian left-additive categories equipped with an operator R of the following type, and satisfying axioms RD.1–RD.7 detailed in [10, Definition 13].

$$\frac{A \xrightarrow{f} B}{A \times B \xrightarrow[\mathsf{R}[f]]{} A}$$

Intuitively, for a morphism $f : A \to B$ we think of its reverse derivative $R[f] : A \times B \to A$ as approximately computing the change of *input* to f required to achieve a given change in *output*. That is, if f is a function, we should have

$$f(x) + \delta_y \approx f(x + \mathsf{R}[f](x, \delta_y))$$

The authors of [10] go on to show that any reverse derivative category also admits a *forward* differential structure: i.e., it is also a Cartesian Differential Category (CDC). This means the existence of a forward differential operator D satisfying various axioms, and having the following type:

$$\frac{A \xrightarrow{f} B}{A \times A \xrightarrow[\mathsf{D}[f]]{} B}$$

In an RDC, the forward differential operator is defined in terms of R as the following string diagram, with $\mathsf{R}^{(n)}$ denoting the n-fold application[1] of R:

$\mathsf{D}[f] \quad :=$ 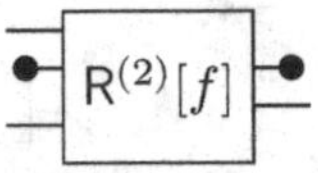

In contrast to the R operator, we think of D as computing a change in *output* from a given change in *input*, whence 'forward' and 'reverse' derivative:

$$f(x + \delta_x) \approx f(x) + \mathsf{D}[f](x, \delta_x)$$

The final pieces we need to state our definition of RDCs are the (cartesian differential) notions of *partial derivative* and *linearity* defined in [10]. Graphically, the partial derivative of $g : A \times B \to C$ with respect to B is defined as follows:

$\mathsf{D}_B[g] \quad :=$ 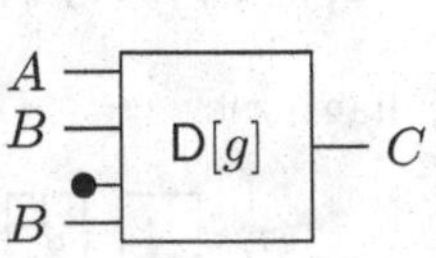

Finally we say that g is *linear in* B when

$\mathsf{D}_B[g] \quad =$ 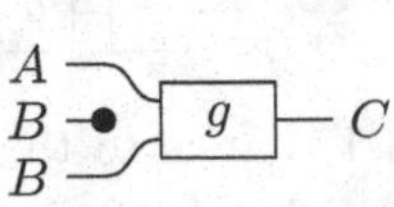

and more generally that $f : A \to B$ is *linear* when

$\mathsf{D}[f] \quad =$ 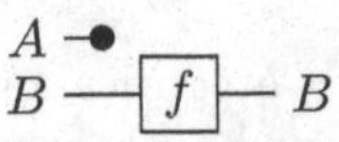

---

1 For example, $R^{(2)}[f]$ denotes the map $R[R[f]]$.

We can now formulate the definition of RDCs. Note that in the following definition and proofs we treat D purely as a syntactic shorthand for its definition in terms of R. We avoid use of CDC axioms to prevent a circular definition, although one can derive them as corollaries of the RDC axioms.

Definition 2. *A* ***Reverse Derivative Category*** *is a cartesian left-additive category equipped with a* reverse differential combinator R*:*

$$\frac{A \xrightarrow{f} B}{A \times B \xrightarrow[\mathsf{R}[f]]{} A}$$

satisfying the following axioms:
*[**ARD.1**] (Structural axioms, equivalent to RD.1, RD.3–5 in [10])*

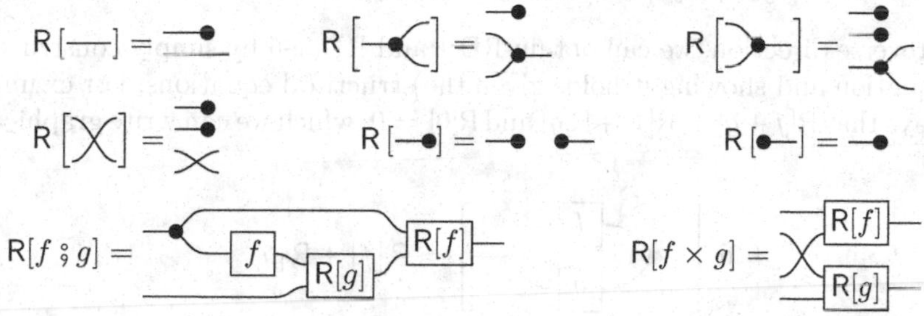

$$\mathsf{R}[f \,;\, g] \qquad \mathsf{R}[f \times g]$$

*[**ARD.2**] (Additivity of change, equivalent to RD.2 in [10])*

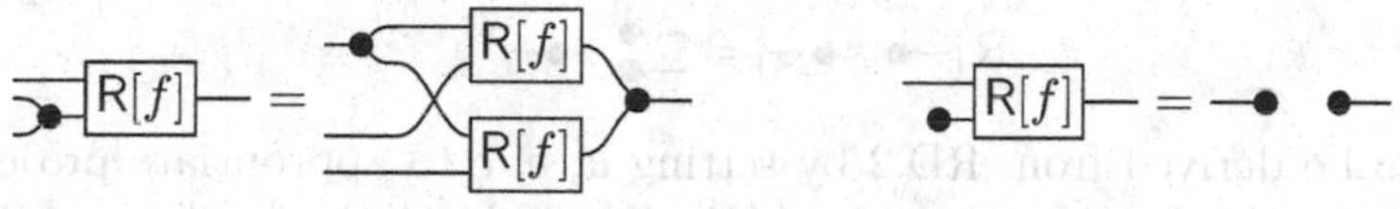

*[**ARD.3**] (Linearity of change, equivalent to RD.6 in [10])*

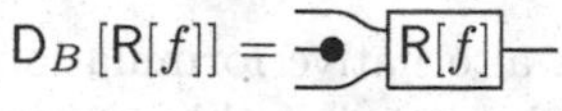

$$\mathsf{D}_B\,[\mathsf{R}[f]] =$$

*[**ARD.4**] (Symmetry of partials, equivalent to RD.7 in [10])*

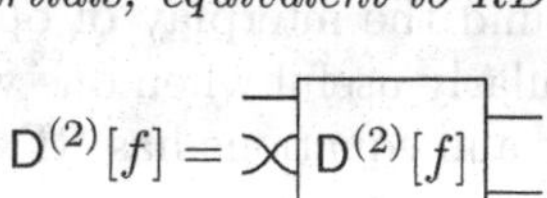

$$\mathsf{D}^{(2)}[f] =$$

Remark 2. Note that we may alternatively write axioms ARD.3 and ARD.4 directly in terms of the R operator by simply expanding the syntactic definition of D.

Note that axioms ARD.1 and ARD.2 are quite different to that of [10], while ARD.3 and ARD.4 are essentially direct restatements in graphical language of RD.6 and RD.7 respectively.

The definition we provide best suits our purposes, although it is different than the standard one provided in [10, Definition 13]. We can readily verify that they are equivalent.

Theorem 1. *Definition 2 is equivalent to [10, Definition 13].*

Proof. Axioms ARD.3–4 are direct statements of axioms RD.6–7, so it suffices to show that we can derive axioms ARD.1–2 from RD.1.5 and vice-versa. The structural axioms ARD.1 follow directly from RD.1 and RD.3–5.

- For R[—] use RD.3 directly.
- For R[X], apply RD.4 to $\langle \pi_1, \pi_0 \rangle$
- For R[>-], apply RD.1 to $\pi_0 + \pi_1$
- For R[•-], apply RD.1 directly.
- For R[-<], apply RD.4 to $\langle \mathsf{id}, \mathsf{id} \rangle$
- For R[-•], apply RD.4 directly.
- For composition $f \,;\, g$, apply RD.5 directly
- For tensor $f \times g$, apply RD.4 to $\langle \pi_0 \,;\, f, \pi_1 \,;\, g \rangle$

In the reverse direction, we can obtain RD.1 and RD.3–5 by simply constructing each equation and showing it holds given the structural equations. For example, RD.1 says that $\mathsf{R}[f+g] = \mathsf{R}[f] + \mathsf{R}[g]$ and $\mathsf{R}[0] = 0$, which we can write graphically as:

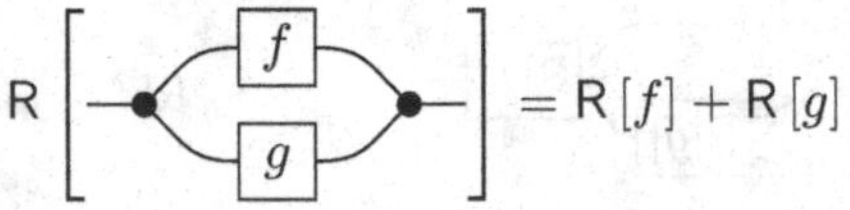

and

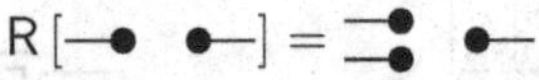

ARD.2 can be derived from RD.2 by setting a, b, c to appropriate projections, and in the reverse direction we can obtain RD.2 simply by applying ARD.2 to its left-hand-side and using naturality of -<.

A main reason to give an alternative formulation of cartesian left-additive and reverse derivative categories is being able to work with a more 'algebraic' definition, which revolves around the interplay of operations >-, •-, -<, and -•. This perspective is particularly useful when one wants to show that the free category on certain generators and equations has RDC structure. We thus recall such free construction, referring to [24, Chapter 2] and [3, Sect. 5] for a more thorough exposition.

Definition 3. *Given a set Obj of* generating objects, *we may consider a set Σ of* generating morphisms $f\colon w \to v$, *where the* arity $w \in Obj^\star$ *and the* coarity $v \in Obj^\star$ *of f are Obj-words. Cartesian left-additive Σ-terms are defined inductively:*

- *Each $f\colon w \to v$ is a Σ-term.*
- *For each $A \in Obj$, the generators (1) and (3) of the cartesian left-additive structure are Σ-terms.*

– *If* $f : w \to v$, $g : v \to u$, *and* $h : w' \to v'$ *are* Σ*-terms, then* $f \mathbin{;} g : w \to u$ *and* $f \otimes h : ww' \to vv'$ *are* Σ*-terms, represented as string diagrams*

$$w \;\text{—}\boxed{f}\text{—}^{v}\text{—}\boxed{g}\text{—}\; u \qquad\qquad \begin{array}{c} w \;\text{—}\boxed{f}\text{—}\; v \\ w' \;\text{—}\boxed{h}\text{—}\; v' \end{array}$$

Let us fix Obj, Σ *and a set* E *of equations between* Σ*-terms. The cartesian left-additive category* $\mathscr{C}$ *freely generated by* (Obj, Σ, E) *is the monoidal category with set of objects* $Obj^\star$ *and morphisms the* Σ*-terms quotiented by the axioms of cartesian left-additive categories and the equations in* E*. The monoidal product in* $\mathscr{C}$ *is given on objects by word concatenation. Identities, monoidal product and sequential composition of morphisms are given by the corresponding* Σ*-terms and their constructors* $f \otimes h$ *and* $f \mathbin{;} g$*.*

One may readily see that $\mathscr{C}$ defined in this way is indeed cartesian left-additive. We say that $\mathscr{C}$ is *presented* by generators (Obj, Σ) and equations E.

3 Reverse Derivatives and Algebraic Presentations

As we will see in Sect. 5, our argument for functional completeness relies on augmenting the algebraic presentation of polynomial circuits with an additional operation. To formulate such result, we first need to better understand how reverse differential combinators may be defined compatibly with the generators and equations presenting a category.

Theorem 2. *Let* $\mathscr{C}$ *be the cartesian left-additive category presented by generators* (Obj, Σ) *and equations* E*. If for each* $s \in \Sigma$ *there is some* $\mathsf{R}[s]$ *which is well-defined (see Remark 3) with respect to* E*, and which satisfies axioms ARD.1–4, then* $\mathscr{C}$ *is a reverse derivative category.*

Proof. Observe that axioms ARD.1 fix the definition of R on composition, tensor product and the cartesian and left-additive structures. It therefore suffices to show that axioms ARD.2–4 are preserved by composition and tensor product. That is, for morphisms f, g of appropriate types, both $f \mathbin{;} g$ and $f \otimes g$ preserve axioms ARD.2–4. Thus, any morphism constructed from generators must also satisfy the axioms ARD.1–4, and $\mathscr{C}$ must be an RDC. Showing that ARD.2–4 are preserved by composition and tensor product can be done graphically, but we omit the proofs here.

Remark 3. In the statement of Theorem 2, strictly speaking $s \in \Sigma$ is just a representative of the equivalence class of Σ-terms (modulo E plus the laws of left-additive cartesian categories) defining a morphism in $\mathscr{C}$. Because of this, we require $\mathsf{R}[s]$ to be 'well-defined', in the sense that if s and t are representatives of the same morphisms of $\mathscr{C}$, then the same should hold for $\mathsf{R}[s]$ and $\mathsf{R}[t]$. In a nutshell, we are allowed to define R directly on Σ-terms, provided our definition is compatible with E and the laws of left-additive cartesian categories.

An immediate consequence of Theorem 2 is that if we have a presentation of an RDC $\mathscr{C}$, we can 'freely extend' it with an additional operation s, a chosen reverse derivative $\mathsf{R}[s]$, and equations E', so long as R is well-defined with respect to E' and the axioms ARD.2–4 hold for $\mathsf{R}[s]$. Essentially, this gives us a simple recipe for adding new 'gadgets' to existing RDCs and ensuring they retain RDC structure.

One particularly useful such 'extension' is the addition of a *multiplication* morphism that distributes over the addition . We define categories with such a morphism as an extension of cartesian left-additive categories as follows:

Definition 4. *A* ***Cartesian Distributive Category*** *is a cartesian left-additive category such that each object A is equipped with a commutative monoid and unit which distributes over the addition . More completely, it is a category having generators*

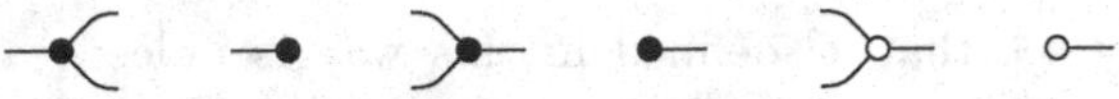

satisfying the cartesianity *equations* (4), *the* left-additivity *equations* (2), *the* multiplicativity *equations*

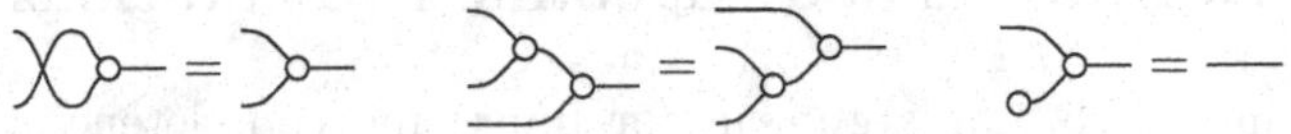

$$(5)$$

and the distributivity *and* annihilation *equations*

$$(6)$$

Just as for cartesian left-additive categories, one may construct cartesian distributive categories freely from a set of objects Obj, a signature Σ, and equations E, the difference being that Σ-term will be constructed using also and , and quotiented also by (5)–(6). The main example of cartesian distributive categories are *Polynomial Circuits*, which we define in Sect. 4 below.

Reverse derivative categories define a reverse differential combinator on a left-additive cartesian structure. As cartesian distributive categories properly extend left-additive ones, it is natural to ask how we may extend the definition of the reverse differential combinator to cover the extra operations and . The following theorem provide a recipe, which we will use in the next section to study RDCs with a cartesian distributive structure. Note that the definition of R^* below is a string diagrammatic version of the reverse derivative combinator defined on POLY in [10].

Theorem 3. *Suppose $\mathscr{C}$ is a left-additive cartesian category presented by (Obj, Σ, E), and assume $\mathscr{C}$ is also an RDC, say with reverse differential combinator R. Then the cartesian distributive category $\mathscr{C}^*$ presented by (Obj, Σ, E), with reverse differential combinator R^* defined as R on the left-additive cartesian structure, and as follows*

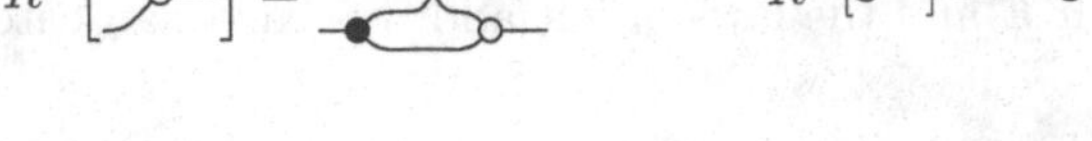

$$R^*[\ ] = \qquad R^*[\ ] = \qquad (7)$$

on the extra distributive structure, is also an RDC.

Proof. It suffices to check that R is well-defined with respect to the additional equations of cartesian distributive categories, and that the new generators and satisfy axioms ARD.2–4.

4 Polynomial Circuits

Our motivating example of cartesian distributive categories is that of *polynomial circuits*, whose morphisms can be thought of as representing polynomials over a commutative semiring. We define them as follows:

Definition 5. *Let S be a commutative semiring. We define* $\mathsf{PolyCirc}_S$ *as the cartesian distributive category presented by (I) one generating object* 1, *(II) for each $s \in S$, a generating morphism* $\boxed{s}$ $: 0 \to 1$, *(III) the 'constant' equations*

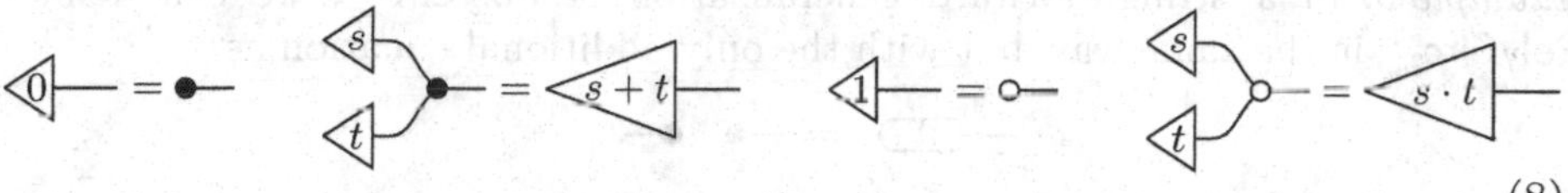

$$0 = \bullet \qquad s, t \mapsto s+t \qquad 1 = \circ \qquad s, t \mapsto s \cdot t \tag{8}$$

for $s, t \in S$, intuitively saying that the generating morphisms respect addition and multiplication of S.

Proposition 1. $\mathsf{PolyCirc}_S$ *is an RDC with* $\mathsf{R}\left[\boxed{s}\right] = \bullet$.

Proof. The type of $\mathsf{R}\left[\boxed{s}\right] : 1 \to 0$ implies that there is only one choice of reverse derivative, namely the unique discard map $\bullet$. Furthermore, R is well-defined with respect to the constant equations (8) for the same reason. Finally, observe that the axioms ARD.2–4 hold for $\mathsf{R}\left[\boxed{s}\right]$, precisely in the same way as for $\mathsf{R}[\bullet]$, and so $\mathsf{PolyCirc}_S$ is an RDC.

Although our Definition 5 of $\mathsf{PolyCirc}_S$ requires that we add an axiom for each possible addition and multiplication of constants, for some significant choices of S an equivalent smaller finite axiomatisation is possible. We demonstrate this with some examples.

Example 1. In the case of $\mathsf{PolyCirc}_{\mathbb{Z}_2}$, the equations of Definition 5 reduce to the single equation

expressing that $x + x = 0$ for both elements of the field $\mathbb{Z}_2$.

Example 2. In the case $\mathsf{PolyCirc}_{\mathbb{N}}$ of the semiring of natural numbers, with the usual addition and multiplication, no extra generating morphisms or equations are actually necessary: all those appearing in Definition 5 may be derived from the cartesian distributive structure. To see why, notice that we may define each constant $s \in S$ as repeated addition:

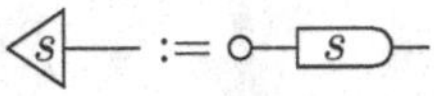

where we define 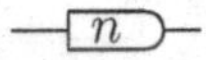 inductively as

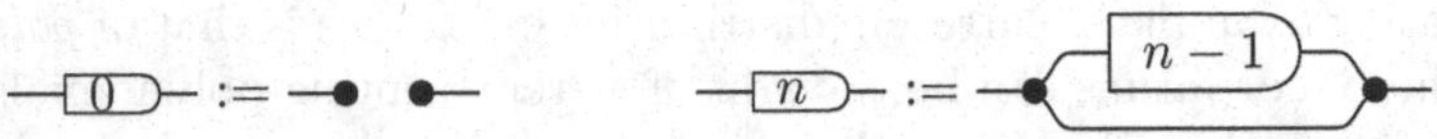

The equations expressing addition and multiplication in $\mathbb{N}$ are then a consequence of those of cartesian distributive categories. In fact, from this observation we have that $\mathsf{PolyCirc}_{\mathbb{N}}$ is the free cartesian distributive category on one generating object.

Example 3. In a straightforward generalization of $\mathsf{PolyCirc}_{\mathbb{Z}_2}$, we can define $\mathsf{PolyCirc}_{\mathbb{Z}_n}$ in the same way, but with the only additional equation as

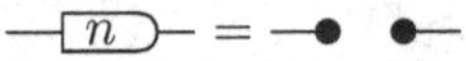

which says algebraically that $(1 + \overset{n}{\ldots} + 1) \cdot x = n \cdot x = 0 \cdot x = 0$.

It is important to note that $\mathsf{PolyCirc}_S$ is isomorphic to the category POLY_S, defined as follows:

Definition 6. POLY_S *is the symmetric monoidal category with objects the natural numbers and arrows $m \to n$ the n-tuples of polynomials in m indeterminates:*

$$\langle p_1(\vec{x}), \ldots, p_n(\vec{x}) \rangle : m \to n$$

with each

$$p_i \in S[x_1, \ldots, x_m]$$

where $S[x_1, \ldots x_m]$ denotes the polynomial ring in m indeterminates over S.

The isomorphism $\mathsf{PolyCirc}_S \cong \mathsf{POLY}_S$ is constructed by using that homsets $\mathsf{PolyCirc}_S(m, n)$ and $\mathsf{POLY}_S(m, n)$ have the structure of the free module over the polynomial ring $S[x_1 \ldots x_m]^n$ which yields a unique module isomorphism between them. We do not prove this isomorphism here, other than to say that it follows by the same argument as presented in [10, Appendix A].

Remark 4. Note in [10] POLY_S is proven to be a reverse derivative category, meaning that we could have derived Proposition 1 as a corollary of the isomorphism $\mathsf{PolyCirc}_S \cong \mathsf{POLY}_S$. We chose to provide a 'native' definition of the reverse differential combinator of $\mathsf{PolyCirc}_S$ because–as we will see shortly–we will need to extend it with an additional generator. The reason for this is to gain the property of 'functional completeness', which will allow us to express any function $S^m \to S^n$. This new derived category will in general no longer be isomorphic to POLY_S, and so we must prove it too is an RDC: we do this straightforwardly using Theorem 2.

Remark 5. When S is a bonafide ring, we may account for its inverse by extending $\mathsf{PolyCirc}_S$ with a 'negate' generating morphism —■—, together with the additional equation —●⟨■⟩●— = —● ●—. Then Theorem 2 suggests us how to extend the reverse differential combinator of $\mathsf{PolyCirc}_S$ to this new category:

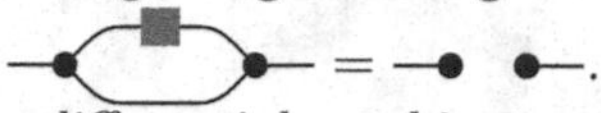

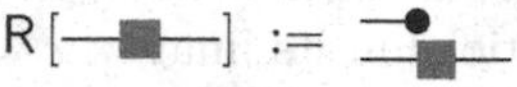

$$\mathsf{R}[—■—] := \ldots$$

5 Functional Completeness

We are now ready to consider the *expressivity* of the model class of polynomial circuits. More concretely, for a given commutative semiring S, we would like to be able to represent any function between sets $S^m \to S^n$ as a string diagram in $\mathsf{PolyCirc}_S$. This property, which we call 'functional completeness', is important for a class of machine learning models to satisfy because it guarantees that we may always construct an appropriate model for a given dataset. It has been studied, for instance, in the context of the various 'universal approximation' theorems for neural networks (see e.g. [16,18]).

To formally define functional completeness, let us fix a finite set S. Recall the cartesian monoidal category FinSet_S, whose objects are natural numbers and a morphism $m \to n$ is a function of type $S^m \to S^n$.

Definition 7. *We say a category $\mathscr{C}$ is* **functionally complete** *with respect to a finite set S when there a full identity-on-objects functor $\mathsf{F} : \mathscr{C} \to \mathsf{FinSet}_S$.*

The intuition for Definition 7 is that we call a category $\mathscr{C}$ 'functionally complete' when it suffices as a syntax for FinSet_S—that is, by fullness of F we may express any morphism in FinSet_S. Note however that two distinct morphisms in $\mathscr{C}$ may represent the same function—F is not necessarily faithful.

In general, $\mathsf{PolyCirc}_S$ is not functionally complete with respect to S. Take for example the boolean semiring $\mathbb{B}$ with multiplication and addition as AND and OR respectively. It is well known [21] that one cannot construct every function of type $\mathbb{B}^m \to \mathbb{B}^n$ from only these operations.

Nonetheless, we claim that in order to make $\mathsf{PolyCirc}_S$ functionally complete it suffices to add to its presentation just one missing ingredient: the 'comparator' operation, which represents the following function:

$$\mathsf{compare}(x, y) = \begin{cases} 1 & \text{if } x = y \\ 0 & \text{otherwise} \end{cases}$$

The following result clarifies the special role played by the comparator.

Theorem 4. *Let S be a finite commutative semiring. A category $\mathscr{C}$ is functionally complete with respect to S iff. there is a monoidal functor $\mathsf{F} : \mathscr{C} \to \mathsf{FinSet}_S$ in whose image are the following functions:*

- *$\langle\rangle \mapsto s$ for each $s \in S$ (constants)*

- $\langle x, y\rangle \mapsto x + y$ *(addition)*
- $\langle x, y\rangle \mapsto x \cdot y$ *(multiplication)*
- compare

Proof. Suppose $\mathscr{C}$ is functionally complete with respect to S, where S is a finite commutative semiring. Then by definition there is a functor $\mathsf{F} : \mathscr{C} \to \mathsf{FinSet}_S$ with each of the required functions in its image.

Now in the reverse direction, we will show that any function can be constructed only from constants, addition, multiplication, and comparison. The idea is that because S is finite, we can simply encode the function table of any function $f : S^m \to S$ as the following expression:

$$x \mapsto \sum_{s \in S^m} \mathsf{compare}(s, x) \cdot f(s) \tag{9}$$

Further, since $\mathscr{C}$ is cartesian, we may decompose any function $f : S^m \to S^n$ into an n-tuple of functions of type $S^m \to S$. More intuitively, for each of the n outputs, we simply look up the appropriate output in the encoded function table.

It follows immediately that $\mathsf{PolyCirc}_S$ is functionally complete with respect to S if and only if one can construct the compare function in terms of constants, additions, and multiplications. We illustrate one such case below.

Example 4. $\mathsf{PolyCirc}_{\mathbb{Z}_p}$ is functionally complete for prime p. To see why, recall Fermat's Little Theorem [12], which states that

$$a^{p-1} \equiv 1 (\mathrm{mod} p)$$

for all $a > 0$. Consequently, we have that

$$(p-1) \cdot a^{p-1} + 1 = \begin{cases} 1 & \text{if } a = 0 \\ 0 & \text{otherwise} \end{cases}$$

We denote this function as $\delta(a) := (p-1) \cdot a^{p-1} + 1$ to evoke the dirac delta 'zero indicator' function. To construct the compare function is now straightforward:

$$\mathsf{compare}(x_1, x_2) = \sum_{s \in S} \delta(x_1 + s) \cdot \delta(x_2 + s)$$

However, as we already observed, it is not possible in general to construct the compare function in terms of multiplication and addition. Therefore, to guarantee functional completeness we must *extend* the category of polynomial circuits with an additional comparison operation.

Definition 8. *We define by* $\mathsf{PolyCirc}_S^{=}$ *as the cartesian distributive category presented by the same objects, operations, and equations of* $\mathsf{PolyCirc}_S$*, with the addition of a 'comparator' operation*

$$\boxed{=} \tag{10}$$

and equations

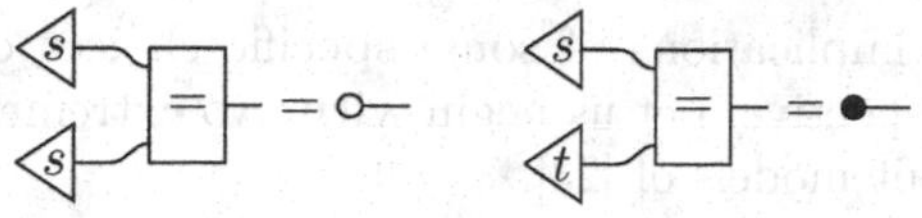

(11)

for $s, t \in S$ *with* $s \neq t$.

To make $\mathsf{PolyCirc}_S^{=}$ a reverse derivative category, we can once again appeal to Theorem 2. However, we must choose an apropriate definition of R[compare] which is well-defined and satisfies axioms ARD.1–4.

A suggestion for this choice comes from the machine learning literature. In particular, the use of the 'straight-through' estimator in quantized neural networks, as in e.g. [4]. Typically, these networks make use of the dirac delta function in the forward pass, but this causes a catastrophic loss of gradient information in the backwards pass since the gradient is zero almost everywhere. To fix this, one uses the *straight-through estimator*, which instead passes through gradients directly from deeper layers to shallower ones.

In terms of reverse derivatives, this amounts to setting $\mathsf{R}[\delta] = \mathsf{R}[\mathsf{id}]$. Of course, we need to define R for the full comparator, not just the zero-indicator function δ, and so we make the following choice:

Theorem 5. $\mathsf{PolyCirc}_S^{=}$ *is an RDC with* R *as for* $\mathsf{PolyCirc}_S$, *and*

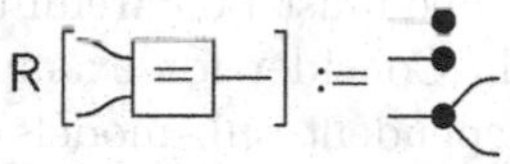

R[ ] :=

Proof. R is well-defined with respect to the equations (11) since both sides of each equation must equal the unique discard morphism. Further, R[ ] satisfies axioms ARD.2–4 in the same way that R[ ] does, and so by Theorem 2 $\mathsf{PolyCirc}_S^{=}$ is a reverse derivative category.

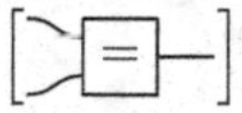

From Theorem 4, we may derive:

Corollary 1. $\mathsf{PolyCirc}_S^{=}$ *is functionally complete with respect to* S.

Finally, note that we recover the dirac delta function by 'capping' one of the comparator's inputs with the zero constant:

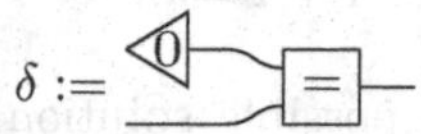

$\delta :=$

whose reverse derivative is equivalent to the 'straight-through' estimator:

R[ ] = = R[—]

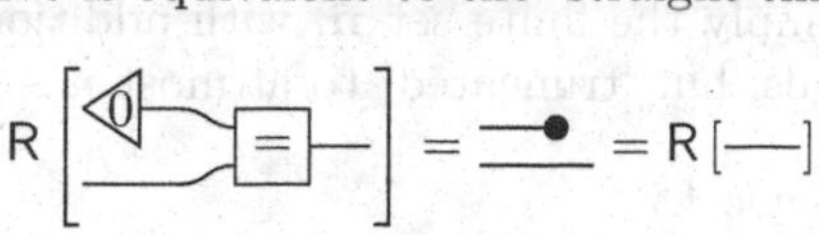

6 Polynomial Circuits in Machine Learning: Case Studies

We now discuss the implications of some specific choices of semiring from a machine learning perspective. Let us begin with two extremes: neural networks, and the boolean circuit models of [23].

Neural Networks. We may think of a neural network as a circuit whose wires carry values in $\mathbb{R}$. Of course, in order to compute with such circuits we must make a finite approximation of the reals–typically using floating-point numbers. However, this approximation introduces two key issues. First, floating point arithmetic is significantly slower than integer arithmetic. Second, the floating point operations of addition and multiplication are not even associative, which introduces problems of *numerical instability*. Although attempts exist to address issues of floating point arithmetic (such as 'posits' [15]), these still do not satisfy the ring axioms; to properly account for these approximations would require additional work.

Boolean Circuits and $\mathbb{Z}_2$. One may note that since we must always eventually deal with finite representations of values, we may as well attempt to define our model class directly in terms of them. This is essentially the idea of [23]: the authors use the category $\mathsf{PolyCirc}_{\mathbb{Z}_2}$ (which they call simply **PolyCirc**) as a model class since it is already functionally complete[2] and admits a reverse derivative operator. However, using a semiring of modular arithmetic in general introduces a different problem: one must be careful to construct models so that gradients do not 'wrap around'. Consider for example the model below, which can be thought of as two independent sub-models f_1 and f_2 using the same parameters[3] but applied to different parts of the input X_1 and X_2

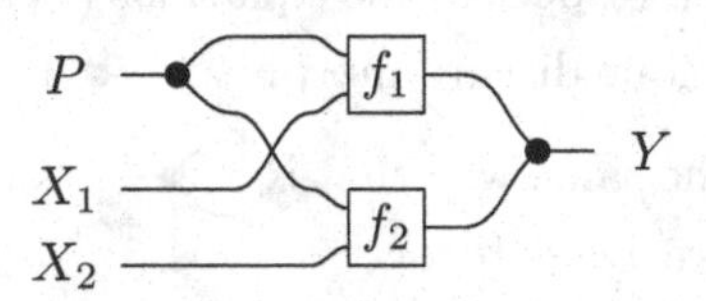

Since $\mathsf{R}[\ ] = \mathsf{R}[\ ]$, when we compute the gradient update for P we will sum the gradients of f_1 and f_2. In the extreme case when the underlying semiring is $\mathbb{Z}_2$, then when the gradients of f_1 and f_2 are both 1, the result will 'wrap around' to 0 and P will not be updated. This is clearly undesirable: here we should prefer that $1 + 1 = 1$ to $1 + 1 = 0$.

Saturating Arithmetic. Another possible solution is to use the semiring Sat_n as a model of *saturating unsigned integer arithmetic* for a given 'precision' n. The underlying set is simply the finite set $\mathbf{\bar{n}}$, with addition and multiplication defined as for the naturals, but 'truncated' to at most $n - 1$. We define Sat_n as

---

[2] We discuss why in Example 4.

[3] This approach is called 'weight-tying' in neural networks literature.

follows, noting that it is equivalent to the semiring $B(n, n-1)$ first defined in [2, Example 3] (see also [14]).

Definition 9. *The semiring* Sat_n *has as addition and multiplication the operations*

$$x_1 + x_2 := \min(n-1, x_1 + x_2) \qquad x_1 \cdot x_2 := \min(n-1, x_1 \cdot x_2)$$

over the set $\bar{\mathrm{n}} := \{0 \ldots n-1\}$

Note that while Sat_n is a commutative semiring, it is certainly *not* a ring: the introduction of inverses means that the associativity axiom of semirings is violated.

Finally, note that for each of these choices of semiring S, in general $\mathsf{PolyCirc}_S$ is not functionally complete. Thus, in order to obtain a model class which is functionally complete and is a reverse derivative category, we must use $\mathsf{PolyCirc}_S^{=}$.

7 Conclusions and Future Work

In this paper, we studied in terms of algebraic presentations categories of polynomial circuits, whose reverse derivative structure makes them suitable for machine learning. Further, we showed how this class of categories is functionally complete for finite number representations, and therefore provides sufficient expressiveness. There remain however a number of opportunities for theoretical and empirical work.

On the empirical side, we plan to use this work combined with data structures and algorithms like that of [22] as the basis for practical machine learning tools. Using these tools, we would like to experimentally verify that models built using semirings like those presented in Sect. 6 can indeed be used to develop novel model architectures for benchmark datasets.

There also remains a number of theoretical avenues for research. First, we want to generalise our approach to functional completeness to the continuous case, and then to more abstract cases such as polynomial circuits over the Burnside semiring. Second, we want to extend the developments of Sect. 3 in order to provide a reverse derivative structure for circuits with notions of feedback and delay, such as the stream functions described in [13].

References

1. Abadi, M., et al.: TensorFlow: large-scale machine learning on heterogeneous systems (2015). https://www.tensorflow.org/
2. Alarcón, F., Anderson, D.: Commutative semirings and their lattices of ideals. Houston J. Math. **20** (1994). https://www.math.uh.edu/~hjm/vol20-4.html
3. Baez, J.C., Coya, B., Rebro, F.: Props in network theory (2017). http://www.tac.mta.ca/tac/volumes/33/25/33-25abs.html

4. Bengio, Y., Léonard, N., Courville, A.: Estimating or propagating gradients through stochastic neurons for conditional computation (2013). https://doi.org/10.48550/ARXIV.1308.3432, https://arxiv.org/abs/1308.3432
5. Blute, R.F., Cockett, J.R.B., Seely, R.A.G.: Cartesian differential categories. Theory App. Categ. **22**, 622–672 (2009). https://emis.univie.ac.at/journals/TAC/volumes/22/23/22-23abs.html
6. Bonchi, F., Gadducci, F., Kissinger, A., Sobocinski, P., Zanasi, F.: String diagram rewrite theory i: rewriting with frobenius structure (2020). https://doi.org/10.48550/ARXIV.2012.01847
7. Bonchi, F., Gadducci, F., Kissinger, A., Sobocinski, P., Zanasi, F.: String diagram rewrite theory ii: Rewriting with symmetric monoidal structure (2021). https://doi.org/10.48550/ARXIV.2104.14686
8. Bonchi, F., Gadducci, F., Kissinger, A., Sobociński, P., Zanasi, F.: String diagram rewrite theory iii: confluence with and without frobenius (2021). https://doi.org/10.48550/ARXIV.2109.06049
9. Choi, J., et al.: Accurate and efficient 2-bit quantized neural networks. In: Talwalkar, A., Smith, V., Zaharia, M. (eds.) Proceedings of Machine Learning and Systems. vol. 1, pp. 348–359 (2019). https://proceedings.mlsys.org/paper/2019/file/006f52e9102a8d3be2fe5614f42ba989-Paper.pdf
10. Cockett, R., Cruttwell, G., Gallagher, J., Lemay, J.S.P., MacAdam, B., Plotkin, G., Pronk, D.: Reverse derivative categories (2019). https://doi.org/10.4230/LIPIcs.CSL.2020.18
11. Cruttwell, G.S.H., Gavranović, B., Ghani, N., Wilson, P., Zanasi, F.: Categorical foundations of gradient-based learning (2021). https://doi.org/10.48550/ARXIV.2103.01931
12. de Fermat, P.: Letter to frénicle de bessy (1640)
13. Ghica, D.R., Kaye, G., Sprunger, D.: Full abstraction for digital circuits (2022). https://doi.org/10.48550/ARXIV.2201.10456
14. Golan, J.S.: Semirings and their Applications. Springer, Dordrecht, Netherlands (2010). https://doi.org/10.1007/978-94-015-9333-5
15. Gustafson, J.L., Yonemoto, I.T.: Beating floating point at its own game: posit arithmetic. Supercomput. Front. Innov. **4**(2), 71–86 (2017). https://doi.org/10.14529/jsfi170206
16. Hornik, K., Stinchcombe, M., White, H.: Multilayer feedforward networks are universal approximators. Neural Netw. **2**(5), 359–366 (1989). https://doi.org/10.1016/0893-6080(89)90020-8
17. Lafont, Y.: Towards an algebraic theory of Boolean circuits. J. Pure Appl. Algebra **184**(2–3), 257–310 (2003). https://doi.org/10.1016/S0022-4049(03)00069-0
18. Leshno, M., et al.: Multilayer feedforward networks with a nonpolynomial activation function can approximate any function. Neural Netw. **6**(6), 861–867 (1993). https://doi.org/10.1016/s0893-6080(05)80131-5
19. Paszke, A., et al.: Pytorch: an imperative style, high-performance deep learning library. In: Advances in Neural Information Processing Systems, vol. 32, pp. 8024–8035. Curran Associates, Inc. (2019)
20. Selinger, P.: A survey of graphical languages for monoidal categories. In: Coecke, B. (eds) New Structures for Physics. Lecture Notes in Physics, vol. 813, pp. 289–355. Springer, Berlin (2010). https://doi.org/10.1007/978-3-642-12821-9_4
21. Wernick, W.: Complete sets of logical functions. Trans. Am. Math. Soc. **51**(1), 117 (1942). https://doi.org/10.2307/1989982

22. Wilson, P., Zanasi, F.: The cost of compositionality: a high-performance implementation of string diagram composition (2021). https://doi.org/10.48550/ARXIV.2105.09257
23. Wilson, P., Zanasi, F.: Reverse derivative ascent: a categorical approach to learning Boolean circuits. Electron. Proc. Theoret. Comput. Sci. **333**, 247–260 (2021). https://doi.org/10.4204/eptcs.333.17
24. Zanasi, F.: Interacting HOPF algebras: the theory of linear systems (2018). https://doi.org/10.48550/ARXIV.1805.03032

Application Domains

A Generic Construction for Crossovers of Graph-Like Structures

Gabriele Taentzer(✉), Stefan John(✉), and Jens Kosiol(✉)

Philipps-Universität Marburg, Marburg, Germany
{taentzer,johns,kosiolje}@mathematik.uni-marburg.de

Abstract. In model-driven optimization (MDO), domain-specific models are used to define and solve optimization problems with evolutionary algorithms. Models are typically evolved using mutations, which can be formally specified as graph transformations. So far, only mutations have been used in MDO to generate new solutions from existing ones; a crossover mechanism has not yet been elaborated. In this paper, we present a generic crossover construction for graph-like structures that can be used to implement crossover operators in MDO. We prove basic properties of our construction and show how it can be used to implement a whole set of crossover operators that have been proposed for specific problems and situations on graphs.

Keywords: Evolutionary Computation · Crossover · Model-driven optimization · Category Theory

1 Introduction

In software development, software engineers often make design decisions in the context of competing constraints ranging from requirements to technology. To efficiently find optimal solutions, Search-Based Software Engineering (SBSE) [16] attempts to formulate software engineering problems as optimization problems that capture the constraints of interest as objectives. By using meta-heuristic search techniques, good solutions can often be found with reasonable effort. Because of their generality, evolutionary algorithms, and in particular genetic algorithms [5,17] that use mutation, crossover, and selection to perform a guided search over the search space, are a technique of particular relevance. According to e.g. [13], the definition of an evolutionary algorithm requires a representation of problem instances and search space elements (i.e., solutions). It also includes a formulated optimization problem that clarifies which of the solutions are *feasible* (i.e., satisfy all constraints of the optimization problem) and best satisfy the objectives. The key ingredients of the optimization process are a procedure for generating a start population of solutions, a mechanism for generating new solutions from existing ones (e.g., by mutation and crossover), a selection mechanism that typically establishes the evolutionary concept of survival of the fittest, and

N. Behr and D. Strüber (Eds.): ICGT 2022, LNCS 13349, pp. 97–117, 2022.
https://doi.org/10.1007/978-3-031-09843-7_6

a condition for stopping evolutionary computations. Selecting these ingredients so that an evolutionary algorithm is effective and efficient is usually a challenge.

Model-driven optimization (MDO) aims at reducing the required level of expertise of users of meta-heuristic techniques. Two main approaches have emerged in MDO: the model-based approach [7,8] performs optimization directly on models, while the rule-based approach [1,4] searches for optimized model transformation sequences. In this paper, we focus on the model-based approach since it tends to be more effective [20] and refer to it as MDO for short. In MDO, optimization problems are specified as models that capture domain-specific information about a problem and its solutions. In that way, users can interact with a domain-specific formulation of their problem, rather than traditional encodings that are typically closer to implementation. While the search space consists of models, the mutation of search space elements is specified by model transformations. In sophisticated evolutionary algorithms, mutations typically perform local changes, while crossovers are used to generate offspring by recombining existing search space elements. For (the model-based approach to) MDO, no crossover mechanism has been worked out yet. This paper fills this research gap and presents a crossover construction for graph-based models.

Several graph-based approaches to crossover have been suggested in the literature, e.g. [27,29]. In most cases, these crossovers are not *generic* (in the sense of different kinds of graphs), but are designed with specific semantics of the underlying graphs in mind. We aim to develop a generic construction of crossovers that can be applied to different kinds of graph-like structures. Moreover, this construction of crossovers is applicable regardless of the semantics of the graphs of interest. We also prove the correctness and completeness of our crossover construction.

The paper is organized as follows: We start with an example MDO problem and discuss a possible crossover in this context in Sect. 2. Section 3 recalls preliminaries. The main contribution of this paper, a pushout-based crossover construction, is presented in Sect. 4. In Sect. 5, we explain how our new crossover construction encompasses important, more specific approaches to crossover (on graph-like structures) that have been suggested in the literature. We close with a discussion of related work and a conclusion in Sects. 6 and 7. All proofs are given in Appendix A.

2 Running Example

The CRA case [6] is an optimization problem from the domain of software design that has recently established itself as an easily understood use case in the context of MDO. Given a software product represented by a set of features (i.e., attributes and methods) and dependency relations between them, the task is to modularize the software by encapsulating its features into classes. Two well-known quality aspects are used to evaluate the quality of solutions: cohesion and coupling. Cohesion rewards classes in which features are highly interdependent, while coupling captures the interdependencies of features that exist between classes. A highly cohesive design with low coupling is considered easy to understand and maintain. Therefore, maximizing cohesion and minimizing coupling are the opposing objectives of the CRA case.

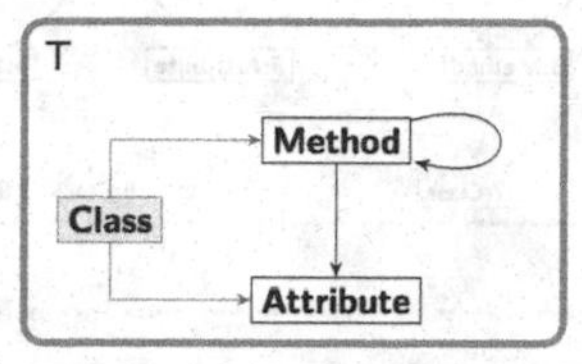

Fig. 1. Type graph of the CRA case. White solid elements specify invariant problem parts, the red colored class element and its relations are solution specific.

The structure of models in the CRA case can be defined by the type graph shown in Fig. 1. A problem instance consists at least of the features and their dependencies. These elements form the invariant part of a concrete problem. Classes (and their relationships), on the other hand, can be added, modified and removed to explore the search space and create new solutions. Typical mutations for the CRA case include small changes like adding or removing a class, assigning a feature to a class, or changing the assignment of a feature from one class to another. Mutation usually does not consider already well optimized substructures that might be worth being shared with other solutions.

In the CRA case, a subset of features, along with their current assignment to classes, contains potentially valuable information. The exchange of this information between two solutions represents a promising crossover as we will see in the following example. Consider solutions $\overline{\mathsf{E}}$ and $\overline{\mathsf{F}}$ in Fig. 2, for a problem instance consisting of four methods and two attributes. Let a crossover choose to recombine them by exchanging their assignment information for the features 1:Method, 2:Attribute and 3:Method. This results in two offspring solutions. Solution $\overline{\mathsf{E}^1\mathsf{F}^2}$ keeps the original assignments of 4:Method, 5:Attribute, and 6:Method as found in solution $\overline{\mathsf{F}}$ and combines them with the assignments of $\overline{\mathsf{E}}$ for the exchanged features. The solution $\overline{\mathsf{E}^2\mathsf{F}^1}$ is constructed in the opposite way.

Note that combining 1:Method, 2:Attribute and 3:Method into one class (as done in solution $\overline{\mathsf{E}}$) seems a reasonable choice. Their pairwise dependencies promote cohesion, while splitting them would lead to coupling. The same is true for the features of class 12: Class in solution $\overline{\mathsf{F}}$. Consequently, the offspring $\overline{\mathsf{E}^1\mathsf{F}^2}$ combines the best of both worlds.

3 Preliminaries: $\mathcal{M}$-Adhesive Categories

In this section, we briefly recall our central formal preliminaries, namely $\mathcal{M}$-*adhesive categories* and $\mathcal{M}$-*effective unions* [12], which provide the setting in which we formulate our contribution. $\mathcal{M}$-adhesive categories with $\mathcal{M}$-effective unions are categories where pushouts along certain monomorphisms interact in a particularly nice way with pullbacks. This is of importance because our construction of crossovers is based on pushouts. Moreover, working in the framework of $\mathcal{M}$-adhesive categories allows us to easily abstract from the concrete choice of graphs used to formalize the models of interest (such as typed, labeled, and attributed graphs). We only use category-theoretic concepts that are common in the context of algebraic graph transformation, and refer to [11,12] for introductions.

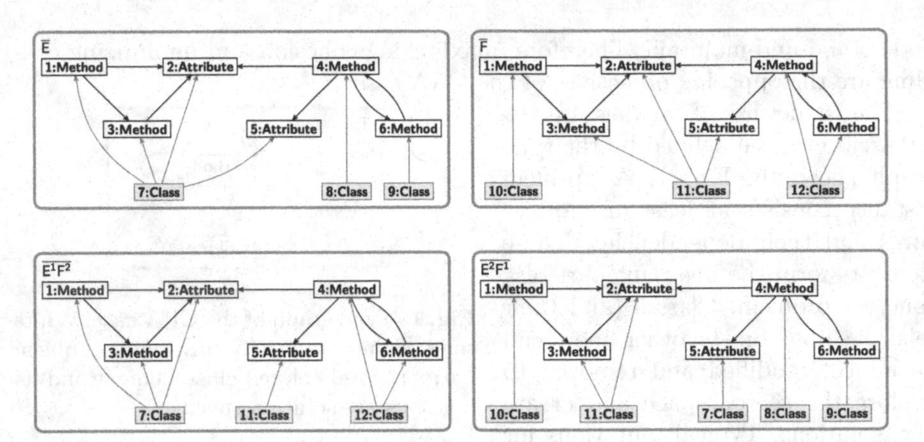

Fig. 2. Example crossover in the CRA case that creates the offspring $\overline{E^1F^2}$ and $\overline{E^2F^1}$ by exchanging the assignments of features 1:Method, 2:Attribute, and 3:Method between the solutions $\overline{E}$ and $\overline{F}$.

Definition 1 ($\mathcal{M}$-adhesive category). *A category $\mathcal{C}$ with a morphism class $\mathcal{M}$ is an* $\mathcal{M}$-adhesive category *if the following properties hold:*

- *$\mathcal{M}$ is a class of monomorphisms closed under isomorphisms (f isomorphism implies that $f \in \mathcal{M}$), composition ($f, g \in \mathcal{M}$ implies $g \circ f \in \mathcal{M}$), and decomposition ($g \circ f, g \in \mathcal{M}$ implies $f \in \mathcal{M}$).*
- *$\mathcal{C}$ has pushouts and pullbacks along $\mathcal{M}$-morphisms, i.e., pushouts and pullbacks where at least one of the given morphisms is in $\mathcal{M}$, and $\mathcal{M}$-morphisms are closed under pushouts and pullbacks, i.e., given a pushout like the left square in Fig. 3a, $m \in \mathcal{M}$ implies $n \in \mathcal{M}$ and, given a pullback, $n \in \mathcal{M}$ implies $m \in \mathcal{M}$.*
- *Pushouts in $\mathcal{C}$ along $\mathcal{M}$-morphisms are* vertical weak van Kampen squares, *i.e., for any commutative cube in $\mathcal{C}$ (as in the right part of Fig. 3a) where we have the pushout with $m \in \mathcal{M}$ in the bottom, $b, c, d \in \mathcal{M}$, and pullbacks as back faces, the top is a pushout if and only if the front faces are pullbacks.*

We speak of $\mathcal{M}$-adhesive categories $(\mathcal{C}, \mathcal{M})$ and indicate arrows from $\mathcal{M}$ as hooked arrows in diagrams. Examples of categories that are $\mathcal{M}$-adhesive include sets with injective functions, graphs with injective graph morphisms and various varieties of graphs with special forms of injective graph morphisms. In particular, typed attributed graphs form an $\mathcal{M}$-adhesive category (where the class $\mathcal{M}$ consists of injective morphisms where the attribute part is an isomorphism).

The existence of $\mathcal{M}$-*effective unions* ensures that the $\mathcal{M}$-subobjects of a given object form a lattice.

Definition 2 ($\mathcal{M}$-effective unions). *An $\mathcal{M}$-adhesive category $(\mathcal{C}, \mathcal{M})$ has* $\mathcal{M}$-effective unions *if for each pushout of a pullback of a pair of $\mathcal{M}$-morphisms the induced mediating morphism belongs to $\mathcal{M}$ as well, i.e., if in each diagram like*

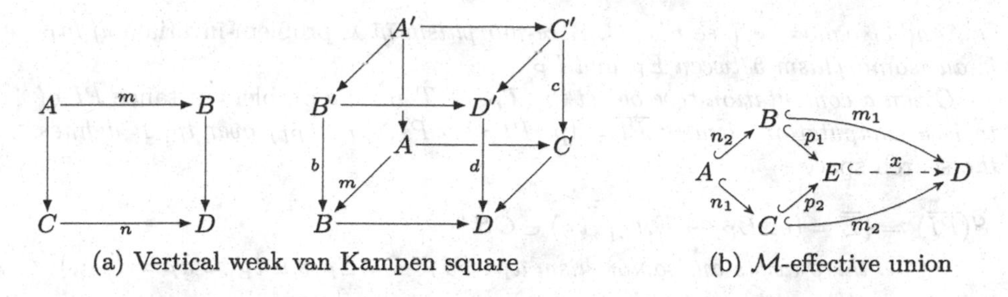

(a) Vertical weak van Kampen square (b) $\mathcal{M}$-effective union

Fig. 3. Defining $\mathcal{M}$-adhesive categories with $\mathcal{M}$-effective unions

the one depicted in Fig. 3b where the outer square is a pullback of $\mathcal{M}$-morphisms and the inner one a pushout, the induced morphism x is an $\mathcal{M}$-morphism.

4 A Pushout-Based Crossover Construction

In this section, we develop our approach to crossover. We start with introducing the objects to which crossover will be applied.

In MDO, optimization problems are defined based on modeling languages, typically specified with meta-models. Various MDO approaches in the literature such as [7,8] have chosen to represent problem instances and solutions by models. Both can contain invariant problem parts as well as solution specific parts, a distinction typically embedded in the associated meta-model. In our formalization, this is reflected in the fact that a *computation element* is given by an object that conforms to a *computation type object*. The type object specifies which parts of a computation element are invariant and which parts contribute to the solution. A concrete problem to be optimized is given by a *problem instance*; every computation element can serve as such. The *search space* of a problem instance includes all computation elements with the same problem object as specified by the given problem instance. In MDO, problem instances and solutions are typically further constrained by additional conditions. We leave this refinement to future work.

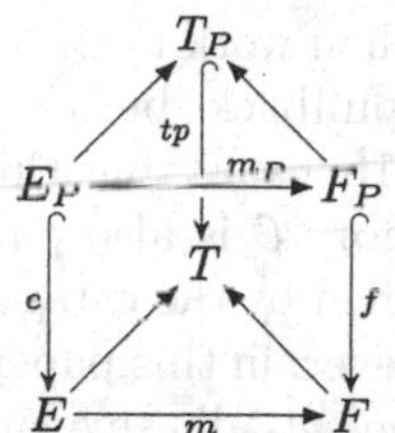

Fig. 4. Computation elements and ce-morphism

Definition 3 (Computation element. Problem instance. Search space). *Let $(\mathcal{C}, \mathcal{M})$ be an $\mathcal{M}$-adhesive category. A* computation type object *in $\mathcal{C}$ is an $\mathcal{M}$-morphism $tp\colon T_P \hookrightarrow T$; T_P is called the* problem type object. *A* computation element $\overline{E} = (e\colon E_P \hookrightarrow E, t_{E_P}, t_E)$ *over tp is an $\mathcal{M}$-morphism e together with* typing morphisms $t_{E_P}\colon E_P \to T_P$ *and $t_E\colon E \to T$ such that the induced square (over tp) is a pullback. The pair (E_P, t_{E_P}) is the* problem object *of $\overline{E}$. If defined, the initial pushout over e yields the* solution part *of $\overline{E}$, written $E \setminus E_P$.*

A computation-element morphism $\overline{m} = (m_P, m)$, *short* ce-morphism, *from computation element $\overline{E}$ to computation element $\overline{F}$ is a pair of morphisms $m_P\colon E_P \to F_P$ and $m\colon E \to F$ that are compatible with typing, i.e., $t_{F_P} \circ m_P =$*

t_{E_P} and $t_F \circ m = t_E$ (see Fig. 4). A ce-morphism $\overline{m}$ is problem-invariant *if m_P is an isomorphism between E_P and F_P.*

Given a computation type object $tp\colon T_P \hookrightarrow T$ in $\mathcal{C}$, a problem instance *$\overline{PI}$ of tp is a computation element $\overline{PI} = (p\colon PI_P \hookrightarrow PI, t_{PI_P}, t_{PI})$ over tp. It defines the* search space

$$S(\overline{PI}) := \{\overline{E} = (e\colon E_P \hookrightarrow E, t_{E_P}, t_E) \in CS \mid \text{there exists an isomorphism } a_P\colon PI_P \xrightarrow{\sim} E_P \text{ s.t. } t_{E_P} \circ a_P = t_{PI_P}\}.$$

Each element of the search space $S(\overline{PI})$ is called solution (object) *for $\overline{PI}$.*

Given a solution $\overline{E}$ for $\overline{PI}$, a subsolution *of $\overline{E}$ is a solution $\overline{E^1}$ from the search space $S(\overline{PI})$ such that there exists a problem-invariant ce-morphism $\overline{s^1}$ from $\overline{E^1}$ to $\overline{E}$ where $s^1 \in \mathcal{M}$.*

Before providing an example, some remarks with respect to the above definition and notation are in order. Since the typing of the problem object of a computation element is defined via a pullback, pullback decomposition implies that a ce-morphism is indeed a pullback square (compare Fig. 4). Thus, in abstract terms, we fix an $\mathcal{M}$-morphism $T_P \hookrightarrow T$ from a given $\mathcal{M}$-adhesive category $\mathcal{C}$. We then work in the category that has pullback squares over $T_P \hookrightarrow T$ as objects and pullbacks between such pullback squares as arrows. The results in [23, Theorem 1] ensure that this category is again $\mathcal{M}$-adhesive, provided that the original category $\mathcal{C}$ is also partial-map adhesive (as defined in [18]); a property that is satisfied by the category of attributed graphs; see, for example, [23, Corollary 1]. However, in this paper it will suffice to consider the arising diagrams as diagrams in the $\mathcal{M}$-adhesive category $\mathcal{C}$.

To shorten the presentation, we often only speak of computation elements $\overline{E}$ and ce-morphisms $\overline{m}$ and use their components (such as E_P, t_{E_P}, or m) freely without introducing them explicitly. Furthermore, we often let the typing be implicit; in particular, we omit it in almost all diagrams. In our examples, we use the category of graphs as the underlying $\mathcal{M}$-adhesive category $\mathcal{C}$. Finally, we specify problem instances in terms of the actual computation elements (and not just in terms of their problem objects) to account for the fact that in practice the problem of interest may be given as part of a (suboptimal) solution.

Example 1. The graph T in Fig. 1 can be viewed as a compact representation of a computation type graph where the black part marks the embedded problem type graph. Similarly, the typed graphs of Fig. 2 are interpreted as computation elements over T, with the black parts typed over the problem type graph; the typing is indicated by the names of the nodes. Since the typing morphisms form pullbacks, these black parts represent the problem graphs of the respective computation elements. Having identical problem graphs, all four graphs belong to the same search space, which can be defined using either of them. This reflects that a user might want to optimize an existing assignment of features to classes, rather than just specifying the features and their interdependencies.

Taking two computation elements (from the same search space) and splitting their solution parts, two offspring solutions are constructed by recombining the resulting subsolutions crosswise. In the following, we formally develop this intuition (based on the category-theoretic concept of pushouts) and prove basic properties of this construction of crossovers. We begin by defining the *split* of a given solution.

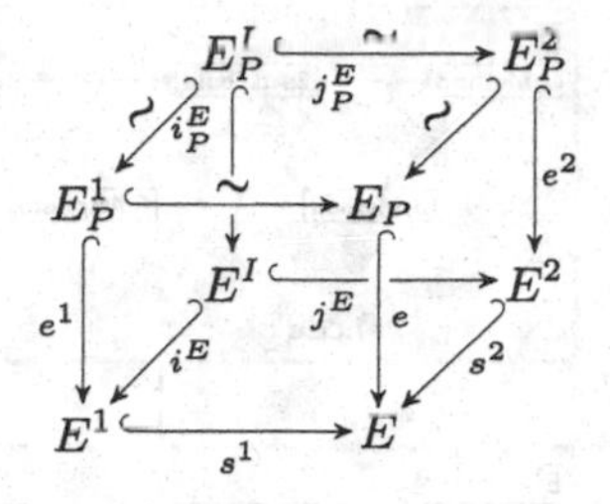

Fig. 5. Split of solution $\overline{E}$

Definition 4 (Split). *Given a problem instance $\overline{PI}$ and a solution $\overline{E}$ for $\overline{PI}$, a* split *of $\overline{E}$ is a commuting cube as depicted in Fig. 5 where the bottom square is a pushout, the vertical squares constitute ce-morphisms, all morphisms come from $\mathcal{M}$, and all problem objects (the objects in the square at the top) are isomorphic to PI_P. The bottom square is called* solution split *and $\overline{E^I}$ is a* split point *of $\overline{E}$. The subsolutions $\overline{E^1}$ and $\overline{E^2}$ of $\overline{E}$ are called* (solution) split objects *of $\overline{E}$.*

A solution can be split in several ways; the central idea is that each solution item of $\overline{E}$ occurs in (at least) one of the solution parts of $\overline{E^1}$ or $\overline{E^2}$. We next present a concrete construction that implements the above declarative definition.

Definition 5 (Split construction). *Given a solution $\overline{E}$, the* split construction *consists of the following steps:*

1. *Choose an $\mathcal{M}$-subobject $s^1 \colon E^1 \hookrightarrow E$ from E (in $\mathcal{C}$) such that when pulling back s^1 along e, the morphism s_P^1 opposite to s^1 is an isomorphism (in particular, $E_P^1 \cong E_P \cong PI_P$, where E_P^1 is the object computed by this pullback). The typing morphisms $t_{E_P^1}$ and t_{E^1} are defined as $t_{E_P} \circ s_P^1$ and $t_E \circ s^1$, respectively.*
2. *Choose another such $\mathcal{M}$-subobject $s^2 \colon E^2 \hookrightarrow E$ from E such that s^1, s^2 are jointly epi (again, typing is defined by composition).*
3. *Complete the cube by constructing pullbacks. That is, determine E^I as the pullback of s^1 and s^2, E_P^I as the pullback of the isomorphisms at the top of the cube, and $e^I \colon E_P^I \hookrightarrow E^I$ as the morphism that is induced by the universal property of the bottom pullback. Again, when considered as computation element, the typing of $\overline{E^I}$ is defined by composition.*

Remark 1. While in general categories the above construction need not be constructive, it is when the underlying category is one of the familiar categories of graphs (being, e.g. typed, labeled, or attributed). Then, the choice of E^1 amounts to extending (an isomorphic copy of) E_P by a choice of solution elements from E; s^1 extends the isomorphism accordingly. Since pullbacks of injective morphisms compute intersections, the pullback of s^1 along e computes the chosen isomorphic copy (up to unique isomorphism). For the choice of E^2, one again extends an isomorphic copy of E_P by a choice of solution elements from E. To ensure that s^1 and s^2 become jointly epi (that is, jointly surjective in our case), one must include at least all solution elements of E not chosen in the construction of E^1.

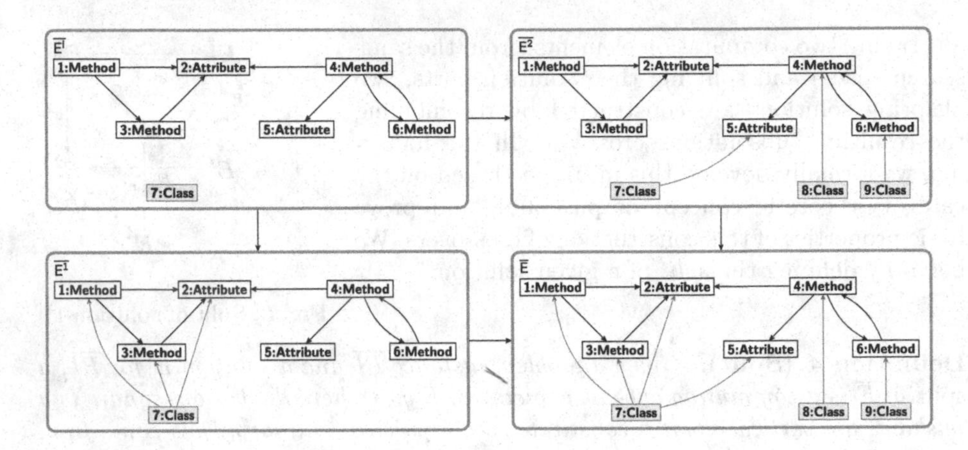

Fig. 6. A split of solution $\overline{\mathsf{E}}$

Example 2. Given the two degrees of freedom for a split, different splits can be constructed from solution $\overline{\mathsf{E}}$ shown in Fig. 2. In steps (1) and (2) we have all possibilities to extend its problem graph $\mathsf{E_P}$ (or an isomorphic copy) with solution parts that yield E^1 and E^2 as long as E^1 and E^2 form graphs and jointly cover E.

A possible split of the solution $\overline{\mathsf{E}}$ is shown in Fig. 6. Here, $\overline{\mathsf{E}}$ is split by first inserting the assignment relations of 1:Method, 2:Attribute, and 3:Method into E^1 along with the associated class 7:Class. The rest of the feature assignments and the necessary classes become part of E^2. The pullback E^I of E^1 and E^2 contains their common solution element 7:Class. To simplify the presentation, the problem graph $\mathsf{E_P}$ is reused in all four graphs. Note that the morphisms in Fig. 6 are indicated by equal numbers in the corresponding nodes. They uniquely induce the mapping of edges. We use these conventions in all of the following examples.

Proposition 1 (Correctness and completeness of split construction). *In an $\mathcal{M}$-adhesive category with $\mathcal{M}$-effective unions, the split construction in Definition 5 is* correct *and* complete*: it always yields a split of the given solution and every possible split can be realized through it. Moreover, for each choice of an $\mathcal{M}$-subobject $s^1 : E^1 \hookrightarrow E$ there exists at least one possible split.*

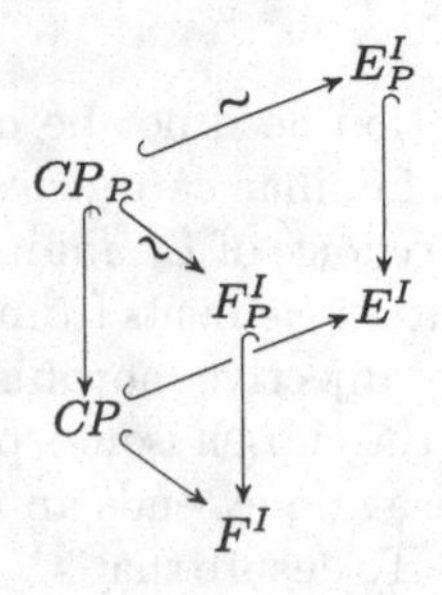

Fig. 7. Crossover point

Given a problem instance $\overline{PI}$ and two solutions $\overline{E}$ and $\overline{F}$ for it, a crossover of $\overline{E}$ and $\overline{F}$ can be performed. Their offspring are basically constructed by recombining solution split objects crosswise. Variations of recombinations are possible, since solution-split objects resulting from solution splits of $\overline{E}$ and $\overline{F}$ can be recombined with more or less overlap. To uniquely determine a crossover of $\overline{E}$ and $\overline{F}$, we define a crossover point that specifies the overlap of their solution split objects.

Definition 6 (Crossover point). *Given a problem instance $\overline{PI}$, two solutions $\overline{E}$ and $\overline{F}$ for $\overline{PI}$, with splits having split points $\overline{E^I}$ and $\overline{F^I}$, respectively, a* crossover point *$\overline{CP}$ is a common subsolution of $\overline{E^I}$ and $\overline{F^I}$. That is, a crossover point is a span of problem-invariant ce-morphisms as depicted in Fig. 7 (with bottom components coming from $\mathcal{M}$).*

We will explain crossover points later along with the crossover operation as such. Next we briefly mention that it is always possible to find a crossover point in a trivial way – the problem object of the given problem instance can always serve as such.

Lemma 1 (Existence of crossover points). *Given a problem instance $\overline{PI} = (p\colon PI_P \hookrightarrow PI, t_{PI_P}, t_{PI})$ over type object tp, two solutions $\overline{E}$ and $\overline{F}$ for $\overline{PI}$, and splits with split points $\overline{E^I}$ and $\overline{F^I}$, respectively, $\overline{CP} := (id : PI_P \hookrightarrow PI_P, t_{PI_P}, tp \circ t_{PI_P})$ is always a crossover point for them. In particular, for each two splits of solutions for the same problem instance there always exists a crossover point.*

Taking two solutions $\overline{E}$ and $\overline{F}$ for a common problem instance and splitting them into subsolutions $\overline{E^1}, \overline{E^2}$ and $\overline{F^1}, \overline{F^2}$, we choose a crossover point for these splits and now define a crossover of these solutions. It basically recombines the subsolutions of $\overline{E}$ and $\overline{F}$ crosswise at the crossover point and yields the computation elements $\overline{E^1F^2}$ and $\overline{E^2F^1}$. We show in Proposition 2 that these two offspring are also solutions to the joint problem instance.

Definition 7 (Crossover). *Let a problem instance $\overline{PI}$, two solutions $\overline{E}$ and $\overline{F}$ for $\overline{PI}$, splits of these two solutions with split objects $\overline{E^1}, \overline{E^2}, \overline{F^1}, \overline{F^2}$ and split points $\overline{E^I}$ and $\overline{F^I}$, respectively, and a* crossover point *$\overline{CP}$ for these splits be given. Then, a* crossover *of solutions $\overline{E}$ and $\overline{F}$ (at $\overline{CP}$ and these splits) yields the two* offspring solutions *$\overline{O_1}$ and $\overline{O_2}$ of $\overline{E}$ and $\overline{F}$ that are shown in Fig. 8 and constructed as follows:*

1. *The ce-morphisms from $\overline{CP}$ to $\overline{E^1}$ and $\overline{E^2}$ are obtained by composing the ce-morphism from $\overline{CP}$ to $\overline{E^I}$ (given by the crossover point) with the ce-morphisms from $\overline{E^I}$ to $\overline{E^1}$ and $\overline{E^2}$ (given by the solution split of $\overline{E}$), respectively. The ce-morphisms from $\overline{CP}$ to $\overline{F^1}$ and $\overline{F^2}$ are obtained analogously.*
2. *The top and bottom squares of the cubes are computed as pushouts (in $\mathcal{C}$) yielding the objects $(E^1F^2)_P$, E^1F^2, $(E^2F^1)_P$, and E^2F^1. The typing morphisms for these objects are obtained from the universal properties of the respective pushout.*
3. *The morphisms $o_1\colon (E^1F^2)_P \hookrightarrow E^1F^2$ and $o_2\colon (E^2F^1)_P \hookrightarrow E^2F^1$ are also induced by the universal property of the pushout squares at the top of the cubes. These morphisms form the objects of $\overline{O_1}$ and $\overline{O_2}$.*

We illustrate the construction before establishing some of its basic properties such as its correctness.

Example 3. A split of solution $\overline{\mathsf{F}}$ (introduced in Fig. 2) is shown in Fig. 9. Again, the split point extends the problem graph by a Class element. Therefore, a

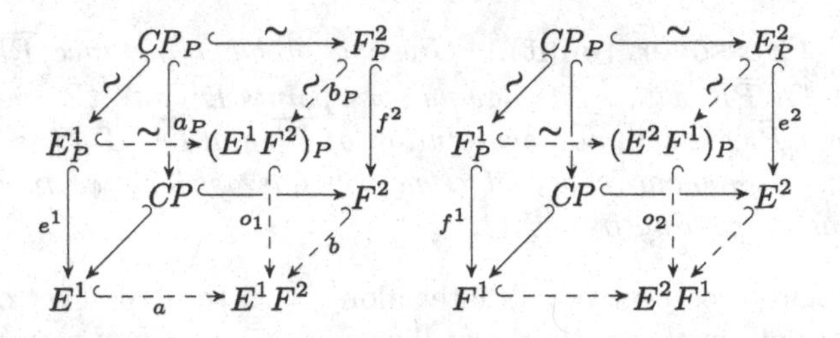

Fig. 8. Crossover of solutions $\overline{E}$ and $\overline{F}$

crossover point for $\overline{\mathsf{E}}$ and $\overline{\mathsf{F}}$ (with the splits given in Figs. 6 and 9) consists either of their common problem graph only, or of this problem graph extended by a single Class. Figure 2 already shows the two offspring graphs that result from applying crossover to $\overline{\mathsf{E}}$ and $\overline{\mathsf{F}}$ where the problem graph is chosen as crossover point. In contrast, adding a Class to the crossover point would merge 7:Class and 11:Class during the recombination and result in the offspring shown in Fig. 10.

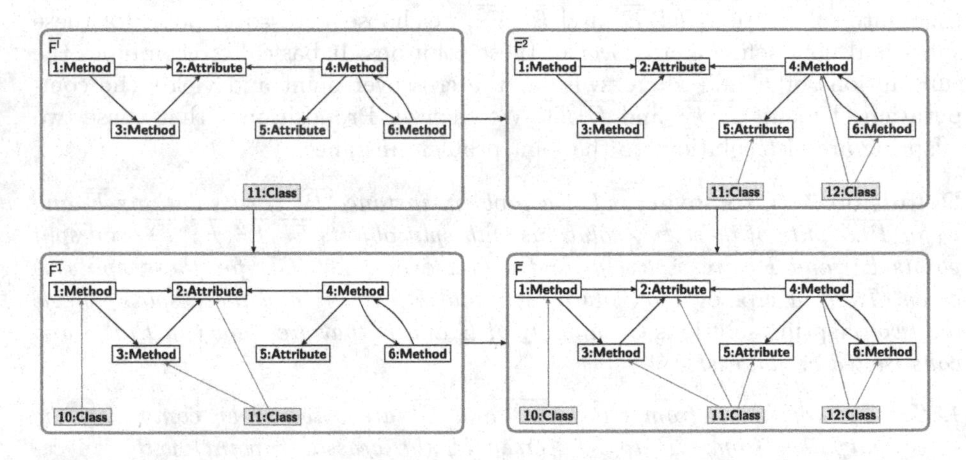

Fig. 9. A split of solution $\overline{\mathsf{F}}$ originally presented in Fig. 2

The next proposition shows that a crossover calculate the offspring correctly, i.e. all offspring calculated represent solutions (for the given problem instance).

Proposition 2 (Correctness of offspring). *Given a problem instance $\overline{PI}$, two solutions $\overline{E}$ and $\overline{F}$ for $\overline{PI}$, splits with split objects $\overline{E^1}, \overline{E^2}, \overline{F^1}, \overline{F^2}$ and split points $\overline{E^I}$ and $\overline{F^I}$, respectively, and a crossover point $\overline{CP}$ for these splits, then there is always a crossover and the two offspring solutions $\overline{O_1}$ and $\overline{O_2}$ are solutions for $\overline{PI}$.*

Next we characterize the expressiveness of the presented crossover construction: Given two solutions $\overline{E}$ and $\overline{F}$, all solutions that can be understood as

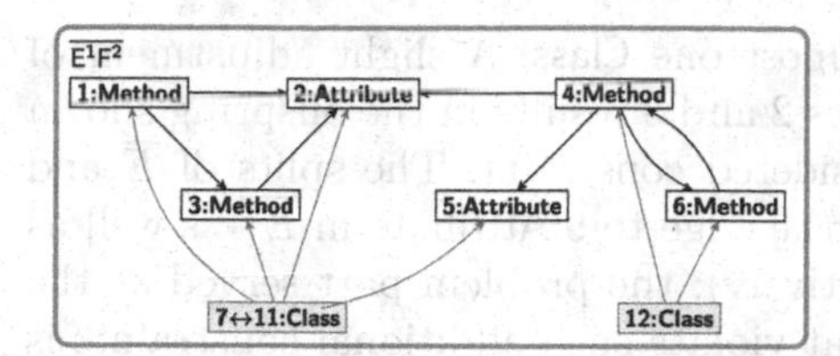

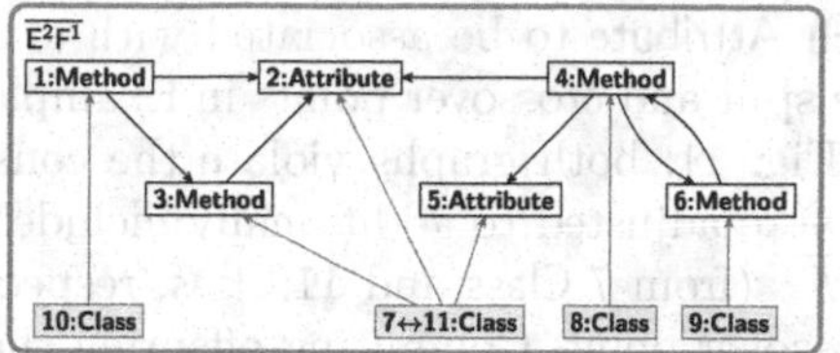

Fig. 10. Two offspring models $\overline{\mathsf{E}^1\mathsf{F}^2}$, $\overline{\mathsf{E}^2\mathsf{F}^1}$, based on the splits of Figs. 6 and 9 and a crossover point containing an additional class

results of splitting $\overline{E}$ and $\overline{F}$ and their recombination can indeed be generated as offspring of the construction in Definition 7 (by different choices of solution splits and crossover points). This is reminiscent of the expressiveness of *uniform crossover* when using arrays of, e.g., bits as genotype [13].

Proposition 3 (Completeness of crossover). *Let the underlying $\mathcal{M}$-adhesive category $\mathcal{C}$ have $\mathcal{M}$-effective unions, and let a problem instance $\overline{PI}$ and solutions $\overline{E}$, $\overline{F}$, and $\overline{O}$ for $\overline{PI}$ be given. The solution $\overline{O}$ can be obtained as offspring from a crossover of $\overline{E}$ and $\overline{F}$ if and only if there are subsolutions $\overline{E^1}$ of $\overline{E}$ and $\overline{F^2}$ of $\overline{F}$ with problem-invariant ce-morphisms $\overline{i}\colon \overline{E^1} \to \overline{O}$ and $\overline{j}\colon \overline{F^2} \to \overline{O}$ such that i and j are jointly epic $\mathcal{M}$-morphisms.*

Discussion. As mentioned earlier, $\mathcal{M}$-adhesive categories include various categories of (typed, labeled, or attributed) graphs that can be used to formalize modeling approaches. In particular, our construction supports crossovers of graphs with inheritance and attribution – concepts that are regularly used in modeling. As for the construction of splits and crossover points, our approach provides several degrees of freedom. In principle, for any implementation of these variation points, the definitions and results in this section are sufficient to complement evolutionary computations in model-based MDO with crossovers. Moreover, our proposed crossover construction is *generic* in the sense that it can be applied to any meta-model; it only needs to be possible to formalize the optimization problem of interest and its search space according to Definition 3. Then, whenever two solution models are chosen for crossover, Proposition 1 ensures that both can be split. Next, Lemma 1 ensures that regardless of which splits are chosen, a crossover point exists for these splits. Finally, Proposition 2 ensures that, for two splits and a crossover point, there is always a crossover that provides solutions of the search space.

Beyond typing, meta-modeling typically employs *integrity constraints* that express further requirements for instances being considered well-formed; *multiplicities* are a typical example. We do not consider such constraints so far. This means that given a meta-model with additional integrity constraints and two of its instance models satisfying these constraints, computing crossover as specified in this work may result in offspring models that violate the constraints. We illustrate this with our running example: In practical applications, the meta-model (type graph) from Fig. 1 would have a constraint requiring each Method and each Attribute to be associated with at most one Class. A slight adjustment of the split and crossover points in Examples 2 and 3 results in the offspring shown in Fig. 11; both graphs violate the considered constraint. The splits of $\overline{E}$ and $\overline{F}$ were adjusted to additionally include the edge to 5:Attribute in $\overline{E^1}$ as well as in $\overline{F^1}$ (from 7:Class and 11:Class, respectively); the problem part served as the crossover point. Computing offspring that violate such additional constraints is not in itself a problem; several methods have been developed in evolutionary algorithm research to deal with this. For example, such *infeasible solutions* can be eliminated by the selection operator, or they can be tolerated (with a reduced fitness assigned to them); after all, even an infeasible solution can lead to a feasible solution of high quality later during the evolutionary computation. However, producing too many infeasible solutions can waste valuable resources and slow down the evolutionary computation process.

Summarizing, we expect evolutionary search to profit most if domain-specific knowledge is used to direct the choices of splits and crossover points, that is, if these choices are adapted to the problem at hand (possibly including the preservation of additional constraints). Thus, while our construction can principally yield problem-agnostic crossovers, it can also (and maybe better) be understood as a *generic* construction that offers a unifying framework for the implementation of specific crossovers on graph-like structures. In the next section, we substantiate the claim that our construction offers such a unifying framework.

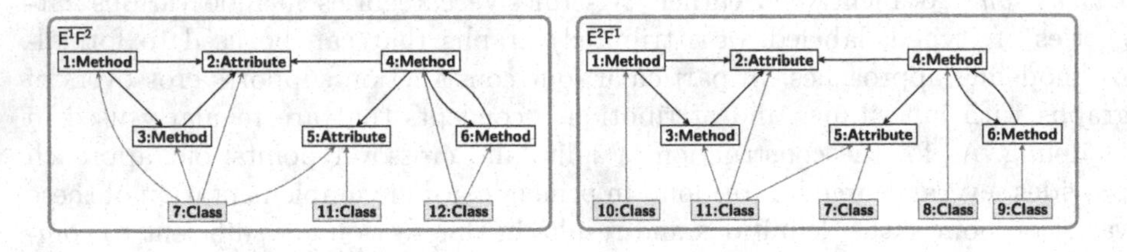

Fig. 11. Offspring violating an integrity constraint

5 Instantiating Existing Approaches to Graph-Based Crossover

In this section, we exemplify how our generic construction includes existing crossover operators that can be applied to graph-like structures. We discuss *uniform*, *k-point* and *subtree crossover*, as these are classic operators that are commonly applied [13,24]. In addition, we consider *horizontal gene transfer* (HGT), which was recently introduced in a setting similar to ours [2].

Uniform and k-Point Crossover are crossover operators commonly used when solutions are encoded as strings (arrays) of bits (or other alphabets) [13]. In k-point crossover, two given parent strings of equal length are split into $k+1$ substrings at k randomly selected crossover points (at equal positions in both strings). The two offspring solutions are obtained by alternately concatenating a substring from each parent, resulting in solutions of the same length as the given parents. In uniform crossover, a new decision is made at each position (according to a given probability) which offspring gets the entry from which parent. This can be understood as k-point crossover with varying k.

Strings can be represented as graphs by simply considering each character of a string as an edge typed or labeled with that character; see, e.g., [30]. Using this representation, our construction of crossovers can be used to implement uniform and k-point crossover. Here, the problem object (graph) is given by the nodes of the graphs (which encode the length of the given strings). The splits are chosen such that (i) the edges are partitioned (disjointly) into the solution splits and (ii) the same partitions are chosen for both parents (i.e., if the first edge of the first parent is included in its first subsolution, the first edge of the second parent is also included in its first subsolution). This partitioning can be done according to the rules of k-point or uniform crossover. The only available crossover point is the set of nodes (i.e. the problem graph), since the edges are distributed disjointly. The calculation of the crossover, i.e.performing the two pushouts, results in two offspring solutions with the same length as the parents.

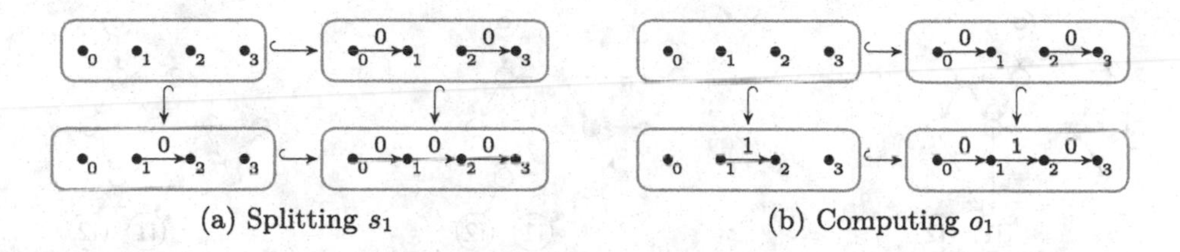

(a) Splitting s_1 (b) Computing o_1

Fig. 12. Implementing classic 2-point crossover

For the k-point crossover, we consider the concrete example of a 2-point crossover of the strings $s_1 : 0|0|0$ and $s_2 : 1|1|1$, where $|$ represents the chosen crossover points. The computed offspring strings are $o_1 : 010$ and $o_2 : 101$. Figure 12 outlines how this calculation is implemented in our approach.

Subtree Crossover is the recombination operator commonly used in genetic programming [24]. In genetic programming, a program is represented by its syntax tree. Such a tree serves as a genotype for an evolutionary computation that aims at finding an (optimal) program for the given task. Given two syntax trees, subtree crossover (randomly) selects and exchanges one subtree from each of them. With our approach, we can implement subtree crossover if we use a little trick in representing the trees: We explicitly encode the edges of the trees as nodes (for a representation of (hyper)edges as special kinds of nodes, see, e.g., their (visual) representation in [31]). The problem tree (graph) is always empty. A split divides a tree into a subtree and the remaining tree, where the node encoding the reference to the subtree is common in both split objects. This node serves as a crossover point to exchange subtrees crosswise at the correct positions. Figure 13 schematically represents a subtree crossover, where R_1 is the root node of the first tree, all ST_i represent subtrees, and nodes of type ref represent edges. Note that representing edges as nodes allows us to split an edge into two parts and distribute it between the two split parts. In this way, we can redirect edges.

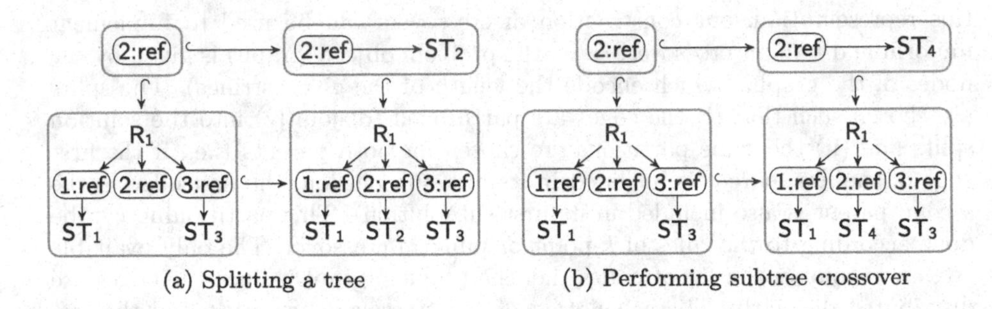

(a) Splitting a tree (b) Performing subtree crossover

Fig. 13. Implementing subtree crossover

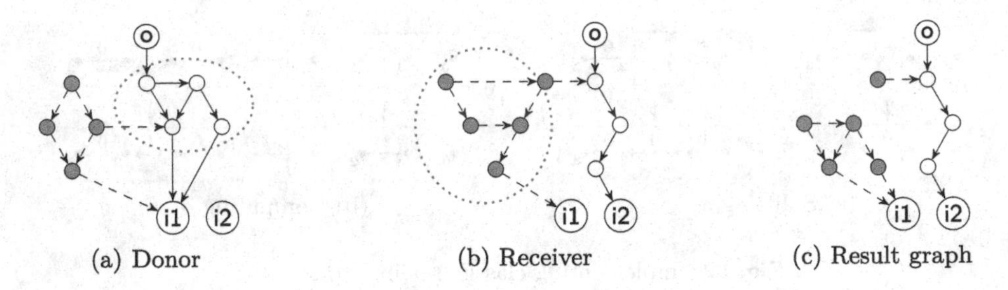

(a) Donor (b) Receiver (c) Result graph

Fig. 14. Example of the horizontal gene transfer (HGT) proposed in [2]. o is the fixed output node. Active nodes are depicted in white, passive nodes are gray. i1 and i2 are input nodes. The marked nodes of the receiver (including outgoing edges) are substituted by the marked parts of the donor.

Horizontal Gene Transfer (HGT) was proposed by Atkinson et al. in [2] as a non-recombinative method for transferring genetic information between individuals. In their work, graphs are used to represent functions (or, with small adaptations, neural networks); the reachability of fixed output nodes determines the *active component* of a graph. As indicated in Fig. 14, HGT takes the active component of one graph (the donor) and copies it to the passive component of another graph (the receiver); to maintain a fixed number of nodes, an appropriate number of passive nodes is deleted from the receiver beforehand. Input nodes representing parameters are identified during that process. In our construction the output and input nodes would be considered the problem part. Choosing the active component as the solution split for the donor, the subgraph that remains after deleting the passive nodes as the solution split of the receiver, and the problem part as crossover point, our approach can compute HGT as a crossover.

6 Related Work

In addition to the approaches presented in detail above in Sect. 5, we now relate our crossover construction to other variants of crossover on graph-like structures. For each approach, we clarify whether it can be simulated by our approach and how expressive it is. We then discuss the crossover variants used so far in MDO.

6.1 Further Approaches for Graph-Based Crossover

The two most general crossover variants on graph-like structures that we are aware of are those proposed by Niehaus [27] and Machado et al. [26]. Niehaus introduces *random crossover* on directed graphs, where a subgraph of one graph is removed and replaced by a subgraph of another graph; in particular, only one offspring is computed. To avoid dangling edges, the exchanged subgraphs must have the same in- and out-degrees with respect to the edges that connect them to the rest of the graph. By using the trick of representing edges as a special kind of node, we can realize this crossover with our approach.

Machado et al. [26] also exchange subgraphs between graphs. The subgraphs are constructed as radii around randomly chosen nodes. To connect the exchanged subgraphs to their new host graphs, a correspondence is established between the nodes that were adjacent to them in their former host graphs. If this correspondence is one-to-one, we can implement this operator in our approach by again representing edges by a special type of node. However, Machado et al. also allow for correspondences that are not one-to-one. To implement this feature, we would need to allow non-injective mappings from the crossover point to the splits in our approach. Unlike this approach, our approach is not limited to choosing subgraphs as radii around randomly chosen nodes.

Other approaches are less general since they depend to a greater extent on the chosen representation or semantics of the graphs used [9,10,19,21,22,28]. In these cases, it does not seem straightforward to apply the proposed crossovers in other contexts. The kind of computations that can be performed using crossover may also be less expressive than those in the approaches already discussed [19,21,22,28,29]. We can implement the crossovers proposed in [9,10,19,28,29] in our approach, often by representing edges as a special type of node. The approach by Kantschik and Banzhaf [22] cannot be implemented for reasons similar to those discussed for [26]. Furthermore, we cannot implement the *subgraph crossover* proposed in [21], because this approach allows random insertion of new edges into an offspring and these edges do not come from any parent.

In summary, our generic approach to crossover on graph-like structures encompasses most of the approaches proposed for more specific situations. Our approach allows more general exchanges of subgraphs than most of the approaches discussed. Moreover, our Theorem 3 is the first result (that we know of) that formally clarifies the expressiveness of the proposed crossover. We have identified two reasons why our approach is not able to encompass an existing approach: First, crossover could cause two (or more) edges that targeted different nodes in their original graph to target the same node in their new context.

Second, elements that do not originate from either parent are reintroduced in the offspring. However, both kinds of changes can be realized in our approach by the subsequent application of mutation operators. We could also solve the first problem by allowing non-injective mappings from crossover points to the splits when performing crossover. However, this would complicate the theory we can provide for our construction: Pushouts along any two morphisms need not exist in $\mathcal{M}$-adhesive categories, and even if the necessary pushouts did exist, ensuring that the computed results come from the search space under consideration (i.e., represent an $\mathcal{M}$-morphism) would only be possible for certain morphisms.

6.2 Crossover in MDO

In the rule-based approach to MDO, the solutions are represented as sequences of model transformations [1,4]. This allows traditional crossovers (e.g., k-point crossover, uniform crossover) to be applied seamlessly. However, they have been shown to be disruptive because the transformations can depend on each other [20] and repair strategies must be used to mitigate this problem. As for the effects of crossover in the rule-based approach, no theoretical results are available. To date, neither a formal basis nor alternatives to traditional crossover have been developed in this context.

Burton et al. were the first to perform optimization directly on models as search space elements [8]. Their specific use case allows for the adaptation of single-point crossover through model transformations. However, their crossover implementation is not described in detail. Recent applications of the model-based approach neglect crossover and stick to mutation as their only change operator, such as [7]. In [32], Zschaler and Mandow present a generalized view on the model-based approach to MDO and point out the challenge of specifying crossover in such a setting. They briefly discuss model differencing and model merging as related concepts, but do not elaborate on this idea. To our knowledge, this paper presents the first approach to address this issue.

7 Conclusion

There is theoretical and practical evidence that evolutionary algorithms in general benefit from the use of crossover [2,9,19] in the sense that the search for optimal solutions can be more effective and efficient. However, in the absence of suitable crossover approaches for (the model-based approach to) MDO, the effect of crossover in this context has not yet been studied. Our proposed generic crossover construction can serve as a basis to start with.

How existing solutions are split and the selection of common crossover points for such splits are critical design decisions. Which of these decisions are beneficial to the effectiveness and efficiency of an optimization remains to be explored. Apart from the typing of objects, our approach neglects additional constraints of an optimization problem, i.e., crossover may lead to violations of constraints. Whether our approach needs to be refined to guarantee constraint-preserving offspring remains for future work. In addition to theoretical exploration of our approach, an implementation is needed to enable empirical analysis. Additionally, specification concepts need to be elaborated to allow users to conveniently specify different split strategies and crossover points that fit their domain.

Acknowledgements. This work has been partially supported by the German Research Foundation (DFG), grant no. TA 294/19-1. We thank the anonymous reviewers for their insightful comments.

A Proofs

The following lemma is the central ingredient for the proof of Proposition 1 and also used in the one of Theorem 3. For adhesive categories, it has already been stated in the extended version of [14]. Here, we present it in the more general context of $\mathcal{M}$-adhesive categories. Because of that, we need to additionally assume the existence of $\mathcal{M}$-effective unions.

Lemma 2 (Pullbacks as pushouts). *In an $\mathcal{M}$-adhesive category $(\mathcal{C}, \mathcal{M})$ with $\mathcal{M}$-effective unions, let $(e_1, e_2) : L_1, L_2 \hookrightarrow E$ be a pair of jointly epimorphic $\mathcal{M}$-morphisms. Then the pullback of (e_1, e_2) is also a pushout.*

Proof. Given the diagram below, where P arises as pullback of (e_1, e_2), Q as pushout of (p_1, p_2), and the morphism h from the universal property of Q, we show that h is an isomorphism.

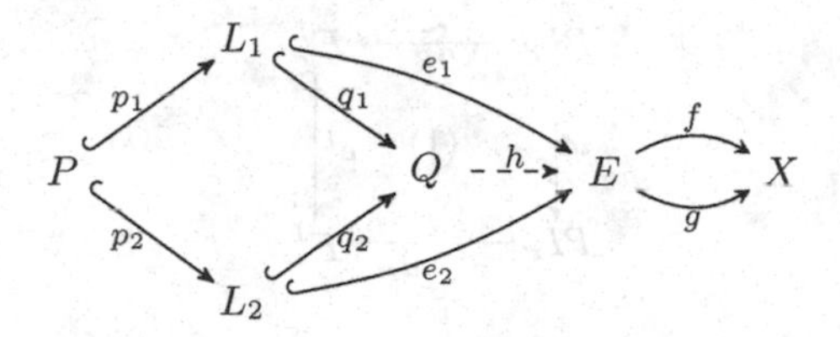

First, since e_1, e_2 are $\mathcal{M}$-morphisms, the morphism h is an $\mathcal{M}$-morphism, assuming $\mathcal{M}$-effective unions. This means that h is a regular monomorphism (compare [25, Lemma 4.8], which is easily seen to also hold in $\mathcal{M}$-adhesive categories).

Secondly, given two morphisms $f, g : E \to X$ with $f \circ h = g \circ h$, it follows that $f \circ h \circ q_1 = g \circ h \circ q_1$ which implies $f \circ e_1 = g \circ e_1$; analogously, $f \circ e_2 = g \circ e_2$ holds. Since e_1, e_2 are jointly epimorphic, it follows that $f = g$, and h is an epimorphism. Thus, h is epi and regular mono and therefore an isomorphism. □

Proof (of Proposition 1). Given a solution split as depicted in Fig. 5, it is straightforward to realize this split via the split construction. One just chooses the already given morphisms s^1 and s^2. As the bottom square in Fig. 5 is a pushout, s^1 and s^2 are jointly epimorphic. Moreover, in an $\mathcal{M}$-adhesive category that square is also a pullback because $E^I \hookrightarrow E^1$ (or, equally, $E^I \hookrightarrow E^2$) $\in \mathcal{M}$.

To show that the construction always computes a solution split, we have to show that it produces a commuting cube of $\mathcal{M}$-morphisms (with isomorphisms at the top) such that the bottom square is a pushout and the four vertical squares constitute ce-morphisms (i.e., are also pullbacks and are compatible with typing). It is well-known that, in every category, in a cube that is computed via pullbacks as stipulated by our construction, all squares are pullbacks; see, e.g., [3, 5.7 Exercises, 2. (b)]. By closedness of $\mathcal{M}$-morphism under pullbacks, this in turn implies that all morphisms are $\mathcal{M}$-morphisms (because e, s^1, and s^2 are). The two morphisms at the front of the top square are isomorphisms by assumption; the other two become isomorphisms by closedness of isomorphisms under pullback. Finally, in an $\mathcal{M}$-adhesive category with $\mathcal{M}$-effective unions, the pullback of jointly epimorphic $\mathcal{M}$-morphisms is always a pushout (see Lemma 2 above). Therefore, the bottom square (computed as pullback of the jointly epic $\mathcal{M}$-morphisms s^1 and s^2) is a pushout as desired. The typing of $\overline{E^1}$ and $\overline{E^2}$ is compatible with the typing of $\overline{E}$ by definition; moreover, the squares obtained from the typing morphisms are pullbacks by pullback composition.

For the last statement, it suffices to observe that E^2 can always be chosen as E, embedded via the identity morphism (which then leads to $E^I \cong E^1$). □

Proof (of Lemma 1). To prove the statement, we have to show that there exists a ce-morphism (a_P, a) from $\overline{CP} := (id : PI_P \hookrightarrow PI_P, t_{PI_P}, tp \circ t_{PI_P})$ to $\overline{E^I}$ such that a_P is an isomorphism and $a \in \mathcal{M}$; the analogous statement for $\overline{F^I}$ is proved in exactly the same way.

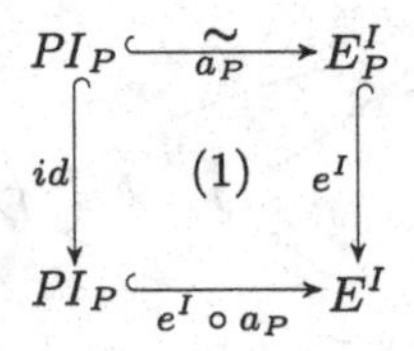

Fig. 15. Showing $\overline{CP}$ to constitute a crossover point

We define such a ce-morphism using the isomorphism a_P with $t_{E^I_P} \circ a_P = t_{PI_P}$ that exists since $\overline{E^I}$ is an element of the search space of $\overline{PI}$. Figure 15 depicts this. The square commutes and $a, e^I \circ a \in \mathcal{M}$ by closedness of $\mathcal{M}$ under isomorphisms and composition. Moreover, using the fact that e^I is a monomorphism, it is also easy to check that the square constitutes a pullback. Finally, using $t_{E^I_P} \circ a_P = t_{PI_P}$ we compute

$$
\begin{aligned}
t_{E^I} \circ e^I \circ a_P &= tp \circ t_{E^I_P} \circ a_P \\
&= tp \circ t_{PI_P}
\end{aligned}
$$

which shows $(a_P, e^I \circ a_P)$ to be type-compatible. □

Proof (of Proposition 2). First, in an $\mathcal{M}$-adhesive category, pushouts along $\mathcal{M}$-morphisms exist. This means that, given two solution splits and a crossover point, crossover is always applicable. Since isomorphisms are closed under pushout, the top squares in the construction consist of isomorphisms only. In particular, $(E^1F^2)_P \cong PI_P \cong (E^2F^1)_P$ (because $E^1_P \cong PI_P \cong E^2_P$ by assumption).

By definition, o_1 is the unique morphism such that

$$o_1 \circ a_P = a \circ e^1 \text{ and } o_1 \circ b_P = b \circ f^2,$$

where (a_P, a) and (b_P, b) denote the ce-morphisms from e^1 and f^2 to o_1 (see Fig. 8). A standard diagram chase (using the facts that the top squares in Fig. 8 consist of isomorphisms only and that diagrams remain commutative if one replaces isomorphisms by their inverses) then shows that $a \circ e^1 \circ a_P^{-1}$ (or, equally, $b \circ f^2 \circ b_P^{-1}$) exhibits this universal property. Therefore, $o_1 = a \circ e^1 \circ a_P^{-1} \in \mathcal{M}$ as composition of $\mathcal{M}$-morphisms. Again, this uses the fact that $\mathcal{M}$ contains all isomorphisms.

Finally, that the typing morphisms of $\overline{O_1}$ induce even a pullback square over tp (and not merely a commuting one) follows exactly as in the proof of Lemma 2.2 in [15], using the facts that the ambient category $\mathcal{C}$ is $\mathcal{M}$-adhesive and $tp \in \mathcal{M}$. □

Proof (of Proposition 3). Let solution $\overline{O}$ be computed via a crossover from $\overline{E}$ and $\overline{F}$. It is immediately clear from the construction that there exist the two required ce-morphisms $\bar{\imath}$ and $\bar{\jmath}$ such that i, j are jointly epic $\mathcal{M}$-morphisms because the projections of a pushout are jointly epi and $\mathcal{M}$-morphisms are closed under pushout.

For the converse direction, O is jointly covered by E^1 and F^2, which stem from subsolutions $\overline{E^1}$ and $\overline{F^1}$ of $\overline{E}$ and $\overline{F}$ by assumption. If the underlying category has $\mathcal{M}$-effective unions, pulling these morphisms back results in a pushout. Let $\overline{CP}$ be the object resulting from that pullback (exactly as in the proof of Proposition 1). We merely have to show that there exist solution splits of $\overline{E}$ and $\overline{F}$ that split up $\overline{E}$ into $\overline{E^1}$ and some suitable subsolution $\overline{E^2}$ of $\overline{E}$ and $\overline{F}$ into $\overline{F^2}$ and some suitable subsolution $\overline{F^1}$ of $\overline{F}$ for which $\overline{CP}$ can serve as a crossover point. As in (the proof of) the second part of Proposition 1, we can use $\overline{E}$ as $\overline{E^2}$ and, because of the symmetric nature of a solution split, $\overline{F}$ as $\overline{F^1}$ and obtain splits of $\overline{E}$ and $\overline{F}$ with $\overline{E^I} = \overline{E^1}$ and $\overline{F^I} = \overline{F^2}$. Hence, $\overline{CP}$, together with the morphisms that stem from its computation as a pullback, can serve as a crossover point for these splits, and applying the crossover construction computes the given solution $\overline{O}$. □

References

1. Abdeen, H., et al.: Multi-objective optimization in rule-based design space exploration. In: Crnkovic, I., Chechik, M., Grünbacher, P. (eds.) ACM/IEEE International Conference on Automated Software Engineering, ASE 2014, Vasteras, Sweden - 15–19 September 2014, pp. 289–300. ACM (2014). https://doi.org/10.1145/2642937.2643005
2. Atkinson, T., Plump, D., Stepney, S.: Horizontal gene transfer for recombining graphs. Genet. Program. Evolvable Mach. **21**(3), 321–347 (2020). https://doi.org/10.1007/s10710-020-09378-1

3. Awodey, S.: Category Theory, Oxford Logic Guides, 2nd edn. vol. 52. Oxford University Press, Oxford (2010)
4. Bill, R., Fleck, M., Troya, J., Mayerhofer, T., Wimmer, M.: A local and global tour on MOMoT. Softw. Syst. Model. **18**(2), 1017–1046 (2019). https://doi.org/10.1007/s10270-017-0644-3
5. Boussaïd, I., Siarry, P., Ahmed-Nacer, M.: A survey on search-based model-driven engineering. Autom. Softw. Eng. **24**(2), 233–294 (2017). https://doi.org/10.1007/s10515-017-0215-4
6. Bowman, M., Briand, L.C., Labiche, Y.: Solving the class responsibility assignment problem in object-oriented analysis with multi-objective genetic algorithms. IEEE Trans. Softw. Eng. **36**(6), 817–837 (2010). https://doi.org/10.1109/TSE.2010.70
7. Burdusel, A., Zschaler, S., John, S.: Automatic generation of atomic multiplicity-preserving search operators for search-based model engineering. Softw. Syst. Model. **20**(6), 1857–1887 (2021). https://doi.org/10.1007/s10270-021-00914-w
8. Burton, F.R., Paige, R.F., Rose, L.M., Kolovos, D.S., Poulding, S., Smith, S.: Solving acquisition problems using model-driven engineering. In: Vallecillo, A., Tolvanen, J.-P., Kindler, E., Störrle, H., Kolovos, D. (eds.) ECMFA 2012. LNCS, vol. 7349, pp. 428–443. Springer, Heidelberg (2012). https://doi.org/10.1007/978-3-642-31491-9_32
9. Doerr, B., Happ, E., Klein, C.: Crossover can provably be useful in evolutionary computation. Theor. Comput. Sci. **425**, 17–33 (2012). https://doi.org/10.1016/j.tcs.2010.10.035
10. Downey, C., Zhang, M., Browne, W.N.: New crossover operators in linear genetic programming for multiclass object classification. In: Pelikan, M., Branke, J. (eds.) Proceedings of the Genetic and Evolutionary Computation Conference, GECCO 2010, Portland, Oregon, USA, 7–11 July 2010. pp. 885–892. ACM (2010). https://doi.org/10.1145/1830483.1830644
11. Ehrig, H., Ehrig, K., Prange, U., Taentzer, G.: Fundamentals of Algebraic Graph Transformation. Monographs in Theoretical Computer Science, Springer, Heidelberg (2006).https://doi.org/10.1007/3-540-31188-2
12. Ehrig, H., Ermel, C., Golas, U., Hermann, F.: Graph and model transformation - general framework and applications. In: Monographs. in Theoretical Computer Science. An EATCS Series, Springer (2015). https://doi.org/10.1007/978-3-662-47980-3
13. Eiben, A.E., Smith, J.E.: Introduction to Evolutionary Computing. NCS, Springer, Heidelberg (2015). https://doi.org/10.1007/978-3-662-44874-8
14. Fritsche, L., Kosiol, J., Schürr, A., Taentzer, G.: Short-cut rules. In: Mazzara, M., Ober, I., Salaün, G. (eds.) STAF 2018. LNCS, vol. 11176, pp. 415–430. Springer, Cham (2018). https://doi.org/10.1007/978-3-030-04771-9_30
15. Garner, R., Lack, S.: On the axioms for adhesive and quasiadhesive categories. Theory Appl. Categor. **27**(3), 27–46 (2012), https://www.emis.de/journals/TAC/volumes/27/3/27-03abs.html
16. Harman, M., Jones, B.F.: Search-based software engineering. Inf. Softw. Technol. **43**(14), 833–839 (2001). https://doi.org/10.1016/S0950-5849(01)00189-6
17. Harman, M., Mansouri, S.A., Zhang, Y.: Search-based software engineering: trends, techniques and applications. ACM Comput. Surv. **45**(1), 11:1–11:61 (2012). https://doi.org/10.1145/2379776.2379787
18. Heindel, T.: Adhesivity with partial maps instead of spans. Fundam. Informaticae **118**(1-2), 1–33 (2012). https://doi.org/10.3233/FI-2012-704

19. Husa, J., Kalkreuth, R.: A comparative study on crossover in cartesian genetic programming. In: Castelli, M., Sekanina, L., Zhang, M., Cagnoni, S., García-Sánchez, P. (eds.) EuroGP 2018. LNCS, vol. 10781, pp. 203–219. Springer, Cham (2018). https://doi.org/10.1007/978-3-319-77553-1_13
20. John, S., Alexandru Burdusel, R.B., Strüber, D., Taentzer, G., Zschaler, S., Wimmer, M.: Searching for optimal models: comparing two encoding approaches. J. Obj. Technol. **18**(3), 6:1–22 (2019). https://doi.org/10.5381/jot.2019.18.3.a6
21. Kalkreuth, R., Rudolph, G., Droschinsky, A.: A new subgraph crossover for Cartesian genetic programming. In: McDermott, J., Castelli, M., Sekanina, L., Haasdijk, E., García-Sánchez, P. (eds.) EuroGP 2017. LNCS, vol. 10196, pp. 294–310. Springer, Cham (2017). https://doi.org/10.1007/978-3-319-55696-3_19
22. Kantschik, W., Banzhaf, W.: Linear-graph GP - a new GP structure. In: Foster, J.A., Lutton, E., Miller, J., Ryan, C., Tettamanzi, A. (eds.) EuroGP 2002. LNCS, vol. 2278, pp. 83–92. Springer, Heidelberg (2002). https://doi.org/10.1007/3-540-45984-7_8
23. Kosiol, J., Fritsche, L., Schürr, A., Taentzer, G.: Double-pushout-rewriting in S-cartesian functor categories: rewriting theory and application to partial triple graphs. J. Log. Algebraic Methods Program. **115**, 100565 (2020). https://doi.org/10.1016/j.jlamp.2020.100565
24. Koza, J.R.: Genetic Programming: On the Programming of Computers by Means of Natural Selection. MIT Press, Complex Adaptive Systems (1992)
25. Lack, S., Sobociński, P.: Adhesive and quasiadhesive categories. RAIRO Theor. Informat. Appl. **39**(3), 511–545 (2005). https://doi.org/10.1051/ita:2005028
26. Machado, P., Nunes, H., Romero, J.: Graph-based evolution of visual languages. In: Di Chio, C., Brabazon, A., Di Caro, G.A., Ebner, M., Farooq, M., Fink, A., Grahl, J., Greenfield, G., Machado, P., O'Neill, M., Tarantino, E., Urquhart, N. (eds.) EvoApplications 2010. LNCS, vol. 6025, pp. 271–280. Springer, Heidelberg (2010). https://doi.org/10.1007/978-3-642-12242-2_28
27. Niehaus, J.: Graphbasierte Genetische Programmierung. Ph.D. thesis, Technical University of Dortmund (2003). http://hdl.handle.net/2003/2744
28. Nobile, M.S., Besozzi, D., Cazzaniga, P., Mauri, G.: The foundation of evolutionary Petri Nets. In: Balbo, G., Heiner, M. (eds.) Proceedings of the International Workshop on Biological Processes & Petri Nets, Milano, Italy, CEUR 24 June 2013. http://ceur-ws.org/Vol-988/paper6.pdf
29. Pereira, F.B., Machado, P., Costa, E., Cardoso, A.: Graph based crossover - a case study with the busy beaver problem. In: Banzhaf, W., Daida, J.M., Eiben, A.E., Garzon, M.H., Honavar, V. (eds.) Proceedings of the 1st Annual Conference on Genetic and Evolutionary Computation, GECCO 1999, Vol. 2, pp. 1149–1155. Morgan Kaufmann Publishers Inc., San Francisco (1999)
30. Plump, D.: Termination of graph rewriting is undecidable. Fundam. Informaticae **33**(2), 201–209 (1998). https://doi.org/10.3233/FI-1998-33204
31. Plump, D.: Term graph rewriting. In: Ehrig, H., Engels, G., Kreowski, H.J., Rozenberg, G. (eds.) Handbook of Graph Grammars and Computing by Graph Transformation, vol. 2, pp. 3–61. World Scientific (1999). https://doi.org/10.1142/9789812815149_0001
32. Zschaler, S., Mandow, L.: Towards model-based Optimisation: using domain knowledge explicitly. In: Milazzo, P., Varró, D., Wimmer, M. (eds.) STAF 2016. LNCS, vol. 9946, pp. 317–329. Springer, Cham (2016). https://doi.org/10.1007/978-3-319-50230-4_24

Towards Development with Multi-version Models: Detecting Merge Conflicts and Checking Well-Formedness

Matthias Barkowsky(✉) and Holger Giese

Hasso-Plattner Institute at the University of Potsdam,
Prof.-Dr.-Helmert-Str. 2-3, 14482 Potsdam, Germany
{matthias.barkowsky,holger.giese}@hpi.de

Abstract. Developing complex software requires that multiple views and versions of the software can be developed in parallel and merged as supported by views and managed by version control systems. In this context, this paper considers permanent monitoring of merging and related consistency problems at the level of models and abstract syntax. The presented approach introduces multi-version models based on typed graphs that permit to store changes and multiple versions in one graph in a compact form and allow (1) to study well-formedness for all versions without the need to extract each version individually, (2) to report all possible merge conflicts without the need to merge all pairs of versions, and (3) to report all violations of well-formedness conditions that will result for merges of any two versions independent of any merge decisions without the need to merge all pairs of versions. Thereby, the approach aims to permit early and frequent conflict detection while developing in parallel. The paper defines the related concepts and algorithms operating on multi-version models, proves their correctness w.r.t. the usually employed three-way-merge, and reports on preliminary experiments concerning the scalability.

1 Introduction

Developing complex software nowadays requires that multiple views and versions of the software can be developed in parallel and merged as supported by views and managed by version control systems [12]. For complex software, living with inconsistencies at least temporarily is inevitable, as enforcing consistency may lead to loss of important information [11] and is hence neither always possible nor desirable. However, working with multiple versions in parallel and changing each version on its own for longer periods of time can introduce substantial conflicts that are difficult and expensive to resolve. Therefore, it is necessary to manage consistency when combining views and versions using merge approaches [12,20].

This paper considers permanent monitoring of merging and related consistency problems at the level of models and abstract syntax. This aims to permit

This work was developed mainly in the course of the project modular and incremental Global Model Management (project number 336677879) funded by the DFG.

N. Behr and D. Strüber (Eds.): ICGT 2022, LNCS 13349, pp. 118–136, 2022.
https://doi.org/10.1007/978-3-031-09843-7_7

early and frequent conflict detection while developing in parallel, as suggested in approaches to detect conflicts early and to enable collaboration to manage conflicts and their risks [4].

The presented approach therefore introduces multi-version models based on typed graphs, which permit to store changes and multiple versions in one graph in a compact form and allow to study the different versions and their merge combinations. The following capabilities are considered: (1) Study well-formedness for all versions at once without the need to extract and explicitly consider each version individually. (2) Report all possible merge conflicts that may result for merges of any two versions without the need to extract and explicitly merge all pairs of versions. (3) Report all violations of well-formedness conditions that will result for merges of any two versions independent of any merge decisions without the need to extract and explicitly merge all pairs of versions.

The approach thus promises to support early conflict detection and collaboration for managing conflicts and their risks, while not having to decide how to later merge conflicting versions. The technique also aims for a better scalability in case there are many versions that are considered in parallel.

Furthermore, the developed multi-version models permit to study the phenomena of versions, merging, and well-formedness conditions in the unifying framework of typed graphs. This enables us to (a) formulate algorithms that can obtain several analysis results without the need to consider a specific version, merge of a pair of versions, or strategy for conflict resolution and (b) prove that the algorithms compute the same results as if we would explicitly consider all specific versions, merges of pairs of versions, or strategies for conflict resolution.

The paper defines the related concepts and algorithms operating on multi-version models, proves their correctness w.r.t. the usually employed three-way-merge, and reports on first experiments concerning the scalability. In Sect. 2, we summarize the preliminaries of the presented approach, including basic definitions for typed graphs, well-formedness conditions, and graph modifications. Then, as a baseline, single-version models in the form of typed graphs with well-formedness conditions are defined in Sect. 3, before multi-version models are introduced in Sect. 4. Determining all merge conflicts and checking well-formedness for all merge results based on multi-version models is then considered in Sect. 5. Results of first experiments for our prototypical implementation of the algorithms are presented in Sect. 6. A summary of related work is given in Sect. 7. Finally, the conclusions of the paper and an outlook of planned future work are presented in Sect. 8.

2 Preliminaries

We briefly reiterate the basic concepts of graphs, graph modifications, and well-formedness conditions used in the remainder of the paper.

A graph $G = (V^G, E^G, s^G, t^G)$ consists of a set of nodes V^G, a set of edges E^G and two functions $s^G : E^G \rightarrow V^G$ and $t^G : E^G \rightarrow V^G$ assigning each edge its source and target, respectively. We assume that graph elements have identities and source and target of an edge are invariant if an edge is part of multiple graphs, that is, for two graphs G and H and an edge $e \in E^G \cap E^H$, it holds that $s^G(e) = s^H(e)$ and $t^G(e) = t^H(e)$. This also implies that, in the context of this paper, $(V^G = V^H \wedge E^G = E^H) \rightarrow (G = H)$.

A graph morphism $m : G \rightarrow H$ is given by a pair of functions $m^V : V^G \rightarrow V^H$ and $m^E : E^G \rightarrow E^H$ that map elements from G to elements from H such that $s^H \circ m^E = m^V \circ s^G$ and $t^H \circ m^E = m^V \circ t^G$ [9].

A graph G can be typed over a type graph TG via a typing morphism $type : G \rightarrow TG$, forming the typed graph $G^T = (G, type^G)$. A typed graph morphism between two typed graphs $G^T = (G, type^G)$ and $H^T = (H, type^H)$ with the same type graph then denotes a graph morphism $m^T : G \rightarrow H$ such that $type^G = type^H \circ m^T$. A (typed) graph morphism m is a monomorphism iff its functions m^V and m^E are injective.

Figure 1 shows an example typed graph M_1 and associated type graph TM from the software development domain. M_1 represents an abstract syntax graph for a program written in an object-oriented language that contains four classes represented by nodes. The type graph also allows representing superclass relationships with edges.

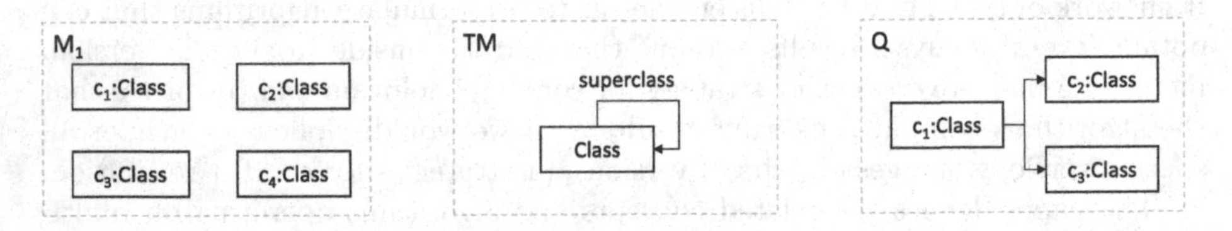

Fig. 1. Example graph, type graph, and violation pattern

The structure of a typed graph G can be restricted by a well-formedness condition ϕ, which in the context of this paper is characterized by a typed graph Q typed over the same type graph. G then satisfies the condition ϕ, denoted $G \models \phi$, iff there exists no monomorphism $m : Q \rightarrow G$. We also call such monomorphisms *matches* and Q the *violation pattern* of ϕ.

Figure 1 shows a violation pattern Q for an example well-formedness constraint that forbids a class having two outgoing superclass relationships.

A graph modification as defined by Taentzer et al. [26] formalizes the difference between two graphs G and H and is characterized by an intermediate graph K and a span of monomorphisms $(G \leftarrow K \rightarrow H)$. In this paper, we assume that the two morphisms are always subgraph inclusions. K then characterizes the subgraph that is preserved through the modification, whereas elements in G that are not in K are deleted and elements in H but not in K are created.

Figure 2 shows an example graph modification from the graph M_1 from Fig. 1 to a new graph M_2, where a superclass edge from class c_1 to class c_3 is created and the class c_4 is deleted. The morphisms are implied by node labels.

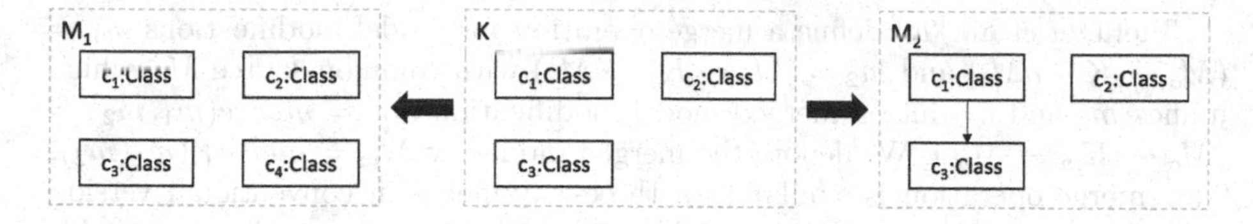

Fig. 2. Example graph modification

Graphs and graph modifications correspond to versions and differences in conventional, line-based version control systems like Git [16], where versions of a development artifact and intermediate differences form a directed acyclic graph.

3 Single-Version Models

In this paper, we consider models in the form of typed graphs that are required to adhere to a set of well-formedness conditions. Effectively, the combination of type graph and well-formedness conditions then acts as a metamodel with potential further constraints. Note that attributes, as usually employed in real-world models, can in this context be modeled as dedicated nodes [17].

For Φ the set of well-formedness conditions, a model M_i is *well-formed* iff $\forall \phi \in \Phi : M_i \models \phi$. We assume $pcheck(M_i, \phi)$ to report all violations to property ϕ with violation pattern Q for model M_i in the form of matches for Q, essentially realizing $\models$ as $pcheck(M_i, \phi) = \emptyset \iff M_i \models \phi$. If violations exist, the model M_i is also called *ill-formed*.

For the notion of models as typed graphs, model modifications correspond to graph modifications as presented in Sect. 2. We say a model modification $(M_i \leftarrow K \rightarrow M_j)$ with subgraph inclusions is *maximally preserving* iff it does not delete and recreate identical elements. Formally, $K = (V^{M_i} \cap V^{M_j}, E^{M_i} \cap E^{M_j}, s^K, t^K)$, where s^K and t^K are uniquely defined assuming invariant edge sources and targets. Consequently, for two models M_i and M_j, the maximally preserving model modification $(M_i \leftarrow K \rightarrow M_j)$ is uniquely defined.

For a set of model modifications $\Delta^{M_{\{1,\ldots,n\}}}$ between models $M_{\{1,\ldots,n\}} = \{M_1, \ldots, M_n\}$, with $\forall (G \leftarrow K \rightarrow H) \in \Delta^{M_{\{1,\ldots,n\}}} : G \in M_{\{1,\ldots,n\}} \wedge H \in M_{\{1,\ldots,n\}}$, we can define the set of predecessors $pre(i) \subset M_{\{1,\ldots,n\}}$ of a version M_i as the set of versions M_j such that there exists a sequence of model modifications $(M_{x_1} \leftarrow K_{x_1} \rightarrow M_{x_2})$, $(M_{x_2} \leftarrow K_{x_2} \rightarrow M_{x_3})$, ..., $(M_{x_{n-1}} \leftarrow K_{x_{n-1}} \rightarrow M_{x_n})$ where $x_1 = j$, $x_n = i$, and $(M_{x_k} \leftarrow K_{x_k} \rightarrow M_{x_{k+1}}) \in \Delta^{M_{\{1,\ldots,n\}}}$ for $1 \leq k < n$.

$\Delta^{M_{\{1,\ldots,n\}}}$ describes a *correct version history* if all morphisms in the individual model modifications are subgraph inclusions, all model modifications are maximally preserving, the *pre* relation is acyclic and there exists a model M_α such that $M_\alpha \in pre(i)$ for all models $M_i \neq M_\alpha$. Effectively, a correct version history describes a directed acyclic graph of model versions $M_{\{1,\ldots,n\}}$ that are derived from an original model M_α via the model modifications in $\Delta^{M_{\{1,\ldots,n\}}}$, and therefore closely corresponds to the versioning of some development artifact in a conventional version control system.

Taentzer et al. [26] define a merge operation for model modifications $m_1 = (M_c \leftarrow K_i \rightarrow M_i)$ and $m_2 = (M_c \leftarrow K_j \rightarrow M_j)$ with common source M_c, which unifies m_1 and m_2 into a merged model modification $m_m = merge(m_1, m_2) = (M_c \leftarrow K_m \rightarrow M_m)$. We denote the merged model by $M_m = merge_G(m_1, m_2)$. This merge operation is similar to a three-way-merge in conventional version control systems [20], since m_m in the default case (i) preserves an element $x \in M_c$ iff it is preserved by both m_1 and m_2 (ii) deletes an element $x \in M_c$ iff it is deleted by m_1 or m_2 (iii) creates an element $x \in M_m$ iff it is created by m_1 or m_2.

However, according to [26], model modifications can be in conflict in two cases: (i) insert-delete conflict and (ii) delete-delete conflict. Taentzer et al. state that only (i), where one modification creates an edge connected to a node deleted by the other modification, is an actual conflict, which has to be resolved to create a correct merge result. In this case, the merge result may deviate from the default case. Such conflicts will be reported by $mcheck((M_c \leftarrow K_i \rightarrow M_i), (M_c \leftarrow K_j \rightarrow M_j))$ in the form (e, v), where e is an edge created by one of the modifications and v is a node deleted by the other modification.

For a correct version history $\Delta^{M_{\{1,\dots,n\}}}$, we say that two sequences of model modifications $M_c \Rightarrow^* M_i$ and $M_c \Rightarrow^* M_j$ are in conflict iff their corresponding maximally preserving model modifications $(M_c \leftarrow K_{c,i} \rightarrow M_i)$ and $(M_c \leftarrow K_{c,j} \rightarrow M_j)$ are in conflict. In this case, we also say that M_i and M_j are in conflict for the common predecessor M_c.

Insert-delete conflicts can be resolved by equipping the *merge* operation with a manual or automatic strategy for conflict resolution. We consider such a strategy valid if it decides for each conflict whether to either revert the edge creation or the node deletion and always produces a proper merged graph. The approach in [26] effectively proposes an automatic strategy that favors insertion over deletion in order to preserve as many model elements as possible. Therefore, it reverts any deletions of nodes that would lead to insert-delete conflicts.

In contrast, a strategy for conflict resolution may favor deletion over insertion by reverting any creations of edges that would lead to insert-delete conflicts. Specifically, for model modifications $m_1 = (M_c \leftarrow K_i \rightarrow M_i)$ and $m_2 = (M_c \leftarrow K_j \rightarrow M_j)$, the model modification $m_{min} = merge^{min}(m_1, m_2)$, with $merge^{min}$ a merge operation equipped with this strategy, only creates an edge created by m_1 or m_2 if neither its source nor target is deleted by the other modification.

If all well-formedness conditions are specified by simple violation patterns, m_{min} also yields a model where all well-formedness violations are also present in the merge result for any other conflict resolution strategy:

Theorem 1. *For two model modifications $m_1 = (M_c \leftarrow K_i \rightarrow M_i)$ and $m_2 = (M_c \leftarrow K_j \rightarrow M_j)$ and a well-formedness constraint ϕ with violation pattern Q, it holds that*

$$pcheck(merge_G^{min}(m_1, m_2), \phi) = \bigcap_{str \in S} pcheck(merge_G^{str}(m_1, m_2), \phi),$$

with S the set of all valid conflict resolution strategies.

Proof. (Sketch) Follows directly from the fact that $merge_G^{min}(m_1, m_2)$ *is the smallest common subgraph of all graphs produced by the operation merge for any valid conflict resolution strategy.* □

If there are no conflicts in the merged model operations, the *merge* operation produces the same result regardless of the chosen strategy for conflict resolution.

For a correct version history, two model versions M_i and M_j, and the set of versions $P = pre(i) \cap pre(j)$, we define the function

$$pre^C(i,j) = \begin{cases} \emptyset & M_i \in pre(j) \vee M_j \in pre(i) \\ \{M_c \in P \,|\, \forall M_x \in P : M_c \notin pre(x)\} & \text{otherwise} \end{cases},$$

which returns the set of latest common predecessors of M_i and M_j. Note that our definition of pre^C corresponds to the definition of a best common ancestor in conventional version control systems such as Git [16], which is used to compute the base for three-way merges in these systems.

Figure 3 shows an exemplary version history based on the graph M_1 from Fig. 1. The initial graph $M_\alpha = M_1$ contains four classes. The modification m_1 (not to be confused with a morphism) to M_2 creates a superclass edge from c_1 to c_3 and deletes the node c_4. The modification m_2 to graph M_3 creates superclass edges from c_1 to c_2 and from c_4 to c_2. There is an insert-delete conflict between the two modifications, since the modification to M_2 deletes a node that is needed as the source of an edge created by the modification to M_3. Furthermore, the result of the merge of the two modifications would violate the well-formedness constraint with the violation pattern Q from Fig. 1, since without additional modifications, the node c_1 would have two outgoing superclass edges.

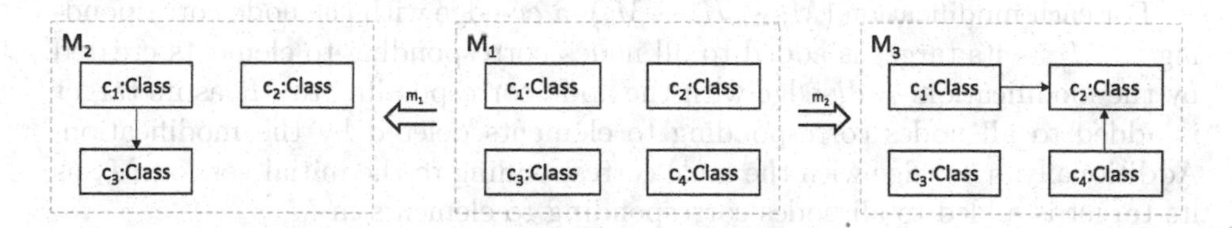

Fig. 3. Example version history

4 Multi-version Models as Typed Graphs

A correct version history $\Delta^{M_{\{1,\ldots,n\}}}$ with model versions $M_{\{1,\ldots,n\}}$ conforming to a type graph TM can be represented by a multi-version model in the form of a single graph that is typed over an adapted type graph.

The adapted type graph TM_{mv} contains a node for each node and edge in TM. It also contains edges connecting each node in TM_{mv} that represents an edge in TM to the nodes representing the edge's source and target in TM. This yields a bijective function $corr_{mv} : V^{TM} \cup E^{TM} \rightarrow V^{TM_{mv}}$, which maps elements from TM to the corresponding node in TM_{mv}, and two bijective functions $corr^s_{mv}, corr^t_{mv} : E^{TM} \rightarrow E^{TM_{mv}}$ mapping edges from TM to the edges in TM_{mv} encoding the source and target relation in TM. In addition, TM_{mv} contains a node *version*, an edge *suc* with source and target *version*, and two edges cv_v and dv_v from each other node $v \in V^{TM_{mv}}$ to the *version* node.

A multi-version model MVM for $\Delta^{M_{\{1,\dots,n\}}}$ is then constructed by an operation *comb* as follows: A subgraph P^M_{mv} encodes structural information about all model versions and is constructed by translating $P^M = \bigcup_{M_i \in M_{\{1,\dots,n\}}} M_i$ to conform to TM_{mv} using an operation $trans_{mv}$. Since source and target functions are invariant in a correct version history, P^M is well-defined.

For each $v \in v^{P^M}$, $trans_{mv}$ creates a node of type $corr_{mv}(v)$ in $V^{P^M_{mv}}$. For each $e \in E^{P^M}$, a node of type $corr_{mv}(e)$ is created. This yields a bijection $origin : P^M_{mv} \rightarrow P^M$ mapping translated elements to their original representation.

In addition, for each edge $e \in E^{P^M}$, an edge of type $corr^s_{mv}(e)$ with source $origin^{-1}(e)$ and target $origin^{-1}(s^{P^M}(e))$ and an edge of type $corr^t_{mv}(e)$ with source $origin^{-1}(e)$ and target $origin^{-1}(t^{P^M}(e))$ are created in $E^{P^M_{mv}}$. Since edge sources and targets are invariant, the corresponding node $v_e = origin^{-1}(e)$ in the end has exactly one edge of type $corr^s_{mv}(e)$ and one of type $corr^t_{mv}(e)$. We thus have two functions $s_{mv} : origin^{-1}(E^{P^M}) \rightarrow E^{P^M_{mv}}$ respectively $t_{mv} : origin^{-1}(E^{P^M}) \rightarrow E^{P^M_{mv}}$ encoding these mappings.

Another, distinct subgraph P^V_{mv} contains versioning information and is constructed as follows: For each $M_i \in M_{\{1,\dots,n\}}$, P^V_{mv} contains a corresponding node of type *version*. For each $(M_i \leftarrow K \rightarrow M_j) \in \Delta^{M_{\{1,\dots,n\}}}$, P^V_{mv} contains an edge of type *suc* from the node representing M_i to the node representing M_j.

For each modification $(M_i \leftarrow K \rightarrow M_j)$, a *cv*-edge with the node corresponding to M_j as its target is added to all nodes corresponding to elements created by the modification. A *dv*-edge with the node corresponding to M_j as its target is added to all nodes corresponding to elements deleted by the modification. Additionally, a *cv* edge with the node corresponding to the initial version M_α as its target is added to all nodes corresponding to elements in M_α.

Since attributes can be encoded by dedicated nodes and assignment edges [17], the construction can be performed analogously for attributed graphs.

For $v \in P^M_{mv}$ and $M_i \in M_{\{1,\dots,n\}}$, we say that v is *mv-present* in M_i, iff for a node m_{cv} connected to v via a *cv* edge, there exists a path from m_{cv} to the node representing M_i via *suc* edges that does not go through a node connected to v via a *dv* edge. We denote the set of versions where v is mv-present by $p(v)$.

A model version M_i can then be derived from MVM via an operation *proj* as follows: Collect all nodes $V_p = \{v_p \in V^{P^M_{mv}} | M_i \in p(v_p)\}$, that is, all nodes that are mv-present in M_i, and translate the induced subgraph into the single-version model M_i with $V^{M_i} = \{origin(v_v) | v_v \in V^{MVM} \wedge corr^{-1}_{mv}(type^{MVM}(v_v)) \in V^{TM}\}$, $E^{M_i} = \{origin(v_e) | v_e \in V^{MVM} \wedge corr^{-1}_{mv}(type^{MVM}(v_e)) \in E^{TM}\}$, $s^{M_i} = origin \circ t^{MVM} \circ s_{mv} \circ origin^{-1}$, and $t^{M_i} = origin \circ t^{MVM} \circ t_{mv} \circ origin^{-1}$.

Correctness

Theorem 2. *For a correct version history $\Delta^{M_{\{1,\ldots,n\}}}$ holds concerning comb and proj:*

$$\forall i \in \{1, \ldots, n\} : M_i = proj(comb(\Delta^{M_{\{1,\ldots,n\}}}), i).$$

Proof. (Sketch) Any element in a version M_i has a corresponding node v in $comb(\Delta^{M_{\{1,\ldots,n\}}})$. By construction, v is connected to a node corresponding to some version M_j via a cv edge, for which there exists a path of suc edges to the node corresponding to M_i. That path does not go through a node connected to v by a dv edge. v is thus mv-present in M_i and hence contained in the projection.

Inclusion of elements in the opposite direction can be shown analogously. Because edge sources and targets are invariant over all graphs, the edges in $comb(M_1, \ldots, M_n)$ correctly encode the source and target functions by construction. Thus, $\forall i \in \{1, \ldots, n\} : M_i = proj(comb(M_1, \ldots, M_n), i)$. □

More detailed proofs for this and other theorems in the paper can be found in the appendix of the preprint version [2].

A maximally preserving model modification ($M_i \leftarrow K \rightarrow M_j$) with $M_i, M_j \in M_{\{1,\ldots,n\}}$ (and thus any model modification in $\Delta^{M_{\{1,\ldots,n\}}}$) can be derived from MVM via $proj^{\Delta}$ as follows: M_i and M_j can be derived via the operation *proj*. K is then the graph containing all elements from $M_i \cap M_j$, with s^K and t^K uniquely defined by the corresponding functions from M_i and M_j and partial identities as morphisms into M_i and M_j.

Theorem 3. *For a correct version history $\Delta^M_{\{1,\ldots,n\}}$ holds concerning comb and $proj^{\Delta}$:*

$$\forall M_i, M_j \in M_{\{1,\ldots,n\}} : m_{i,j} = proj^{\Delta}(comb(\Delta^{M_{\{1,\ldots,n\}}}), i, j),$$

with $m_{i,j}$ the maximally preserving model modification from M_i to M_j.

Proof. Follows trivially from Theorem 2 and the definition of the maximally preserving model modification ($M_i \leftarrow K_{i,j} \rightarrow M_j$). □

Figures 4 and 5 visualize the multi-version model MVM constructed for the example history in Fig. 3 and the associated adapted type graph TM_{mv}. MVM contains a node for each node and edge in the models of the example history, one node of type *version* for each of the graphs M_1, M_2, and M_3, and appropriate edges as created by *comb*.

4.1 Directly Checking Well-Formedness for Multi-version Models

We can use a multi-version model to directly find all well-formedness violations in all individual versions via an operation $pcheck_{mv}$. For a multi-version model MVM with a bijective mapping into a union of original model versions $origin_M$ and a well-formedness constraint ϕ with associated violation pattern Q, $pcheck_{mv}(MVM, \phi)$ works as follows:

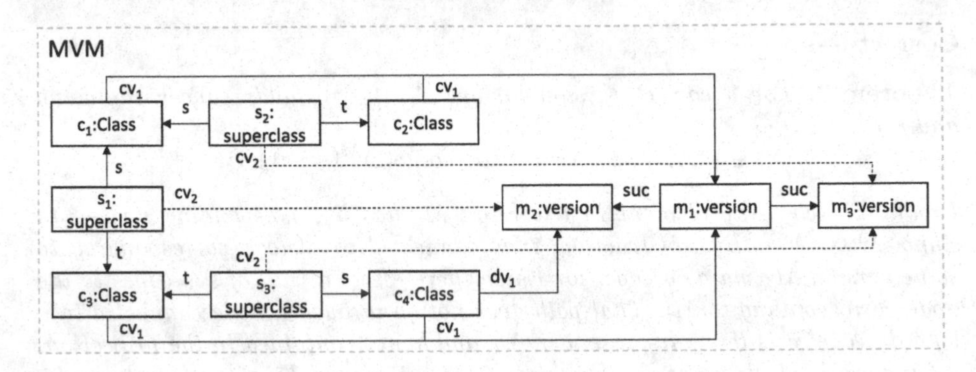

Fig. 4. Multi-version model for the history in Fig. 3

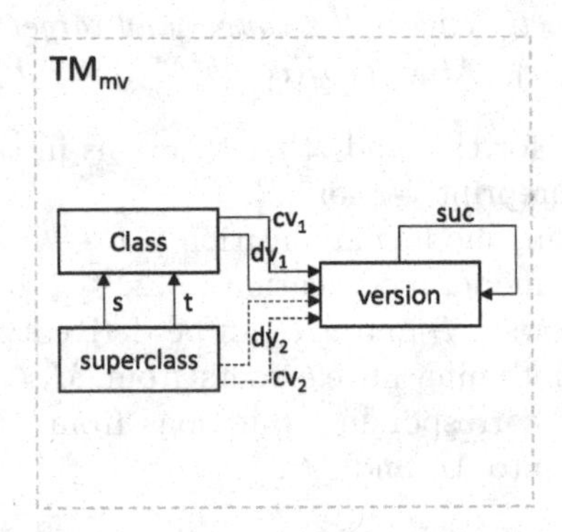

Fig. 5. Adapted type graph for type graph in Fig. 1

First, the graph Q typed over the original type graph is translated into a corresponding graph Q_{mv} typed over the adapted type graph using $trans_{mv}$. This yields a bijective mapping $origin_Q : Q_{mv} \to Q$.

Then, all matches for Q_{mv} in MVM are found. For each such match m_{mv}, $pcheck_{mv}$ computes all versions for which all vertices in the image of the match are mv-present by $P = \bigcap_{v \in V^{Q_{mv}}} p(m_{mv}(v))$. If $P \neq \emptyset$, the match into the original model versions $m = origin_M \circ m_{mv} \circ origin_Q^{-1}$ is constructed and reported as a violation in all versions in P.

Correctness

Theorem 4. *For a well-formedness constraint ϕ with violation pattern Q, a correct version history $\Delta^{M_{\{1,\ldots,n\}}}$, and $MVM = comb(\Delta^M_{\{1,\ldots,n\}})$ holds:*

$$pcheck_{mv}(MVM, \phi) = \biguplus_{i \in \{1,\ldots,n\}} \{(i, m) | m \in pcheck(proj(MVM, i), \phi)\}.$$

Proof. (Sketch) A match $m : Q \to M_i$ for any version M_i has one corresponding match m_{mv} with $m = origin_M \circ m_{mv} \circ origin_Q^{-1}$, where edges created by

$trans_{mv}$ ensure correct connectivity. $P = \bigcap_{v \in V^{Q_{mv}}} p(m_{mv}(v))$ contains exactly the versions containing all elements in $m(Q)$. This yields the stated equality. □

Complexity. The effort for searching all versions $M_{\{1,...,n\}}$ of some version history $\Delta^{M_{\{1,...,n\}}}$ for a pattern Q using *pcheck* is in $O(\sum_{M_i \in M_{\{1,...,n\}}} C(M_i, Q))$, with $C(M_i, Q)$ the effort for finding all matches of Q into M_i.

$P^M_{mv} = trans_{mv}(P^M)$ and $Q_{mv} = trans_{mv}(Q)$ are only different encodings of $P^M = \bigcup_{M_i \in M_{\{1,...,n\}}} M_i$ and Q. Considering computation of the mv-present predicate, the effort for $pcheck_{mv}$ is hence in $O(C(\bigcup_{M_i \in M_{\{1,...,n\}}} M_i, Q) + X \cdot |V^{Q_{mv}}| \cdot |\Delta^{M_{1,...,n}}|)$, with X the number of matches for Q_{mv} into P^M_{mv}.

Discussion. If many elements are shared between individual versions and modifications only perform few changes, the size of the union of all model versions will be small compared to the sum of the sizes of all individual versions. If pattern matching is efficient with respect to the size of the considered model, pattern matching over the union of all model versions will then likely require less effort than matching over each individual version. Intuitively, $pcheck_{mv}$ avoids redundant searches over model parts that are shared between multiple versions and thus saves the related effort. If the number of matches for violation patterns is low, the associated checks performed by $pcheck_{mv}$ will likely be more efficient than the pattern matching over the individual versions.

Overall, $pcheck_{mv}$ will thus likely be more efficient than using *pcheck* in scenarios where pattern matching is efficient, the number of changes between versions is low, and the number of violations in the union of versions is low.

5 Directly Checking Merge Results for Multi-version Models

We can consider multi-version models to directly detect whether (a) merge conflicts exist for any valid pair of encoded model modifications via an operation $mcheck_{mv}$ and (b) any resulting merged model is ill-formed via an operation $pcheck^m_{mv}$, where a pair of model modifications $(M_c \leftarrow K_i \rightarrow M_i)$ and $(M_c \leftarrow K_j \rightarrow M_j)$ is valid iff $M_c \in pre^C(M_i, M_j)$.

5.1 Directly Checking for Merge Conflicts

$mcheck_{mv}$ can be realized for a multi-version model $MVM = comb(\Delta^{M_{\{1,...,n\}}})$ as follows: First, the operation collects all nodes in MVM representing edges that are created by some model modification. This means all nodes $v_e \in V^{MVM}$ where $corr^{-1}_{mv}(type^{MVM}(v)) \in E^{TM}$ connected to a node m_x via a cv edge, where m_x does not correspond to M_α and with TM the original type graph. Then, for each node v_e, we compute the set of versions $P = p(v_e)$ where it is mv-present. If $P \neq p(v_s)$, where $v_s = s^{MVM}(s_{mv}(v_e))$, we then compute a set of versions D

that correspond to nodes reachable via *suc* edges from a node connected to v_s via a *dv* edge without going through nodes connected to v_s via a *cv* edge.

Afterwards, for each pair of versions $M_i \in P$ and $M_j \in D$, we check for each latest common predecessor $M_c \in pre^C(i,j)$ whether $M_c \in p(v_s) \wedge M_c \notin P$. For any triplet of versions (i,j,c) where this is the case, the edge $origin(v_e)$ is then in an insert-delete conflict with its source. To facilitate formalization, this conflict is reported in the normalized form $(min(i,j), max(i,j), c, (origin(v_e), origin(v_s)))$. Insert-delete conflicts with the edge's target are computed analogously.

Correctness

Theorem 5. *For a version history $\Delta^{M_{\{1,\ldots,n\}}}$ and the associated multi-version model $MVM = comb(\Delta^M_{\{1,\ldots,n\}})$ holds:*

$$mcheck_{mv}(MVM) = \biguplus_{(i,j,c)\in Y} \{(i,j,c,m) | m \in mcheck(m_{c,i}, m_{c,j})\},$$

where $Y = \{(i,j,c) \,|\, i,j \in \{1,\ldots,n\} : i < j, c \in \{c | M_c \in pre^C(i,j)\}\}$ and with $m_{c,i} = proj^\Delta(MVM, c, i)$ and $m_{c,j} = proj^\Delta(MVM, c, j)$.

Proof. (Sketch) The collected nodes representing edges correspond to a superset of edges that may be involved in a conflict. The construction of the sets P and D for a collected node v_e ensures that any pair of versions where one may create $e = origin(v_e)$ and the other may delete the source (or target) of e is considered. The condition checked for each common predecessor of a version pair then yields exactly the triplets of versions where e is part of an insert-delete conflict. Because of the normalization of the results of $mcheck_{mv}$, we have the stated equality. □

Complexity. The function pre^C_{mv} can be precomputed in $O(|M_{\{1,\ldots,n\}}|^4)$.

Since information about creation and deletion of elements is not explicitly available in a naïve representation, finding all insert-delete conflicts between two model modifications via *mcheck* has to be done by checking for each edge in either modification's resulting model whether it is created by that modification and its source or target is deleted by the other modification. Since there may exist up to $O(|M_{\{1,\ldots,n\}}|^3)$ possible merges in a version history, in the worst case, this implies effort in $O(|M_{\{1,\ldots,n\}}|^4 + |E^{M_{max}}| \cdot |M_{\{1,\ldots,n\}}|^3)$, where $|E^{M_{max}}|$ is the maximum number of edges present in a single model version.

Created edges can be retrieved efficiently from a multi-version model given appropriate data structures. Computing and checking the required version sets takes $O(|M_{\{1,\ldots,n\}}|^3)$ steps per edge. Therefore, the overall computational complexity of $mcheck_{mv}$ is in $O(|M_{\{1,\ldots,n\}}|^4 + \Delta_+ \cdot |M_{\{1,\ldots,n\}}|^3)$, where Δ_+ is the overall number of elements created in the version history.

Discussion. The efficiency of $mcheck_{mv}$ compared to using $mcheck$ mostly depends on the number of edges created by some model modification compared to the number of edges in the individual versions. If most edges are present in the original model version and are shared between many model versions, $mcheck_{mv}$ will be more efficient. Otherwise, $mcheck_{mv}$ will not achieve a significant improvement and might even perform worse than the operation based on $mcheck$.

Version control systems such as Git typically select a single latest common predecessor as the base for a three way merge [16]. Using a corresponding partial function $pre_1^C : \mathbb{N} \times \mathbb{N} \rightarrow M_{\{1,...,n\}}$ with $pre_1^C(i,j) \in pre^C(i,j)$ if $pre^C(i,j) \neq \emptyset$ and $pre_1^C(i,j) = \perp$ to select a single latest common predecessor of two versions i and j rather than pre^C in $mcheck_{mv}$, by the same logic as used in the proof of correctness, we instead have an analogous equality for pre_1^C. Disregarding the computational effort for precomputing pre_1^C, replacing pre^C by pre_1^C reduces the remaining computational complexity of $mcheck_{mv}$ to $O(\Delta_+ \cdot |M_{\{1,...,n\}}|^2)$.

5.2 Directly Checking Well-Formedness for Merge Results

To find all violations of a well-formedness constraint ϕ characterized by a pattern Q via $pcheck_{mv}^m$ in merge results of a multi-version model MVM, we first translate Q into $Q_{mv} = trans_{mv}$. We then find all matches for Q_{mv} in MVM.

For a match m_{mv} for Q_{mv}, we determine the set of versions $P_v = p(v)$ for each $v \in m_{mv}(V^{Q_{mv}})$. For each pair of versions $M_i \in \arg\min_{P \in \{p(v) | v \in m_{mv}(V^{Q_{mv}})\}} |P|$ and $M_j \in \bigcup_{v \in V^{Q_{mv}}} p(v)$, we check whether $\forall v \in m_{mv}(V^{Q_{mv}}) : M_i \in p(v) \vee M_j \in p(v)$. We then check for each latest common predecessor $M_c \in pre^C(i,j)$ if for all $v \in V^{Q_{mv}}$, it holds that $v \in V^{M_c} \rightarrow (v \in V^{M_i} \wedge v \in V^{M_j})$, that is, v is not deleted in M_i or M_j. If this is the case, the match m into $\bigcup_{M_x \in M_{\{1,...,n\}}} M_x$ corresponding to m_{mv} represents a violation in $merge^{min}((M_c \leftarrow K_i \rightarrow M_i), (M_c \leftarrow K_j \rightarrow M_j))$. We report results in the normalized form $(min(i,j), max(i,j), c, m)$.

Correctness

Theorem 6. *Given a well-formedness constraint ϕ, a correct version history $\Delta^{M_{\{1,...,n\}}}$, and the multi-version model $MVM = comb(\Delta^M_{\{1,...,n\}})$, it holds that:*

$$pcheck_{mv}^m(MVM, \phi) = \biguplus_{(i,j,c) \in Y} \{(i,j,c,m) | m \in pcheck(M_{i,j,c}^{min}, \phi)\},$$

where $Y = \{(i,j,c) \,|\, i,j \in \{1,\ldots,n\} : i < j, c \in \{c | M_c \in pre^C(i,j)\}\}$ and $M_{i,j,c}^{min} = merge_G^{min}(proj^\Delta(MVM,c,i), proj^\Delta(MVM,c,j))$.

Proof. (Sketch) For two versions M_i, M_j with latest common predecessor M_c, a match $m : Q \rightarrow merge_G^{min}(proj^\Delta(MVM,c,i), proj^\Delta(MVM,c,j))$ has one corresponding match $m_{mv} : trans_{mv}(Q) \rightarrow MVM$ by construction, where the edges created by $trans_{mv}$ ensure the correct connectivity. The set of version pairs considered by $pcheck_{mv}^m$ contains all version pairs such that each matched element *is contained in at least one of the versions. The condition checked for every latest common predecessor ensures that only version triplets are reported where the merge result also contains all matched elements if there are no merge conflicts. Since* $merge^{min}$ *resolves conflicts by prioritizing deletion and, as ensured by the check, no matched node is deleted by the merge, conflict resolution cannot invalidate the match or create new matches. We thus have the stated equality.* □

By Theorem 1 and Theorem 6, we also have that $pcheck^{m}_{mv}$ yields the set of violations that cannot be avoided by any conflict resolution strategy:

Corollary 1. *Given a well-formedness constraint* ϕ*, a correct version history* $\Delta^{M_{\{1,...,n\}}}$*, and the multi-version model* $MVM = comb(\Delta^{M_{\{1,...,n\}}})$*, it holds that:*

$$pcheck^{m}_{mv}(MVM, \phi) = \biguplus_{(i,j,c) \in Y} \bigcap_{str \in S} \{(i, j, c, m) | m \in mcheck(M^{str}_{i,j,c}, \phi)\},$$

where $Y = \{(i, j, c) \mid i, j \in \{1, \ldots, n\} : i < j, c \in \{c | M_c \in pre^{C}(i, j)\}\}$ *and* $M^{str}_{i,j,c} = merge^{str}_{G}(proj^{\Delta}(MVM, c, i), proj^{\Delta}(MVM, c, j))$*, and with* S *the set of all valid conflict resolution strategies.*

Complexity. The function pre^{C}_{mv} can be precomputed in $O(|M_{\{1,...,n\}}|^4)$.

With $C(M_i, Q)$ the effort for finding all matches of Q into M_i, finding violations characterized by a pattern Q in all results of a set of possible merges Y using *pcheck* takes effort in $O(O(|M_{\{1,...,n\}}|^4 + \sum_{(m_1,m_2) \in Y} C(merge^{min}_{G}(m_1, m_2), Q))$.

The computation and checking of version triplets for a match in $pcheck^{m}_{mv}$ takes effort in $O(|M_{\{1,...,n\}}|^3)$. For X matches for Q_{mv}, the effort for $pcheck^{m}_{mv}$ is thus in $O(|M_{\{1,...,n\}}|^4 + C(\bigcup_{M_i \in M_{\{1,...,n\}}} M_i, Q) + X \cdot |V^{Q_{mv}}| \cdot |M_{\{1,...,n\}}|^3)$.

Discussion. By the same argumentation as for $pcheck_{mv}$, $pcheck^{m}_{mv}$ will likely be more efficient than the corresponding operation using *pcheck* in scenarios where pattern matching is efficient, the number of changes between versions is low, and the number of violations in the union of model versions is low.

Using some partial function $pre^{C}_{1} : \mathbb{N} \times \mathbb{N} \rightarrow M_{\{1,...,n\}}$ to select a single latest common predecessor rather than pre^{C} in $pcheck^{m}_{mv}$, by the same logic as in the proof of correctness, we have an analogous equality for pre^{C}_{1}. Disregarding the effort for precomputing pre^{C}_{1}, replacing pre^{C} by pre^{C}_{1} reduces the remaining complexity of $pcheck^{m}_{mv}$ to $O(C(\bigcup_{M_i \in M_{\{1,...,n\}}} M_i, Q) + X \cdot |V^{Q_{mv}}| \cdot |M_{\{1,...,n\}}|^2)$.

6 Evaluation

For an initial empirical evaluation of the performance and scalability of the presented operations, we experiment with an application scenario from the software development domain. Therefore, we extract abstract syntax graphs from a small previous research project (**rete**) and a larger open source project (**henshin** [1]) written in Java using the EMF-based [10] MoDisco tool [5]. We store the extracted models in a graph format and fold each of the projects into a multi-version model, using a mapping strategy based on hierarchy and element names.

We then run implementations of the presented operations for conflict detection and well-formedness checking based on multi-version models (**MVM**) and baseline implementations using corresponding single-version models (**SVM**).[1] We consider three well-formedness constraints: uniqueness of a class's superclass, uniqueness of a method's return type, and consistency of an overriden method's return type. We employ our own EMF-based tool [14] for pattern matching.

Figure 6 shows the measured execution times for the operations $pcheck_{mv}$, $mcheck_{mv}$, and $pcheck^{m}_{mv}$ and related single-version-model-based operations over the example models. The execution times for $pcheck_{mv}$ and $pcheck^{m}_{mv}$ correspond to the combined pattern matching time for all considered well-formedness constraints. All reported times exclude the time for computing any merge results required by SVM and the time required to precompute the pre^C function, since it is required by both the MVM and the SVM implementation. Precomputing pre^C took about 5 ms for the smaller project and about 3.5 s for the larger project.

For the tasks related to well-formedness checking, the MVM variant performs better (up to factor 50) than SVM. Since there are only few to no matches for the violation patterns of the considered constraints, the MVM implementation only performs few of the potentially expensive checks over the version graph, while avoiding most of the redundancy in the pattern matching of SVM.

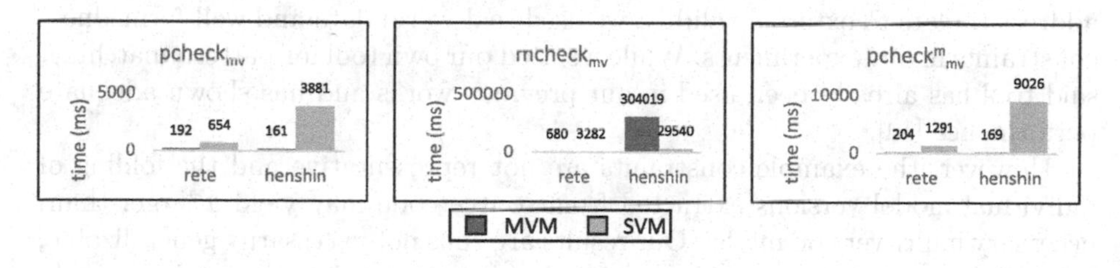

Fig. 6. Measurement results for $pcheck_{mv}$, $mcheck_{mv}$, and $pcheck^{m}_{mv}$

For conflict detection, MVM performs better than SVM for the smaller project (factor 5), but has a substantially higher execution time for the larger project (factor 10). The reason for the bad performance is that most edges are not present in the initial model version. In fact, the number of edges created throughout the version history is much higher than the number of edges in any individual version. Furthermore, in contrast to the solution using *mcheck*, the operation $mcheck_{mv}$ considers versions where the source or target of an edge

[1] All experiments were executed on a Linux SMP Debian 4.19.67-2 machine with Intel Xeon E5-2630 CPU (2.3 GHz clock rate) and 386 GB system memory running OpenJDK version 1.8.0_242. Reported execution times correspond to the minimum of at least five runs of the respective experiment. Memory measurements were obtained in a single run using the native Java library. Our implementation and datasets are available under https://github.com/hpi-sam/multi-version-models.

is *not* present. Due to the high number of versions in the project and because many elements are only present in few versions, this leads to the processing of large version sets, which deteriorates the performance of MVM in this scenario.

The memory consumption of the multi-version models and their representations as collections of single-version models is displayed in Fig. 7. For both projects, the representation as a multi-version model affords a more compact representation compared to a naïve encoding (factor 30 for the larger project).

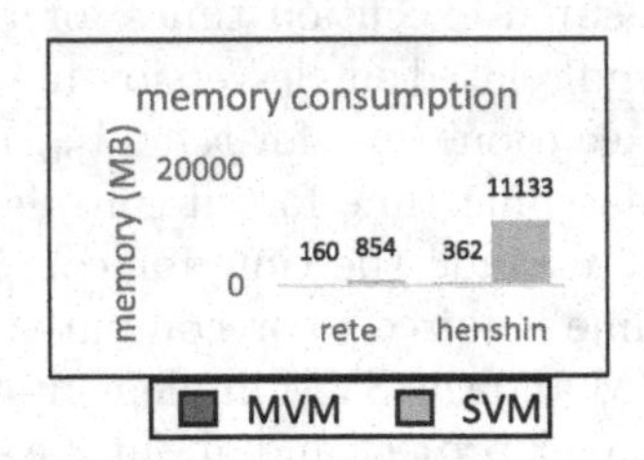

Fig. 7. Measurement results for memory consumption

Threats to Validity. Unexpected JVM behavior poses a threat to internal validity, which we tried to mitigate by performing multiple runs of each experiment measuring execution time and profiling time spent on garbage collection. To address threats to external validity, we used real-world data and well-formedness constraints in our experiments. While we used our own tool for pattern matching, said tool has already been used in our previous works and has shown adequate performance [14].

However, the example constraints are not representative and the folding of individual model versions extracted from source code may yield a larger-than-necessary multi-version model. Our results are thus not necessarily generalizable, but instead constitute an early conceptual evaluation of the presented approach.

7 Related Work

While most practical version control systems operate on text documents [20], versioning and merging of models has also been subject to extensive research.

There already exist several formal and semi-formal approaches to model merging, which compute the result of a three-way-merge of model modifications [26,27]. Notably, the approach by Taentzer et al. [26] represents a formally defined solution that works on the level of graphs, which is why for our approach, we build on their notion of model merging. In their work, Taentzer et al. also consider checking of well-formedness constraints by constructing a tentative merge result over which the check is executed. While this allows their approach to handle arbitrary constraints rather than just simple graph patterns, the check has to be executed for each individual merge.

Some approaches consider detection of merge conflicts [19] or model inconsistencies [3] based on the analysis of sequences of primitive changes. However, these approaches do not consider the case of multiple versions and pairwise merges and naturally do not employ a graph-based definition of inconsistencies.

For the more general problem of model versioning, both formal solutions [8,24] and tool implementations [18,21] have been introduced. Similar to our approach, some of these techniques are based on a joint representation of multiple model versions [21,24]. However, to the best of our knowledge, joint conflict detection or well-formedness checking for all merges at once is not considered.

Model repositories such as Hawk [13] allow storing the evolution of models over time and enable the execution of queries equipped with temporal operators. Folding and joint querying of the temporal evolution of graphs has also been studied in previous work of our group [15,25]. However, these solutions focus on sequences of graph modifications without diverging branches and hence do not consider merging.

The presented encoding of different model versions in a unified multi-version model bears similarity to so-called 150% models from software product lines [6]. A 150% model represents different configurations of a software system as a single unified model, where annotations determine the presence of individual model elements in certain configurations. The derivation of a model instance for a specific configuration from a 150% model then corresponds to the projection from a multi-version model to a specific model version. A realization of 150% models in the context of model-driven engineering is presented in [23].

Westfechtel and Greiner [28] present a solution for propagating presence information from a unified encoding of multiple product line configurations along model transformations. While their approach bears some similarity to the collective well-formedness checking in our solution, the technique in [28] focuses on product lines and hence does not consider version histories and merging.

[7] introduces a new semantics for OCL in the context of software product lines, which allows the collective checking of well-formedness constraints over a unified encoding of product line configurations. However, the application of this approach to model versioning would require a translation of version graphs and model modifications to an encoding of valid configurations and presence annotations. This seems nontrivial, especially if the compression of version histories achieved by multi-version models is to be preserved. However, by relying on OCL as a specification language, the approach in [7] allows a much higher expressiveness when formulating well-formedness conditions compared to simple graph patterns. Adopting some of the ideas in [7] may therefore enable lifting our definition of well-formedness to more expressive formalisms in future work.

A solution to conflict detection for features in software product lines is presented in [22]. In [22], product variability is encoded by so-called delta modules, which represent operations for extending a basic version of the software by certain features and are thus similar to model modifications. The approach checks for syntactic conflicts via pair-wise comparison of delta-modules and thus relates to detection of merge conflicts in the context of model merging. The approach in [22] also considers the case where a third delta module fixes conflicts between two other modules. Considering merges of more than two versions could also be an interesting direction for future work in the context of multi-version models.

8 Conclusion

In this paper, we have presented an approach for encoding a model's version history as a single typed graph. Based on this representation, we have introduced operations for finding merge conflicts and violations of well-formedness conditions in the form of graph patterns in the entire history and related merge results. We have conducted an initial empirical evaluation, which demonstrates potential benefits of the approach, but also highlights shortcomings in unfavorable scenarios.

In future work, we plan to address these shortcomings by studying how to compress the version graph or restrict the set of considered versions to those most relevant to users. We also plan to explore how such a restriction may allow the pruning of superfluous elements from a multi-version model and thereby prevent performance degradation as more versions are introduced. Furthermore, we want to investigate how to lift our notion of well-formedness constraints to more expressive formalisms such as nested graph conditions and develop an incremental version of the approach. Finally, we will extend our empirical evaluation to better characterize our technique's performance.

References

1. Arendt, T., Biermann, E., Jurack, S., Krause, C., Taentzer, G.: Henshin: advanced concepts and tools for in-place EMF model transformations. In: Petriu, D.C., Rouquette, N., Haugen, Ø. (eds.) MODELS 2010. LNCS, vol. 6394, pp. 121–135. Springer, Heidelberg (2010). https://doi.org/10.1007/978-3-642-16145-2_9
2. Barkowsky, M., Giese, H.: Towards Development with Multi-Version Models: Detecting Merge Conflicts and Checking Well-Formedness. arXiv preprint (2022). https://doi.org/10.48550/arXiv.2205.04198
3. Blanc, X., Mougenot, A., Mounier, I., Mens, T.: Incremental detection of model inconsistencies based on model operations. In: van Eck, P., Gordijn, J., Wieringa, R. (eds.) CAiSE 2009. LNCS, vol. 5565, pp. 32–46. Springer, Heidelberg (2009). https://doi.org/10.1007/978-3-642-02144-2_8
4. Brun, Y., Holmes, R., Ernst, M.D., Notkin, D.: Early detection of collaboration conflicts and risks. IEEE Trans. Softw. Eng. **39**(10), 1358–1375 (2013). https://doi.org/10.1109/TSE.2013.28
5. Bruneliere, H., Cabot, J., Jouault, F., Madiot, F.: MoDisco: a generic and extensible framework for model driven reverse engineering. In: Proceedings of the IEEE/ACM International Conference on Automated Software Engineering (2010). https://doi.org/10.1145/1858996.1859032
6. Czarnecki, K., Antkiewicz, M.: Mapping features to models: a template approach based on superimposed variants. In: Glück, R., Lowry, M. (eds.) GPCE 2005. LNCS, vol. 3676, pp. 422–437. Springer, Heidelberg (2005). https://doi.org/10.1007/11561347_28

7. Czarnecki, K., Pietroszek, K.: Verifying feature-based model templates against well-formedness OCL constraints. In: Proceedings of the 5th International Conference on Generative Programming and Component Engineering, GPCE 2006, pp. 211–220. Association for Computing Machinery, New York (2006). https://doi.org/10.1145/1173706.1173738
8. Diskin, Z., Czarnecki, K., Antkiewicz, M.: Model-versioning-in-the-large: algebraic foundations and the tile notation. In: 2009 ICSE Workshop on Comparison and Versioning of Software Models, pp. 7–12. IEEE (2009). https://doi.org/10.1109/CVSM.2009.5071715
9. Ehrig, H., Ehrig, K., Prange, U., Taentzer, G.: Fundamentals of Algebraic Graph Transformation. EATCS, Springer, Heidelberg (2006). https://doi.org/10.1007/3-540-31188-2
10. EMF. https://www.eclipse.org/modeling/emf/. Accessed 23 Feb 2022
11. Finkelstein, A.C.W., Gabbay, D., Hunter, A., Kramer, J., Nuseibeh, B.: Inconsistency handling in multiperspective specifications. IEEE Trans. Software Eng. **20**(8), 569–578 (1994). https://doi.org/10.1109/32.3106670
12. Frühauf, K., Zeller, A.: Software configuration management: state of the art, state of the practice. In: SCM 1999. LNCS, vol. 1675, pp. 217–227. Springer, Heidelberg (1999). https://doi.org/10.1007/3-540-48253-9_15
13. García-Domínguez, A., Bencomo, N., Parra-Ullauri, J.M., García-Paucar, L.H.: Querying and annotating model histories with time-aware patterns. In: 2019 ACM/IEEE 22nd International Conference on Model Driven Engineering Languages and Systems (MODELS), pp. 194–204. IEEE (2019). https://doi.org/10.1109/MODELS.2019.000-2
14. Giese, H., Hildebrandt, S., Seibel, A.: Improved flexibility and scalability by interpreting story diagrams. In: Electronic Communications of the EASST, vol. 18 (2009). https://doi.org/10.14279/tuj.eceasst.18.268
15. Giese, H., Maximova, M., Sakizloglou, L., Schneider, S.: Metric temporal graph logic over typed attributed graphs. In: Hähnle, R., van der Aalst, W. (eds.) FASE 2019. LNCS, vol. 11424, pp. 282–298. Springer, Cham (2019). https://doi.org/10.1007/978-3-030-16722-6_16
16. Git. https://git-scm.com/. Accessed 23 Feb 2022
17. Heckel, R., Küster, J.M., Taentzer, G.: Confluence of typed attributed graph transformation systems. In: Corradini, A., Ehrig, H., Kreowski, H.-J., Rozenberg, G. (eds.) ICGT 2002. LNCS, vol. 2505, pp. 161–176. Springer, Heidelberg (2002). https://doi.org/10.1007/3-540-45832-8_14
18. Koegel, M., Helming, J.: EMFStore: a model repository for EMF models. In: Proceedings of the 32nd ACM/IEEE International Conference on Software Engineering-Volume 2, pp. 307–308 (2010). https://doi.org/10.1145/1810295.1810364
19. Küster, J.M., Gerth, C., Engels, G.: Dependent and conflicting change operations of process models. In: Paige, R.F., Hartman, A., Rensink, A. (eds.) ECMDA-FA 2009. LNCS, vol. 5562, pp. 158–173. Springer, Heidelberg (2009). https://doi.org/10.1007/978-3-642-02674-4_12
20. Mens, T.: A state-of-the-art survey on software merging. IEEE Trans. Softw. Eng. **28**(5), 449–462 (2002). https://doi.org/10.1109/TSE.2002.1000449
21. Murta, L., Corrêa, C., Prudêncio, J.G., Werner, C.: Towards odyssey-VCS 2: improvements over a UML-based version control system. In: Proceedings of the 2008 International Workshop on Comparison and Versioning of Software Models (2008). https://doi.org/10.1145/1370152.1370159

22. Pietsch, C., Kelter, U., Kehrer, T.: From pairwise to family-based generic analysis of delta-oriented model-based SPLs, pp. 13–24. Association for Computing Machinery, New York (2021). https://doi.org/10.1145/3461001.3471150
23. Reuling, D., Pietsch, C., Kelter, U., Kehrer, T.: Towards projectional editing for model-based SPLs. In: Proceedings of the 14th International Working Conference on Variability Modelling of Software-Intensive Systems, VAMOS 2020. Association for Computing Machinery, New York (2020). https://doi.org/10.1145/3377024.3377030
24. Rutle, A., Rossini, A., Lamo, Y., Wolter, U.: A category-theoretical approach to the formalisation of version control in MDE. In: Chechik, M., Wirsing, M. (eds.) FASE 2009. LNCS, vol. 5503, pp. 64–78. Springer, Heidelberg (2009). https://doi.org/10.1007/978-3-642-00593-0_5
25. Sakizloglou, L., Ghahremani, S., Barkowsky, M., Giese, H.: Incremental execution of temporal graph queries over runtime models with history and its applications. Softw. Syst. Model. 1–41 (2021). https://doi.org/10.1007/s10270-021-00950-6
26. Taentzer, G., Ermel, C., Langer, P., Wimmer, M.: A fundamental approach to model versioning based on graph modifications: from theory to implementation. Softw. Syst. Model. **13**(1), 239–272 (2012). https://doi.org/10.1007/s10270-012-0248-x
27. Westfechtel, B.: A formal approach to three-way merging of EMF models. In: Proceedings of the 1st International Workshop on Model Comparison in Practice (2010). https://doi.org/10.1145/1826147.1826155
28. Westfechtel, B., Greiner, S.: Extending single- to multi-variant model transformations by trace-based propagation of variability annotations. Softw. Syst. Model. **19**(4), 853–888 (2020). https://doi.org/10.1007/s10270-020-00791-9

Visual Smart Contracts for DAML

Reiko Heckel[1(✉)], Zobia Erum[1], Nitia Rahmi[2], and Albert Pul[1]

[1] School of Computing and Mathematical Sciences, University of Leicester, Leicester, UK
{rh122,ze19}@le.ac.uk, pulalbert2@gmail.com
[2] PT Bank Rakyat Indonesia (Persero) Tbk, Central Jakarta, Indonesia
nitiarahmi@gmail.com

Abstract. The Digital Asset Modelling Language (DAML) enables low-code development of smart contract applications. Starting from a high-level but textual notation, DAML thus implements the lower end of a model-driven development process, from a platform-specific level to implementations on a range of blockchain platforms. Existing approaches for modelling smart contracts support a domain-oriented, conceptual view but do not link to the same technology-specific level.

We develop a notation based on class diagrams and visual contracts that map directly to DAML smart contracts. The approach is grounded in an operational semantics in terms of graph transformation that accounts for the more complex behavioural features of DAML, such as its role-based access control and the order of contract execution and archival. The models, with their mappings to DAML and their operational semantics, are introduced via the Doodle case study from a DAML tutorial and validated through testing the graph transformation system against the DAML code using the Groove model checker.

Keywords: smart contracts · DAML · model-based development · UML · visual contracts · graph transformation · Groove

1 Introduction

Smart contracts are transactions that automate workflows or document legally relevant events and actions, reducing the need for trusted intermediaries to prevent malicious or accidental deviations from agreed protocols [18]. Since the adoption of smart contracts for transactions on blockchains, an increasing number of platforms are emerging (listed e.g., in [12]), supporting a variety of concepts and languages for different application domains and business models.

This diversity causes familiar challenges, both strategic (which platform to adopt for maturity, long-term stability and support) and short-term (how to find or train qualified developers for specific technologies, integrate with specific technology stacks used, etc.). Moreover, the use of smart contracts to automate workflows demands input from domain experts not trained in the languages employed to program them.

Some of these challenges are addressed by low-code domain-specific languages. The Distributed Asset Modelling Language (DAML) [3] supports a

N. Behr and D. Strüber (Eds.): ICGT 2022, LNCS 13349, pp. 137–154, 2022.
https://doi.org/10.1007/978-3-031-09843-7_8

model-driven approach to smart contact development. It provides primitives for data management, contract creation, execution and archival with fine-grained role-based access control, and supports cross-platform deployment by mapping to a range of smart contract platforms, such as the Ethereum family [19], but also traditional databases. While this supports platform-independence, it does so at programming level, like Java running on virtual machines on different operating systems.

In this sense, DAML code can be seen as a technology-specific model, analogous to object-oriented class diagrams that map to OO languages such as Java, C#, C++, etc., hence supporting the lower part of a model-driven approach to smart contract development: the mapping from technology-specific models (DAML code) to implementation. However, it lacks

1. alignment with object- and component-oriented concepts, to support a seamless design process across applications using smart contracts in conjunction with other technologies and languages;
2. visualisation in mainstream modelling language, to allow the use of common notation across and between projects;
3. comprehensive operational semantics, to give formal explanations of features such as the order of contract execution and archival, which impacts on parties' ability to access different versions of a contract.

The first problem arises from a technology-oriented domain-specific language that supports the concepts of a family of target platforms (e.g. blockchains and smart contracts) but risks creating process silos by forcing early design decisions on how certain requirements should be implemented (e.g. what data to store on a blockchain vs. a centralised database; what functionality to provide through smart contracts vs. a traditional API). By providing a common notation and semantics to address 2, we can defer such decisions, increasing the potential for reuse of conceptual models.

Starting from requirements and domain models, other authors have addressed the conceptual design of smart contract applications using familiar UML diagrams, however without mapping to the technology-specific level (see Sect. 2). This means that we lack a clear understanding of how a conceptual model should be implemented using the concepts of the technological space or, vice versa, how the concepts and behaviours in that space can be expressed precisely and semantically correctly in our designs.

In this paper, we address these shortcomings by providing the link between the upper part of a model-based approach to smart contract development and the technology-specific level. In particular, we develop a mapping between DAML and the object-oriented concepts and notations in UML based on a semantic understanding of DAML's behaviour in terms of formal, visual and executable graph transformation systems [7]. The executable semantics supports DAML's complex operational model and, in the future, will enable formal analysis in particular of access control properties.

Our visual modelling approach relies on

- class diagrams, as standard object-oriented notation for data and operations, augmented by features for modelling DAML parties [3] and their roles and access rights in relation to the contracts;
- visual contracts modelling pre- and postconditions of operations in an object-based notation [4], with annotations for contract creation, update and archival, and nested executions.

Together, these constituents form an integrated DAML model subject to consistency requirements that ensure the model can be mapped to a graph transformation system as operational semantics. In the following section we outline this approach in more detail and discuss the state of the art and related work in modelling blockchains and smart contracts. Then, Sect. 3 describes how the structure and operations of smart contracts are represented in class diagrams, Sect. 4 captures pre- and postconditions of operations as visual contracts. In Sect. 5 we introduce the operational semantics in terms of graph transformations before Sect. 6 shows the application of Groove to analyse properties of the model and Sect. 7 concludes the paper.

2 Visual Models for Smart Contracts

Most current approaches to visual modelling of blockchains and smart contacts are based on combinations of entity-relationship, UML and BPMN diagrams. [17] aims to develop modelling standards at a conceptual, technology-independent level based on the idea of implementing smart contracts in object-oriented languages. [14] proposes the use of agile methodology, such as user stories, and data models in the form of UML diagrams to design blockchain applications. The short paper [6] uses class diagrams to describe data and application structure and sequence diagrams for communication. [20] demonstrates the use of two requirements-level frameworks, i* and UML use case and sequence diagrams. These proposals target a high-level view of applications focusing on functionality and processes without mapping to the technology level.

[21] proposes UML diagrams to model architecture and business processes of blockchain applications. [13] elaborates this by an integrated approach to model-driven engineering of blockchain applications, including business process modelling and management and the generation of registries for the digital assets the processes interact with. Their focus is on asset management, rather then a general approach to DAML smart-contract modelling we are going to propose.

There are tools to visualise structural features of smart contracts. [1] generates class diagrams from Solidity code, where a Solidity variable is an attribute and a function an operation. The DAML tool set can generate a form of call graph from the code. Neither support visualisation of functional behaviour.

A common obstacle to a more formal and comprehensive use of visual models is the lack of a semantic model integrating structural, functional and process views. One such integrated model has been provided in the form of visual contracts (VCs) for the modelling of service-oriented systems and component interfaces [4]. However, VCs do not support nested operations (the invocation of operations from inside others), nor any specific features of smart contracts.

Our approach is based on visual contracts, extending them with the required features, thus creating an integrated DAML model with operational semantics in graph transformations. In particular we formalise

- class diagrams as structural and data model, by attributed type graphs;
- visual contracts as functional model, by graph transformation rules;

In the following sections we first describe the notations used and then the semantic model.

3 Templates as Classes

In this section, we introduce DAML and establish a link with object-oriented concepts by "reverse engineering" DAML code into UML class diagrams. A class can be seen as a template for creating an object with certain features. In DAML, a *template* describes the features of a *contract*, including its ownership and access rights, attributes, and operations called *choices*. A contract is a transaction that is created from a template, like an object from a class, by a committed transaction. It remains active until it is archived. A contract is immutable. To update its field we create a new version of the contract and archive the old one [3].

To represent a template with its access rights as a class, we need the concept of a *party*. If referenced by a template's attributes, a party can be declared as *signatory* (that can create and archive a contract), *controller* (with execution rights), or *observer* (with read access to the contract and able to observe its creation and archival).

In class diagrams we model parties by an special type Party. The specific roles of parties in relation to a contract are specified using derived attributes /signatory, /controller, /observers and /maintainer.

A *contract key* is like the primary key in a database table, providing a way to identify the contract based on its attributes. The key is stable from one version of the contract to the next, in contrast to the *contract id*, which is specific to one version. The contract key contains at least one *maintainer*, i.e., a signatory of the contract that owns the key [3].

We use the Doodle case study [11] from the introductory DAML papers [9,10] to illustrate DAML's features and their representation in visual models. A doodle is a voting system to schedule meetings, where an *organiser* invites *voters* to vote on a set of *options*, recording their preferences in *voting slots*. Everyone can vote at most once for each option, and votes are visible to all.

The DAML code [11] has one data structure and two contract templates: As shown in the listing below, a `VotingSlot` record represents the data about an option, including the vote count and the list of parties who voted for it. The line `deriving (Eq, Show)` states that record equality is based on identity and they support serialisation. A `Doodle` contract (created by the organiser as signatory) offers choices (i.e., operations invoked by the specified controllers) to add and remove voters and issue invites (organiser), and to cast votes (voters). Individual `DoodleInvite` contracts will allow voters to access the `CastVote` choice.

```
1  data VotingSlot = VotingSlot
2    with
3      count : Int
4      voted : [Party]
5        deriving (Eq, Show)
6
7  template Doodle
8    with
9      name: Text
10     organizer : Party
11     voters : [Party]
12     options : [Text]
13     votes : TextMap VotingSlot
14     open: Bool
15   where
16     signatory organizer
17     observer voters
18     ensure (unique voters) && (unique options)
19     key (organizer , name): (Party, Text)
20     maintainer (fst key)
21     ...
```

We illustrate the mapping by creating a class diagram from the DAML code. The result is shown in Fig. 1. A template maps to a class with the same name, and template parameters become its attributes or associations. In particular, role declarations in DAML templates become derived attributes of type Party in the respective classes. In the `Doodle` template, the organiser is the signatory and the voters are observers.

Attributes of native type such as `Text` and `Bool` are shown as common class attributes while attributes referring to contracts or records are associations. DAML uses assembly types such as lists and maps, which require special treatment. E.g., attribute `votes` is of type `TextMap VotingSlot`, which we represent as a class `VotingSlot` with an attribute option: Text. Each instance vs:VotingSlot represents a pair ⟨index, object⟩ of the map with index = vs.option and object = vs. An association with a ≪list≫ stereotype represents a list-valued attribute of the contract. Attributes and associations that jointly form the `key` or are individually declared `unique` are prefixed with corresponding stereotypes. E.g., `(organizer, name)` jointly form a key for a Doodle contract, and `voters` is a unique list (without repetition of elements).

A choice is shown as a method with a stereotype indicating whether it is *post-consuming, pre-consuming* or *non-consuming.* For example, `AddVoter` is a consuming choice: when executing it we archive the contract and replace it with a new version. By default, the choice is executed after the archival of the contract. This is pre-consuming mode, in contrast to a post-consuming choice where the current version of the contract is archived after executing the choice.

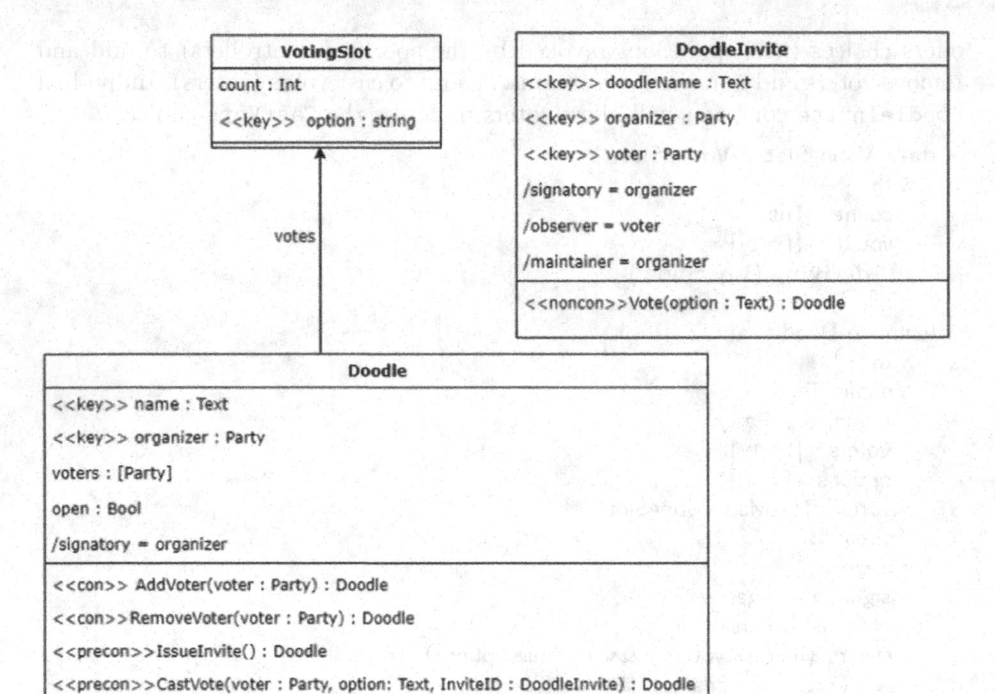

Fig. 1. Class diagram for the Doodle case study

```
1  template Doodle
2  ...
3   choice AddVoter : ContractId Doodle
4       with
5          voter : Party
6        controller organizer
7          do
8            assertMsg
9                  "this doodle has been opened for voting,
10                   cannot add voters"
11               (not open)
12           create this with voters = voter :: voters
13
14      choice RemoveVoter : ContractId Doodle
15        with
16          voter : Party
17        controller organizer
18          do
19            assertMsg
20                 "this doodle has been opened for voting,
21                   cannot remove voters"
22               (not open)
23           create this with voters = DA.List.delete voter voters
```

```
24    preconsuming choice IssueInvites : ContractId Doodle
25      controller organizer
26        do
27          assertMsg
28              "this doodle has been opened for voting,
29                  cannot issue any more invites"
30              (not open)
31          DA.Traversable.mapA
32              (\voter -> create DoodleInvite with
33                doodleName = this.name,
34                organizer = this.organizer, voter = voter)
35              voters
36          -- archive self
37          create this with open = True
38
39    preconsuming choice CastVote: ContractId Doodle
40      with
41        voter: Party
42        option: Text
43        inviteId : ContractId DoodleInvite
44      controller voter
45        do
46          invite <- fetch inviteId
47          assertMsg
48              "this invite was issued for a different doodle"
49              (invite.doodleName == name)
50          assertMsg
51              "the voter casting the vote does not match the voter
52              who received the invite"
53              (invite.voter == voter)
54          assertMsg
55              "the organizer who issued the invite is not the one
56              who created this doodle"
57              (invite.organizer == organizer)
58          assertMsg "this doodle not is open" open
59          assertMsg "voters is not one of the invited voters"
60                          (elem voter voters)
61          assertMsg "this is not a valid option " (elem option options)
62          let
63            crtVotes = fromOptional
64              (VotingSlot with count = 0, voted = [])
65              (DA.TextMap.lookup option this.votes)
66            updatedVotes = DA.TextMap.insert option
67              (VotingSlot with count = crtVotes.count + 1,
68                voted = voter :: crtVotes.voted)
69              this.votes
70          assertMsg
71              "each voter is only allowed to cast one vote per option"
72              (notElem voter crtVotes.voted)
73          create this with votes = updatedVotes
```

There are four choices in the Doodle template, which we represent as class methods treating their `with` fields as input parameters and using stereotypes ≪con≫, ≪noncon≫ and ≪precon≫ to indicate their mode of execution.

Table 1 summarises our mapping from DAML templates to class diagrams. Recall that data structures are special contracts, without choices, used to declare complex data types and their access rights.

4 Choices as Visual Contracts

The behaviour of an operation can be specified as a visual contract (VC) over the class diagram [2,4]. VCs are model-level representations of the design-by-contract paradigm [15] specifying pre- and postconditions of operations.

The operations defined by a DAML template are its choices. Based on our mapping of the structural features of DAML code to class diagrams, we model DAML choices by VCs.

VCs are derived from the choice's interface, its `with` and `do` blocks, see Fig. 2. Fields declared in the `with` block are input parameters and we specify the controller who can execute the choice. We indicate creation and deletion by corresponding constraints new, delete as well as green and blue colour [7]. Note that deleting a contract in a VC represents archival, because contracts are never deleted in DAML.

Table 1. Mapping DAML templates to class diagrams

DAML template	**Class diagram**
template	class
data	class
with (template parameters)	class attributes or associations
party type	Party
signatory, maintainer	/signatory, /maintainer
observer, controller	/observer, /controller
choice name	method name
list type A []	cardinality / stereotype ≪list≫
TextMap A	A <<key>> index : Text
Mode of choice: preconsuming (post)consuming nonconsuming	Stereotype left of method name: ≪precon≫ ≪con≫ ≪noncon≫
with (block)	input parameter
return-type of choice	return-type of method

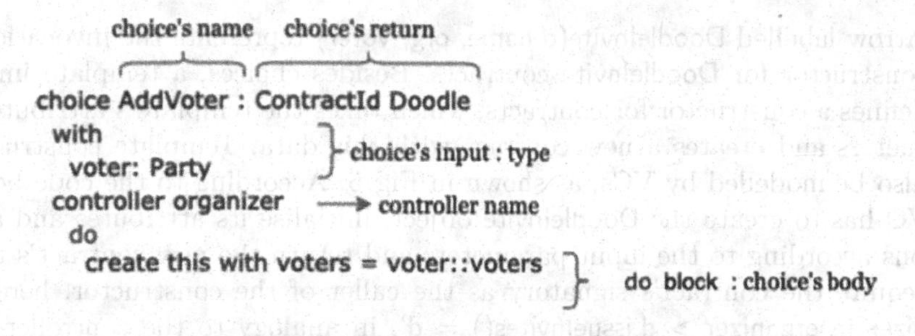

Fig. 2. Components of `AddVoter` choice: organizer > d.AddVoter(voter) = d'

organizer > d.AddVoter(voter) = d'

d → d' : Doodle {new}
open = false voters := voter::voters

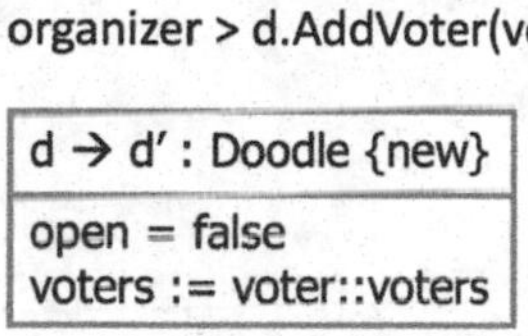

Fig. 3. Choice `AddVoter` as visual contract

In the VC in Fig. 3 we indicate this in the label of the diagram: organizer > d.AddVoter(voter) = d' means that organizer is the controller. The execution of the choice archives the current version d and replaces it with a new version d' indicated by object id d → d'. The choice adds the given `voter` to list `voters`.

`AddVoter` is a consuming choice by default since no qualifier is given left of the choice name. Hence, the contract will be archived before the body of the choice is executed. The id d → d' represents the creation a new contract d' and the archival of the old one d.

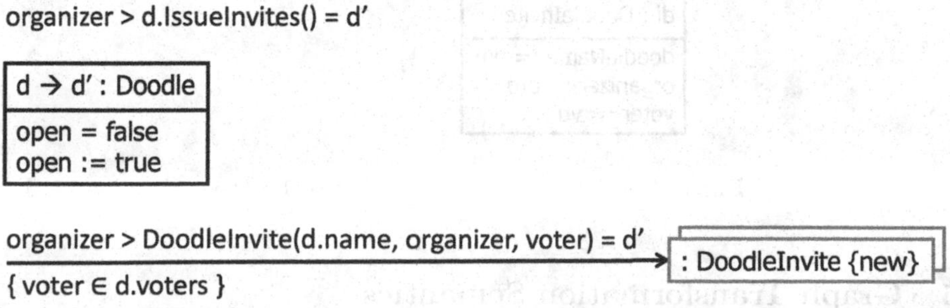

organizer > d.IssueInvites() = d'

d → d' : Doodle
open = false open := true

organizer > DoodleInvite(d.name, organizer, voter) = d' → : DoodleInvite {new}

{ voter ∈ d.voters }

Fig. 4. Choice `IssueInvites` as visual contract

In Fig. 4 we show the VC for the IssueInvites choice, which creates a new DoodleInvite contract for each party in the voters list. The stack notation is used to represent the set of new invites to be generated, one for each voter ∈ voters.

The arrow labelled DoodleInvite(d.name, org, voter) represents the invocation of the constructor for DoodleInvite contracts. Besides choices, a template implicitly defines a constructor for contracts, which takes the template's attributes as parameters and creates a new contract with this data. Template constructors can also be modelled by VCs, as shown in Fig. 5. According to the code below, the VC has to create the DoodleInvite object, initialise its attributes and associations according to the input parameters and return the new contract's id di. We require the contract's signatory as the caller of the constructor, here the organizer in organizer > d.issueInvites() = d', in analogy to the controller of a choice.

```
1  template DoodleInvite
2    with
3      doodleName: Text
4      organizer : Party
5      voter : Party
6    where
7      signatory organizer
8      observer voter
9      key (organizer, voter, doodleName) : (Party, Party, Text)
10     ...
```

Note that input parameters and returns of constructors derive from the class's attributes, associations and name, along with the creation and initialisation actions required. Therefore, this default constructor VC is wholly derivable from the class definition and would not have to be defined explicitly. It is included here for illustrating the principle and to show the possibility of defining customised constructors.

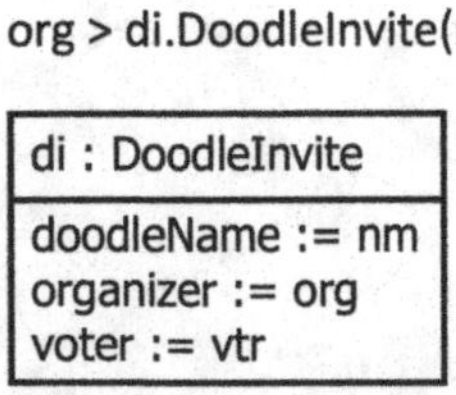

org > di.DoodleInvite(nm, org, vtr) = di

di : DoodleInvite
doodleName := nm organizer := org voter := vtr

Fig. 5. Constructor for template `DoodleInvite`

5 Graph Transformation Semantics

In this section we translate the class diagram and VCs introduced above to a typed graph transformation system. This provides them with operational semantics for the analysis of state invariants and reachability properties expressed in temporal logics, which we will cover in the next section. Since we will use the Groove model checker [5], we present type graph and rules in Groove notation.

Disregarding operations, a class diagram can be seen as an attributed type graph with inheritance [7]. Such a type graph defines a set of attributed instance graphs representing object structures, the possible data states of our system.

A type graph derived from the class diagram in Fig. 1 is shown in Fig. 6. The node and edge types in the centre of the type graph, shown in grey and black, derive directly from the class diagram. They represent the DAML templates with their attributes. Note that we introduce a single node type Party to represent attributes of this type as edge types.

The blue node and edge types in the left and right margins are runtime structures for calls to choices and constructors. Each call is represented by a call node with attributes and edges representing the call's parameters including the caller (the controller of the choice or the signatory of the constructor), the this contract executing the call, input parameters and return. In the case of recursive calls we also include a call edge from the calling to the called operation. For example, where IssueInvites calls DoodleInvite, this is represented by a call edge from IssueInvitesCall to DoodleInviteCall.

Visual contracts [4] are formalised as (sets of) attributed graph transformation rules over operation signatures and conforming to the type graph. They

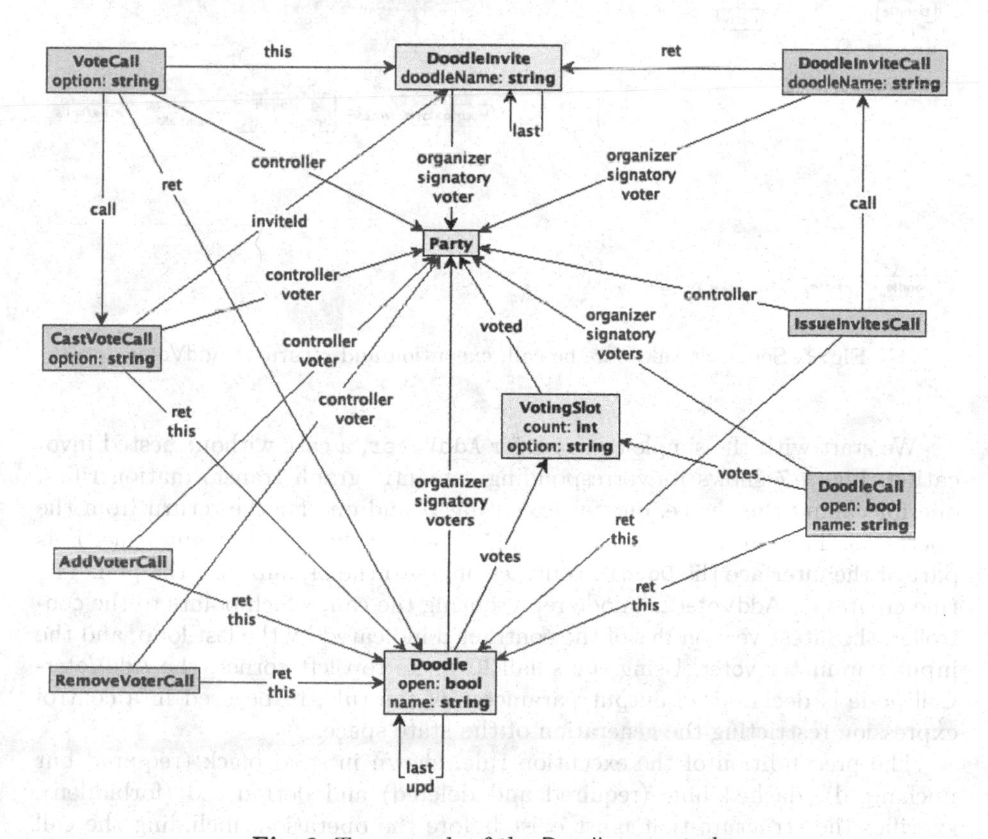

Fig. 6. Type graph of the Doodle case study

capture the pre- and postconditions of the operations and provide an executable model where operations are represented as transformations over instances of the type graph.

The model in [4] supports a view of operations in an interface acting over the data state of a component or service. However, it does not allow to specify the invocation of one operation from inside another one, such as the execution of `DoodleInvite::Vote` calling `Doodle::CastVote`. Visual contracts also do not account for the specific features of smart contracts in DAML, such as contract archival and update, or the different effects of pre-, post- or non-consuming choices. In this section we address these features by showing how the straightforward interpretation of VCs as graph transformation rules can be extended to smart visual contracts for DAML.

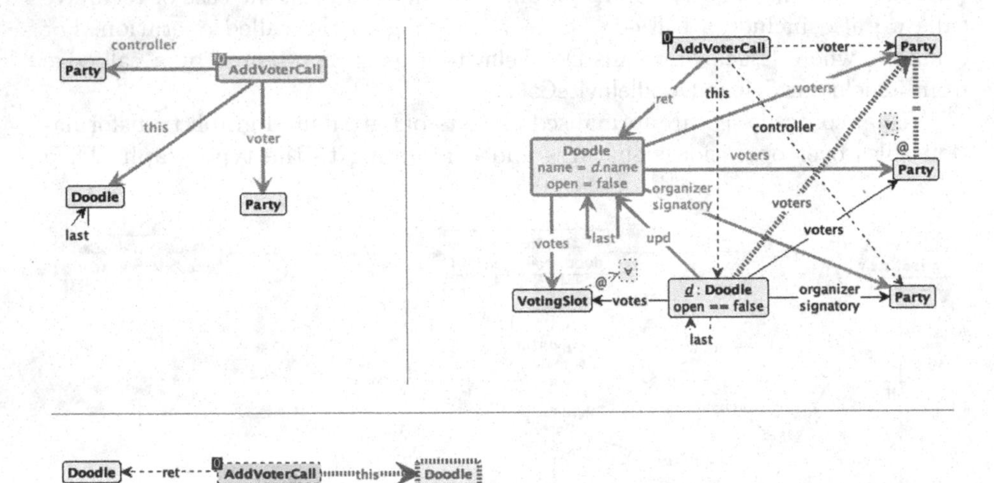

Fig. 7. Semantic rules for the call, execution and return of AddVoter

We start with the simple contract for `AddVoter`, a case without nested invocation. Figure 7 shows its corresponding semantic graph transformation rules, one for calling the choice, one for executing it and one for the return from the operation. The call rule reflects that `AddVoter` can be called at any time. It is part of the interface the `Doodle` contract offers to the organizer of the poll. The rule creates an AddVoterCall node representing the call, which points to the controller, the latest version this of the contract d indicated by the last loop, and the input parameter voter. Using the small !0 in its top left corner, the AddVoterCall node is declared an output parameter of the rule, to be used in a control expression restricting the generation of the state space.

The precondition of the execution rule, shown in solid black (required but unchanged), dashed blue (required and deleted) and dotted red (forbidden), specifies the structure that must exist before the operation, including the call node. The postcondition, which specifies the changes to the graph, is shown in solid black (unchanged) and dashed blue (deleted). Attribute values can be tested in conditions and updated using assignments. The AddVoterCall node is a rule parameter as indicated by the small 0 in its the top left.

In the execution rule, the call node is unlinked from its input parameters and linked to its return, the new version d' of the contract. This is created and linked to the old version, and the last marker is moved to this new version. Data and links are copied from the old to the new version. To facilitate this for an unlimited number of voters, the Party node referred to by the voters edge from d is universally quantified using the @-labelled edge to the ∀ node. That means, new voters edges from the new Doodle node will be created for all Party nodes with an incoming voters edge from d. Any updates, such as the addition of the link to the voter, are applied to the new version d'. This is because the choice is preconsuming, i.e., the contract is archived before changes are applied.

The return rule just deletes the call node. In this case, this could have been done in the same rule because there are no nested calls, but declaring the call node as a rule parameter, and with the help of the Groove recipe below, we ensure that the sequence of call, execution and return rules is executed atomically, avoiding intermediate states.

```
// atomic execution of AddVoter rules
recipe removeVoter (out node callNode) {
      callRemoveVoter(out callNode); // call rule
      execRemoveVoter(callNode);     // execution rule
      retRemoveVoter(callNode);      // return rule
}
```

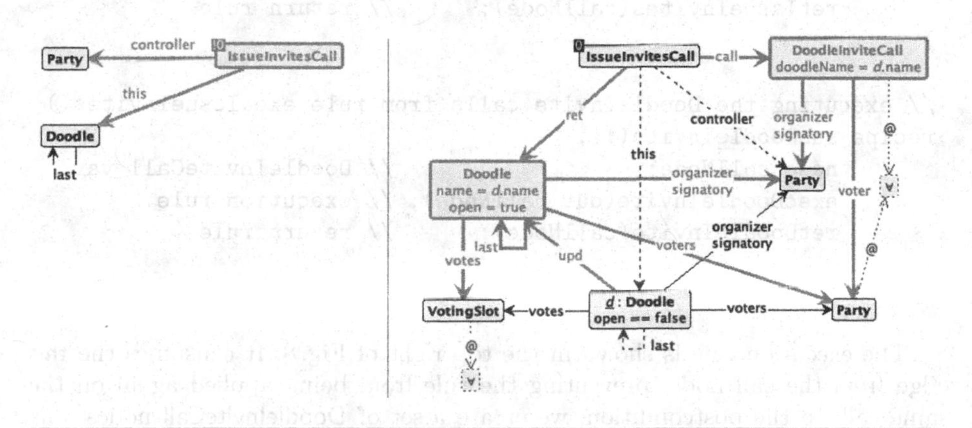

Fig. 8. Semantic rules for call, execution and return of IssueInvites

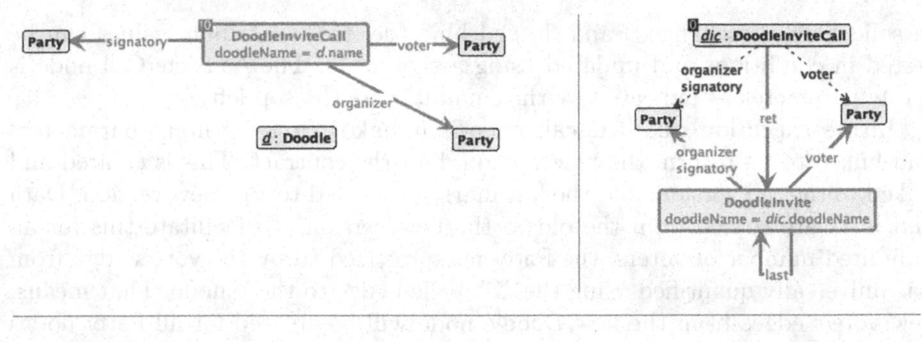

Fig. 9. Semantic rules for call, execution and return of DoodleInvite

Choice `IssueInvites` features an invocation of another operation, the constructor for `DoodleInvite` contracts. Again, the IssueInvites VC is translated into three separate graph transformation rules, for call, execution and return. The recipe is below, utilising a sub-recipe for executing the calls to the DoodleInvite constructor.

```
// atomic execution of IssueInvites rules
recipe issueInvites (out node callNode) {
      callIssueInvites(out callNode); // call rule
      execIssueInvites(callNode);     // execution rule
      #subDoodleInvite();             // as long as possible
      retIssueInvites(callNode);      // return rule
}

// executing the DoodleInvite calls from rule execIssueInvites()
recipe subDoodleInvite(){.
      node callNode;                  // DoodleInviteCall var
      execDoodleInvite(out callNode); // execution rule
      retDoodleInvite(callNode);      // return rule
}
```

The execution rule is shown in the top right of Fig. 8. It consumes the this edge from the call node, preventing the rule from being applied again on the same call. In the postcondition, we create a set of DoodleInviteCall nodes, one for each Party node linked to by a voters edge. This is achieved by a universally qualified subrule denoted by a ∀ node with @-labelled edges from all nodes that are part of that rule. The rules for the DoodleInvite constructor are given in Fig. 9. The return rule for IssueInvites is shown in the bottom of Fig. 8. It uses a negative application condition (in bold red, dotted) to check that there are no constructor calls that haven't returned yet, and deletes its own call node, enabling any other return rules waiting for this call to IssueInvites to finish.

Note that, since archival takes place before executing the choice code, the contract is archived before calling the constructors, which are invoked with the new version of the `Doodle` contract as context. While the overall effect is the same, the distinction is important for the visibility of invites. Quoting from [10]:

> "If the choice were not preconsuming, the version of the Doodle contract on which you exercise the choice would still exist when the contracts are created. [This] ...means that parties which are neither signatories or observers of the VotingInvite contract, will see a copy of it once, when the contract is created."

In particular, once a contract is archived we can no longer use its contract id to access it. In our semantic rules this is reflected by the restriction that we cannot have a contract on the left-hand side that is not tagged by a last edge.

The separation of execution and return rules allows us to model also post-consuming behaviour, where the contract is archived in the return rule, after executing the choice and receiving the returns from any invoked choices.

6 Validation

We analysed the model in Groove [5] both to validate the soundness of the overall approach to mapping smart VCs into graph transformation systems and to experiment with model checking to analyse different types of properties.

First, we tested the semantic graph transformation system against the DAML code. We used the Groove control expression below to specify a range of scenarios and then executed them in Groove. We implemented selected scenarios as DAML test scripts and compared the results.

```
// creating Doodle for existing DoodleCall
node cnDoodle;                          // var for DoodleCall
execDoodle(out cnDoodle);               // execute Doodle constructor
retDoodle(cnDoodle);                    // Doodle constructor return

// adding voters
node cnAddVoter;                        // var for AddVoterCall node
addVoter(out cnAddVoter)*;              // add any subset of parties

// issuing invites
node cnIssueInvites;                    // variable for IssueInvitesCall
issueInvites(out cnIssueInvites);// execute issueInvites recipe

// voting
node cnVote;                            // var for VoteCall
vote(out cnVote)*;                      // any subset of voters vote
```

Tests were executed on a state space of 516 states and 1908 transitions generated from a start graph with a single DoodleCall node, three Party nodes and two VotingSlot nodes. Recipes as shown in Sect. 5 were used to prevent interleaving on intermediate states, significantly reducing the size of the state space. One notable result was a mismatch of how constraints are handled. For example, the DAML code `ensure (unique voters)` guarantees that the same party cannot occur twice in the list of voters. This is checked at runtime and when trying to add the same party twice using `AddVoter`, an exception is thrown. To achieve the same behaviour in our semantics, we added to the execution rule in the top right of Fig. 7 a negative application condition ensuring this constraint by stating the absence of a voters link from d to the top-right Party node. This is an example of a constraint-guaranteeing negative application condition, which can be constructed automatically from the rule and the constraint [8,16].

Then, we analysed safety properties deriving from constraints declared in the class diagram, such as key properties for DoodleInvite (doodleName, organizer, voter are jointly unique), Doodle (name, organizer are jointly unique), and VotingSlot (option is unique), and uniqueness of edges (voters edges represent a collection with unique entries). From the logic of the problem domain we derive requirements such as: A party can vote at most once for each voting slot.

We formalised these constraints as property rules (without effect) expressing the forbidden patterns in their precondition, and verified them in Groove as a CTL formula AG (!propNotDoodleInviteKey & !propNotDoodleKey & !propNotVotingSlotKey & !propNotUniqueVoters & !propNotUniqueVote). In addition, we checked the lifeness property, that it is always possible to reach a state where voting is not enabled anymore (because all invited parties have voted for all possible voting slots), written in CTL as AF !execVote.

This shows that model checking is feasible on graph transformation models derived from visual smart contracts, with many of the properties defined directly by the constraints in the class diagram and other safety and lifeness properties derived from requirements of the problem domain.

A full version of the model including all VCs, transformation and property rules is available at https://www.cs.le.ac.uk/people/rh122/mdd4daml both as pdf and graph grammar in Groove, which we are planning to explore in the future.

7 Conclusion

To support the development of smart contract applications in DAML, we proposed an integrated modelling approach consisting of DAML-specific class diagrams and visual contracts. We established a mapping between DAML and such integrated models, which can be used in both forward and reverse engineering, and demonstrated its use by a case study. We defined operational semantics in terms of graph transformation systems with control expressions in Groove, discussed how they capture the specific behavioural features of DAML's access control mechanisms, and demonstrated the possibility of analysing these semantic models.

The mappings in our case study from smart contracts to smart VCs and from smart VCs to graph transformation rules are designed to be generalisable, but have not been formalised and automated. Apart from the usefulness of visual smart contracts for modelling smart contract applications, the automation of these mappings is essential for a range of model-based development activities, including formal analysis, simulation and model checking, and model-based testing, which we are planning to explore in the future.

References

1. Addison, N.: Solidity 2 UML (2021). https://github.com/naddison36/sol2uml. Accessed 1 Mar 2022
2. Alshanqiti, A., Heckel, R., Kehrer, T.: Visual contract extractor: a tool for reverse engineering visual contracts using dynamic analysis. In: Proceedings of the 31st IEEE/ACM International Conference on Automated Software Engineering, pp. 816–821. Association for Computing Machinery, New York, NY, USA (2016). https://doi.org/10.1145/2970276.2970287
3. Digital Asset: DAML documentation (2021). https://docs.daml.com/index.html. Accessed 1 Mar 2022
4. Engels, G., Lohmann, M., Sauer, S., Heckel, R.: Model-driven monitoring: an application of graph transformation for design by contract. In: Corradini, A., Ehrig, H., Montanari, U., Ribeiro, L., Rozenberg, G. (eds.) ICGT 2006. LNCS, vol. 4178, pp. 336–350. Springer, Heidelberg (2006). https://doi.org/10.1007/11841883_24
5. Ghamarian, A.H., de Mol, M., Rensink, A., Zambon, E., Zimakova, M.: Modelling and analysis using GROOVE. Int. J. Softw. Tools Technol. Transf. **14**(1), 15–40 (2012). https://doi.org/10.1007/s10009-011-0186-x
6. Haelterman, N.: How to model smart contracts within software projects, February 2018. https://blog.fundrequest.io/how-to-model-smart-contracts-within-software-projects-ef1e298b21e6. Accessed 1 Mar 2022
7. Heckel, R., Taentzer, G.: Graph Transformation for Software Engineers - With Applications to Model-Based Development and Domain-Specific Language Engineering. Springer, Cham (2020), https://doi.org/10.1007/978-3-030-43916-3
8. Heckel, R., Wagner, A.: Ensuring consistency of conditional graph rewriting - a constructive approach. Electron. Notes Theor. Comput. Sci. **2**, 118–126 (1995). https://doi.org/10.1016/S1571-0661(05)80188-4
9. Kirschner, E.: A doodle in DAML - Part 1. Medium.com, October 2020. https://entzik.medium.com/a-doodle-in-daml-part-1-d2ef18bbf7e8. Accessed 1 Mar 2022
10. Kirschner, E.: A doodle in DAML - Part 2. Medium.com (November 2020), https://entzik.medium.com/a-doodle-in-daml-part-2-910614d94c62. Accessed 1 Mar 2022
11. Kirschner, E.: Github, November 2020. https://github.com/entzik/daml-examples/blob/master/doodle/daml/Com/Thekirschners/Daml/Doodle.daml. Accessed 1 Mar 2022
12. Lawton, G.: Task markers, May 2021. https://searchcio.techtarget.com/feature/Top-9-blockchain-platforms-to-consider. Accessed 1 Mar 2022
13. Lu, Q., et al.: Integrated model-driven engineering of blockchain applications for business processes and asset management. Softw. Pract. Exp. **51**(5), 1059–1079 (2021). https://doi.org/10.1002/spe.2931

14. Marchesi, M., Marchesi, L., Tonelli, R.: An agile software engineering method to design blockchain applications. In: Proceedings of the 14th Central and Eastern European Software Engineering Conference Russia. CEE-SECR 2018, Association for Computing Machinery, New York, NY, USA (2018). https://doi.org/10.1145/3290621.3290627
15. Meyer, B.: Applying 'design by contract'. Computer **25**(10), 40–51 (1992)
16. Nassar, N., Kosiol, J., Arendt, T., Taentzer, G.: Constructing optimized constraint-preserving application conditions for model transformation rules. J. Log. Algebraic Methods Program. **114**, 100564 (2020). https://doi.org/10.1016/j.jlamp.2020.100564
17. Rocha, H., Ducasse, S.: Preliminary steps towards modeling blockchain oriented software. In: 1st IEEE/ACM International Workshop on Emerging Trends in Software Engineering for Blockchain, WETSEB@ICSE 2018, Gothenburg, Sweden, May 27–June 3, 2018, pp. 52–57. ACM (2018). https://ieeexplore.ieee.org/document/8445060
18. Szabo, N.: Formalizing and securing relationships on public networks. First Monday **2**(9) (1997). https://firstmonday.org/ojs/index.php/fm/article/view/548. Accessed 1 Mar 2022
19. tintinweb: Solidity visual developer, June 2021. https://marketplace.visualstudio.com/items?itemName=tintinweb.solidity-visual-auditor. Accessed 1 Mar 2022
20. Vingerhouts, A.S., Heng, S., Wautelet, Y.: Organizational modeling for blockchain oriented software engineering with extended-i* and UML. In: PoEM Workshops (2020)
21. Xu, X., Weber, I., Staples, M.: Architecture for Blockchain Applications. Springer, Cham (2019). https://doi.org/10.1007/978-3-030-03035-3

Computational Category-Theoretic Rewriting

Kristopher Brown[1(✉)], Evan Patterson[2], Tyler Hanks[1], and James Fairbanks[1]

[1] University of Florida, Gainesville, FL, USA
{kristopher.brown,t.hanks,fairbanksj}@ufl.edu
[2] Topos Institute, Berkeley, CA, USA
evan@topos.institute

Abstract. We demonstrate how category theory provides specifications that can efficiently be implemented via imperative algorithms and apply this to the field of graph transformation. By examples, we show how this paradigm of software development makes it easy to quickly write correct and performant code. We provide a modern implementation of graph rewriting techniques at the level of abstraction of finitely-presented $\mathcal{C}$-sets and clarify the connections between $\mathcal{C}$-sets and the typed graphs supported in existing rewriting software. We emphasize that our open-source library is extensible: by taking new categorical constructions (such as slice categories, structured cospans, and distributed graphs) and relating their limits and colimits to those of their underlying categories, users inherit efficient algorithms for pushout complements and (final) pullback complements. This allows one to perform double-, single-, and sesqui-pushout rewriting over a broad class of data structures. Graph transformation researchers, scientists, and engineers can then use this library to computationally manipulate rewriting systems and apply them to their domains of interest.

Keywords: Double pushout rewriting · category theory · graph rewriting

1 Introduction and Motivation

Term rewriting is a foundational technique in computer algebra systems, programming language theory, and symbolic approaches to artificial intelligence. While classical term rewriting is concerned with tree-shaped terms in a logical theory, the field of graph rewriting extends these techniques to more general shapes of terms, typically simple graphs, digraphs, multigraphs, or typed graphs. Major areas of graph rewriting are graph *languages* (rewriting defines a graph grammar), graph *relations* (rewriting is a relation between input and output graphs), and graph *transition systems* (rewriting evolves a system in time) [14].

When considering the development of software for graph rewriting, it is important to distinguish between studying rewriting systems as mathematical

N. Behr and D. Strüber (Eds.): ICGT 2022, LNCS 13349, pp. 155–172, 2022.
https://doi.org/10.1007/978-3-031-09843-7_9

objects and building applications on top of rewriting as infrastructure. The former topic can answer inquiries into confluence, termination, reachability, and whether certain invariants are preserved by rewriting systems. In contrast, we will focus on answering questions that involve the application of concretely specified rewrite systems to particular data.

Category theory is a powerful tool for developing rewriting software, as the numerous and heterogeneous applications and techniques of rewriting are elegantly unified by categorical concepts. Furthermore, the semantics of categorical treatments of graph rewriting are captured by universal properties of limits and colimits, which are easier to reason about than operational characterizations of rewriting. This is an instance of a broader paradigm of *computational applied category theory*, which begins by modeling the domain of interest with category theory, such as using monoidal categories and string diagrams to model processes. One is then free (but not required) to implement the needed categorical structures in a conventional programming language, where the lack of a restrictive type system facilitates a fast software development cycle and enables algorithmic efficiency. For example, arrays can be used to represent finite sets, and union-find data structures can compute equivalence classes.

Our approach takes the domain of interest modeled by category theory to be the field of graph transformation. This was first suggested by Minas and Schneider [20] and is distinguished from existing tools by working at a higher level of abstraction and developing rewriting capabilities within a broader framework of categorical constructions. While current software tools are connected to category theory through their theoretical grounding in adhesive categories [17], they are specialized to graphs in their implementation.

Connection to ExACT. An orthogonal technique of applying category theory to rewriting software development encodes category theory into the type system of the program itself. This strategy, sometimes called *executable applied category theory* (ExACT), allows type checking to provide static guarantees about the correctness of rewriting constructions. At present, it is not feasible to execute provably-correct programs on large problems, as they generally have poor performance [24]. Translation-based approaches offer an alternative to proof assistants by encoding graph rewriting into first-order logic and computing answers with SMT solvers, which likewise suffer from scalability concerns when used as an engine to compute rewrites at scale [14]. We distinguish computational applied category theory from this paradigm by analogy to the distinction between computational linear algebra and formalizations of linear algebra, a distinction visualized in Fig. 1. One area in which these paradigms can interact is through making the testing of unverified software more robust: extracted programs from formalized proofs can serve as a test oracle and a basis for generating test cases [25].

Structure of the Paper. We will first introduce $\mathcal{C}$-sets and typed graphs, the latter of which has been the focus of preexisting graph rewriting software. Our first contribution is to elucidate the subtle relationships between these two mathematical constructs, and we argue on theoretical and performance grounds that $\mathcal{C}$-sets are more directly applicable to many problems where typed graphs are

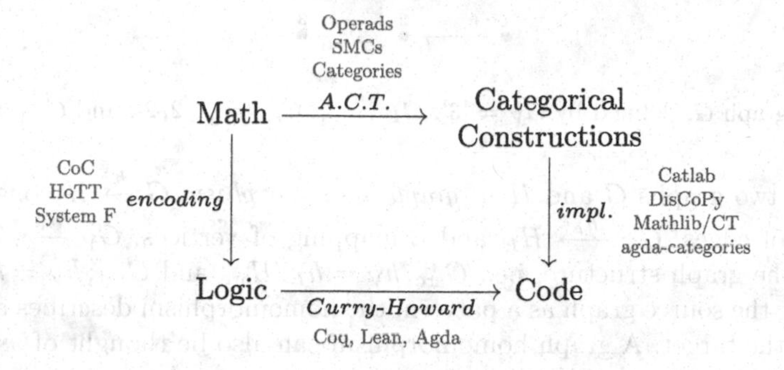

Fig. 1. Two broad strategies for computational category theory. Applied category theory is used to represent the program's *subject matter* in the upper path, while category theory is encoded in the program's *structure* or *type system* in the lower path. This is not a commutative diagram.

currently applied. Our next contribution draws from previous theoretical work of Löwe, who developed theory for DPO and SPO of $\mathcal{C}$-sets [19]. We present the first software implementation of this rewriting on $\mathcal{C}$-sets and extend it with algorithms for SqPO and homomorphism finding. Our last contribution also draws from preexisting theoretical work of Minas and Scheider as mentioned above - we describe a modern realization of computational applied category theory and show how this paradigm allowed for these rewriting techniques to be 1.) efficient, 2.) programmed at a high level, closely matching the mathematical specification, and 3.) extensible to new categories. We lastly outline extensions of rewriting beyond $\mathcal{C}$-sets, which highlight the flexibility of our technique. A supplemental notebook[1] is provided with code to accompany all figures. We assume familiarity with the basic concepts of categories, functors, and natural transformations.

2 Important Categories in Computational Graph Transformation

2.1 Graphs and Their Homomorphisms

We take graphs to be finite, directed multigraphs. Thus, a graph G is specified by two finite sets, G_E and G_V, giving its edges and vertices, and two functions $G_{\text{src}}, G_{\text{tgt}} : G_E \to G_V$, defining the source and target vertex of each edge.

We can compactly represent sets and functions by working in the skeleton of **FinSet**, where a natural number n is identified with the set $[n] := \{1, ..., n\}$. A function $f : [n] \to [m]$ can be compactly written as a list $[x_1, x_2, ..., x_n]$, such that f sends the element $i \in [n]$ to the element $x_i \in [m]$. This leads to the edge list representation of graphs, which are encoded as two natural numbers and two lists of natural numbers (Fig. 2).

[1] https://nbviewer.org/github/kris-brown/Computational-Category-Theoretic-Rewriting/blob/main/Computational_Category_Theoretic_Rewriting.ipynb.

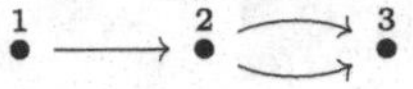

$$\overset{1}{\bullet} \longrightarrow \overset{2}{\bullet} \rightrightarrows \overset{3}{\bullet}$$

Fig. 2. A graph G, defined by $G_V = [3]$, $G_E = [3]$, $G_{\mathrm{src}} = [1,2,2]$, and $G_{\mathrm{tgt}} = [2,3,3]$.

Given two graphs G and H, a *graph homomorphism* $G \xrightarrow{h} H$ consists of a mapping of edges, $G_E \xrightarrow{h_E} H_E$ and a mapping of vertices, $G_V \xrightarrow{h_V} H_V$, that preserve the graph structure, i.e., $G_{\mathrm{src}}; h_V = h_E; H_{\mathrm{src}}$ and $G_{\mathrm{tgt}}; h_V = h_E; H_{\mathrm{tgt}}$. Regarding the source graph as a pattern, the homomorphism describes a pattern match in the target. A graph homomorphism can also be thought of as a typed graph, in which the vertices and edges of G are assigned types from H. For a fixed typing graph X, typed graphs and type-preserving graph homomorphisms form a category, namely the slice category $\mathbf{Grph}/X$ [9].

2.2 $\mathcal{C}$-Sets and Their Homomorphisms

Graphs are a special case of a class of structures called $\mathcal{C}$-sets.[2] Consider the category $\mathcal{C}$ freely generated by the graph $E \overset{s}{\underset{t}{\rightrightarrows}} V$. A $\mathcal{C}$-set is a functor from the category $\mathcal{C}$ to **Set**, which by definition assigns to each object a set and to each arrow a function from the domain set to the codomain set. For this choice of $\mathcal{C}$, the category of $\mathcal{C}$-sets is isomorphic to the category of directed multigraphs. Importantly, we recover the definition of graph homomorphisms between graphs G and H as a natural transformation of functors G and H.

The category $\mathcal{C}$ is called the *indexing category* or *schema*, and the functor category $[\mathcal{C}, \mathbf{Set}]$ is referred to as $\mathcal{C}$-**Set** or the category of *instances*, *models*, or *databases*. Given a $\mathcal{C}$-set X, the set that X sends a component $c \in \mathrm{Ob}\ \mathcal{C}$ to is denoted by X_c. Likewise, the finite function X sends a morphism $f \in \mathrm{Hom}_{\mathcal{C}}(a, b)$ to is denoted by X_f. We often restrict to $[\mathcal{C}, \mathbf{FinSet}]$ for computations.

In addition to graphs, **Set** itself can be thought of as a $\mathcal{C}$-set, where the schema $\mathcal{C}$ is the terminal category **1**. We can change $\mathcal{C}$ in other ways to obtain new data structures, as illustrated in Fig. 3. $\mathcal{C}$-sets can also be extended with a notion of *attributes* to incorporate non-combinatorial data [21,27], such as symbolic labels or real-valued weights. For simplicity of presentation, we focus on $\mathcal{C}$-sets without attributes in our examples.

2.3 Relationships Between $\mathcal{C}$-Sets and Typed Graphs

One reason to prefer modeling certain domains using typed graphs or $\mathcal{C}$-sets rather than graphs is that the domain of interest has regularities that we wish to enforce *by construction*, rather than checking that these properties hold of inputs at runtime and verifying that every rewrite rule preserves them. There are close connections but also important differences between modeling with typed graphs or with $\mathcal{C}$-sets.

---

[2] $\mathcal{C}$-sets are also called *copresheaves* on $\mathcal{C}$ or *presheaves* on $\mathcal{C}^{op}$, and are what Löwe studied as *graph structures* or *unary algebras*.

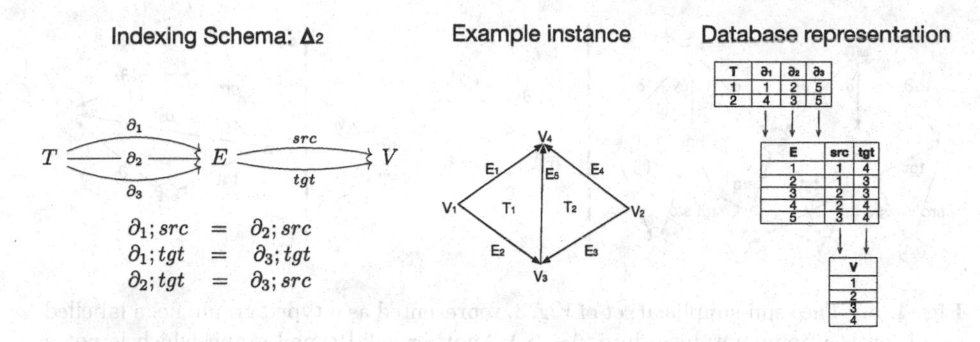

Fig. 3. The schema of two-dimensional semi-simplicial sets, Δ_2, and an example semi-simplicial set, i.e. an object of Δ_2-**Set**. The equations enforce the connectivity of edges to be a triangle. Note that MacLane defines Δ as our Δ^{op}.

Every $\mathcal{C}$-set instance X can be functorially transformed into a typed graph. One first applies the category of elements construction, $\int X : \mathcal{C}\textbf{-Set} \to \textbf{Cat}/\mathcal{C}$, to produce a functor into $\mathcal{C}$. Then the underlying graph functor $\textbf{Cat} \to \textbf{Grph}$ can be applied to this morphism in **Cat** to produce a graph typed by $\mathcal{C}$, i.e., a graph homomorphism into the underlying graph of $\mathcal{C}$. Figure 4a shows a concrete example. However, a graph typed by $\mathcal{C}$ is only a $\mathcal{C}$-set under special conditions. The class of $\mathcal{C}$-typed graphs representable as $\mathcal{C}$-set instances are those that satisfy the path equations of $\mathcal{C}$ and are, moreover, *discrete opfibrations* over $\mathcal{C}$. Discrete opfibrations are defined in full generality in Eq. 1.[3]

$$\begin{aligned}&\text{Given a functor } F : \mathcal{E} \to \mathcal{C} : \text{for all } x \xrightarrow{\phi} y \in \operatorname{Hom}\mathcal{C}, \text{ and for all } e_x \in F^{-1}(x),\\ &\qquad\text{there exists a unique } e_x \xrightarrow{e_\phi} e_y \in \operatorname{Hom}\mathcal{E} \text{ such that } F(e_\phi) = \phi \qquad (1)\end{aligned}$$

However, there is a sense in which every typed graph is a $\mathcal{C}$-set: there exists a schema $\mathcal{X}$ such that $\mathcal{X}$-**Set** is equivalent to **Grph**$/X$. By the fundamental theorem of presheaf toposes [15], $\mathcal{X}$ is the category of elements of the graph X, viewed as a $\mathcal{C}$-set on the schema for graphs. Note this procedure of creating a schema to represent objects of a slice category works beyond graphs, which we use to develop a framework of subtype hierarchies for $\mathcal{C}$-sets, as demonstrated in Fig. 5.

Because every typed graph category is equivalent to a $\mathcal{C}$-set category but not the converse, $\mathcal{C}$-sets are a more general class of structures. The $\mathcal{C}$-set categories equivalent to typed graph categories are those whose instances represent sets and *relations*, in contrast with the general expressive power of $\mathcal{C}$-sets to represent sets and *functions*. Concretely for some edge $a \xrightarrow{f} b$ in a type graph X, graphs typed over X can have zero, one, or many f edges for each vertex of type a, while $\mathcal{C}$-sets

[3] When specialized to typed graphs, $\mathcal{E} \xrightarrow{F} \mathcal{C}$ is a graph homomorphism and the graphs are regarded as their path categories.

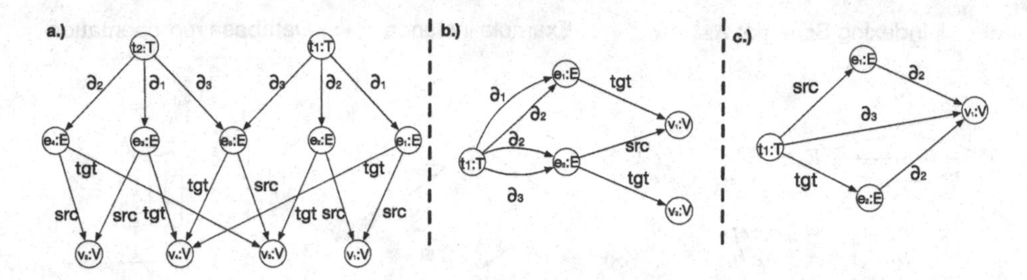

Fig. 4. a.) The semi-simplicial set of Fig. 3, represented as a typed graph, i.e. a labelled graph with a homomorphism into Δ_2. **b.)** Another valid typed graph which is not a $\mathcal{C}$-set for three independent reasons: 1.) T_1 has multiple edges assigned for ∂_2, 2.) e_1 has no vertices assigned for src, and 3.) the last equation of Δ_2 is not satisfied. **c.)** A labelled graph which is not well-typed with respect to Δ_2, i.e. no labelled graph homomorphism exists into Δ_2.

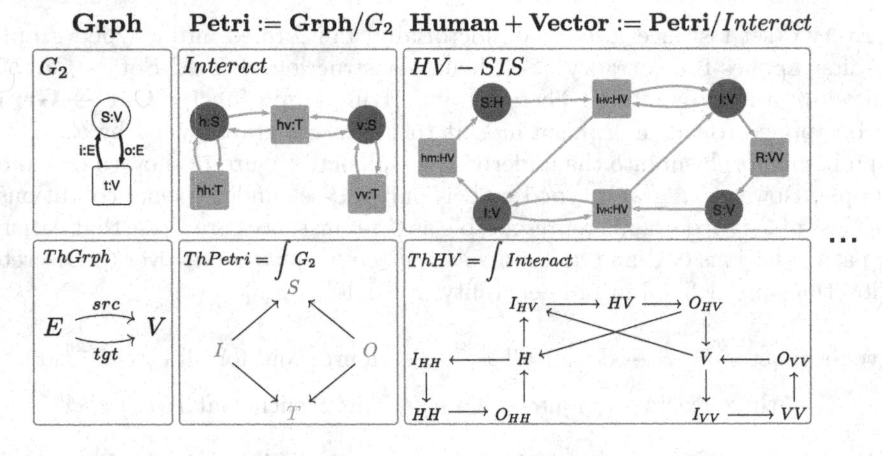

Fig. 5. Beginning with a theory of graphs, we derive a theory of whole-grain Petri nets (or bipartite graphs) by considering two distinct kinds of vertices (states and transitions) and two kinds of edges (inputs and outputs). *ThPetri* is constructed the category of elements of G_2. Then, taking a slice in **Petri** over an instance, *Interact*, which asserts three kinds of transitions and two kinds of states, we define a type system encoding certain domain knowledge about host-vector interactions, such as the impossibility of a transition which converts a host into a vector. As an example of subtyping, we can interpret hosts as a type of state, implying they are also a type of vertex. This process can be repeated, such as considering SIS disease dynamics for both hosts and vectors. Note that for ease of visualization, $\mathcal{C}$-set components at the apex of a span of morphisms (e.g. E, I, O) are represented as directed edges.

come with a restriction of there being exactly one such edge. While functions can represent relations via spans, the converse is not true.

There are practical consequences for this in graph rewriting software, if one is using typed graph rewriting to model a domain that truly has functional relationships. Because rewrite rules could take one out of the class of discrete opfibrations, as in Fig. 4b, this becomes a property that one has to verify of inputs and check all rewrite rules preserve. Typed graph rewriting software can allow declaring these constraints and enforce them, but this becomes an additional engineering task outside of the underlying theory. In contrast, $\mathcal{C}$-sets are discrete opfibrations by construction.

Path equations are another common means of modeling a domain that are not represented in the theory of typed graph rewriting. This means, for example, that the equation $\partial_1; tgt = \partial_2; src$ in a semi-simplicial set must be checked of all runtime inputs as well as confirmed to be preserved by each rewrite rule. This property is not straightforward to guarantee in the case of sesqui-pushout rewriting. As an upcoming example will demonstrate, it is not sufficient to just check that one's rewrite rule satisfies the path equalities: the rewriting itself must take path equalities into account in order to compute the correct result.

Furthermore, there are performance improvements made possible by working with $\mathcal{C}$-sets, rather than typed graphs. Borrowing terminology from relational databases, we first note that data in a $\mathcal{C}$-set is organized into distinct tables, so queries over triangles of a semi-simplicial set do not have to consider vertices or edges, for example. Secondly, the uniqueness of foreign keys allows them to be indexed, which is crucial to performance when performing queries that require table joins. This mirrors the well-known performance differences between queries of data organized in relational databases versus knowledge graphs [5]. We compare both representations within the same rewriting tool in a single benchmark experiment, described in Fig. 6. This preliminary benchmark evaluates the performance of a single rewrite on semi-simplicial sets in a planar network of tessellated triangles. The rewrite locates a pair of triangles sharing an edge (i.e. a quadrilateral with an internal diagonal edge) and replaces them with a quadrilateral containing the opposite internal diagonal edge. We also chart the performance of finding all quadrilateral instances (homomorphisms) in variously sized grids. The results in Fig. 6 demonstrate a lower memory footprint as well as improved rewrite and match searching for $\mathcal{C}$-sets.

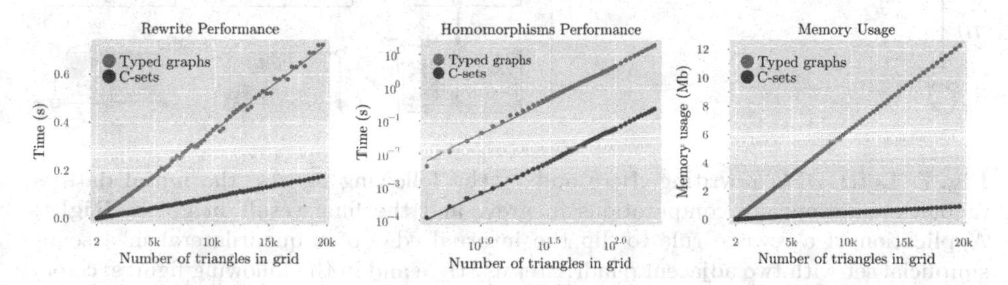

Fig. 6. Semisimplicial set edge flip benchmark results. Time was measured on an AMD EPYC 75F3 Milan 3.0 GHz Core with 4 GB of allocated RAM.

3 Category-Theoretic Rewriting

Pushout Complements. Given a pair of arrows $A \xrightarrow{f} B \xrightarrow{g} C$, one constructs a pushout *complement* by finding a pair of morphisms $A \to D \to C$ such that the resulting square is a pushout. While any category of $\mathcal{C}$-sets has pushouts, pushout complements are more subtle because they are not guaranteed to exist or be unique [4]. These are both desirable properties to have when using the pushout complement in rewriting, so we will demand that identification and dangling conditions (Eqs. 2–3 [19]) hold, which guarantee its existence, and that the first morphism f be monic, which forces it to be unique [18].

$$\begin{gathered}\forall X \in \mathrm{Ob}\,\mathcal{C}, \forall x_1, x_2 \in B_X : \\ g_X(x_1) = g_X(x_2) \implies x_1 = x_2 \vee \{x_1, x_2\} \subseteq f_X(A_X)\end{gathered} \tag{2}$$

$$\begin{gathered}\forall \phi : X \to Y \in \mathrm{Hom}\,\mathcal{C}, \forall x \in C_X : \\ \phi(x) \in g_Y(B_Y - f_Y(A_Y)) \implies x \in g_X(B_X - f_X(A_X))\end{gathered} \tag{3}$$

DPO, SPO, SqPO. The double-pushout (DPO) algorithm [10] formalizes a notion of rewriting a portion of a $\mathcal{C}$-set, visualized in Fig. 7. The morphism m is called the *match* morphism. The meaning of L is to provide a pattern that m will match to a sub-$\mathcal{C}$-set in G, the target of rewriting. R represents the $\mathcal{C}$-set which will be substituted back in for the matched pattern to yield the rewritten $\mathcal{C}$-set, and I indicates what fragment of L is preserved in the rewrite and its relation to R. To perform a rewrite, first, a pushout complement computes K, the original $\mathcal{C}$-set with deletions applied. Second, the final rewritten $\mathcal{C}$-set is computed via pushout along r and i.

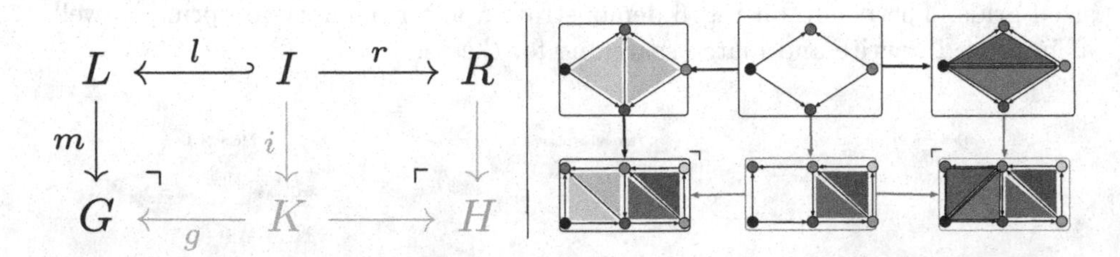

Fig. 7. Left: DPO rewriting. Here and in the following figures, the initial data is in black, intermediate computations in grey, and the final result in green. **Right:** Application of a rewrite rule to flip the internal edge of a quadrilateral in a semi-simplicial set with two adjacent quadrilaterals. Here and in the following figures, colors are used to represent homomorphism data. (Color figure online)

Single-pushout (SPO) rewriting [19] generalizes DPO rewriting, as every DPO transformation can be expressed as a SPO transformation. The additional expressivity allows us to delete in an unknown context, as demonstrated in Fig. 8.

The name comes from the construction being a single pushout in the category of *partial* $\mathcal{C}$-set morphisms, $\mathcal{C}$-**Par**. A partial $\mathcal{C}$-set morphism is a span $L \overset{l}{\hookleftarrow} I \overset{r}{\rightarrow} R$ where l is monic. Lastly, sesqui-pushout (SqPO) rewriting [8] is the most recently developed of the three rewriting paradigms we discuss here. It is defined in terms of the notions of partial map classifiers and final pushout complements, and it further generalizes SPO by allowing both deletion and addition in an unknown context, as demonstrated in Fig. 9.

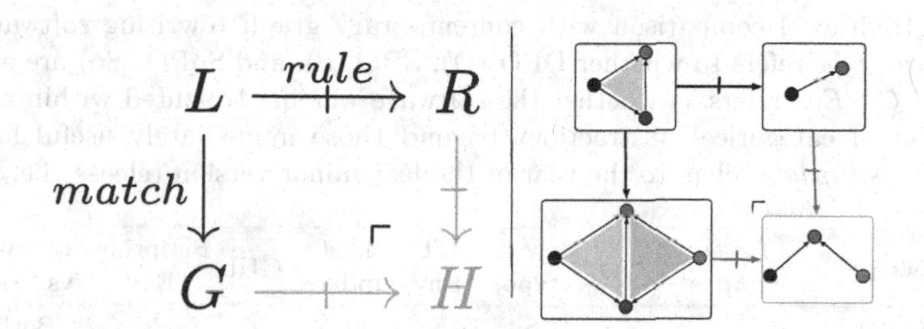

Fig. 8. Left: SPO rewriting **Right:** An instance of deletion in an unknown context.

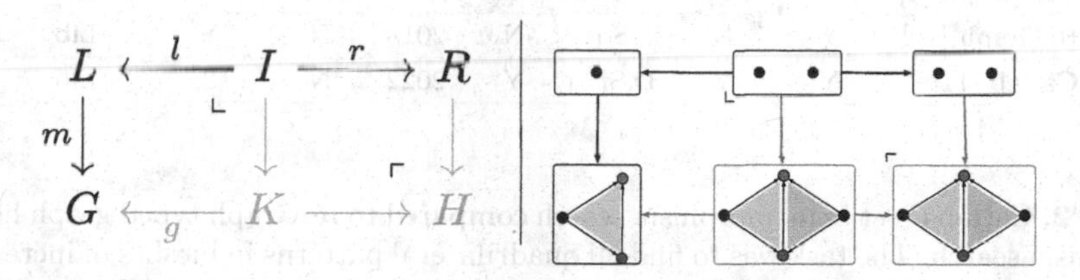

Fig. 9. Left: SqPO rewriting **Right:** an instance of creation in an unknown context. Note that there are multiple possible pushout complements because l is not monic, but performing DPO using any of these would leave the original graph unchanged. Also note that enforcing the Δ_2 equations (in Fig. 3) when computing the partial object classifier affects the results: without equations, there are four resulting 'triangle' objects, although two of these clearly do not form triangles.

4 Design and Implementation of Generic Categorical Rewriting

Within the paradigm of computational category theory, Catlab.jl is an open source framework for applied category theory at the center of an ecosystem of software packages called AlgebraicJulia [11,21]. We have recently extended Catlab to support the categorical rewriting paradigms described above for $\mathcal{C}$-sets on finitely presented schemas $\mathcal{C}$. This class of structures balances expressivity and efficiency of manipulation, given that $\mathcal{C}$-sets are representable in the concrete language of relational databases [27], modulo equations in $\mathcal{C}$. In Catlab, each $\mathcal{C}$-set is automatically specialized to an efficient Julia data type; for example, when specialized to graphs, Catlab's implementation of $\mathcal{C}$-sets, performs competitively against libraries optimized for graphs [21]. Catlab now occupies a unique point in the space of rewriting software tools (Table 1). For performance in pattern matching (often the typical bottleneck of rewriting), Catlab outperforms ReGraph, the nearest alternative in terms of expressive capabilities (SqPO) and usability (Table 2).

Table 1. High-level comparison with contemporary graph rewriting software packages. *Rewrite type* refers to whether DPO (D), SPO (S), and SqPO (Sq) are explicitly supported. *CT Env* refers to whether the software was implemented within a general environment of categorical abstractions beyond those immediately useful for graph rewriting. *Last update* refers to the year of the last minor version release (i.e. X.Y.0).

Software	Typed Graphs	$\mathcal{C}$-sets	Rewrite type	CT Env	Last update	GUI	Scripting Env	Library vs. App
AGG [29]	Y	N	S	N	2017	Y	N	Both
Groove [23]	Y	N	S	N	2021	Y	N	App
Kappa [13]	N	N		N	2021	Y	Y	App
VeriGraph [1]	Y	N	D	Y	2017	N	Y	Lib
ReGraph [12]	Y	N	Sq	N	2018	N	Y	Lib
Catlab [11]	Y	Y	D,S,Sq	Y	2022	N	Y	Lib

Table 2. Catlab $\mathcal{C}$-set homomorphism search compared to ReGraph typed graph homomorphism search. The task was to find all quadrilateral patterns in meshes of increasing size. Tests were conducted on a single AMD EPYC 75F3 Milan 3.0 GHz Core with 4GB of RAM.

Mesh size	Catlab (s)	ReGraph (s)
2 by 2	1.2×10^{-4}	5.3×10^{-3}
2 by 3	2.7×10^{-4}	8.0
2 by 4	4.7×10^{-4}	1313.3
2 by 5	6.7×10^{-4}	44979.8

The development of Catlab has emphasized the separation of syntax and semantics when modeling a domain. This facilitates writing generic code, as diverse applications can share syntactic features, e.g. representability through string diagrams and hierarchical operad composition, with different semantic interpretations of that syntax for diverse applications. One result of this is that library code becomes very reusable and interconnected, such that new features can be built from the composition of old parts with minimal additions, which reduces both developer time and the surface area for new bugs.

This point is underscored by the developer experience of implementing the above rewriting algorithms: because colimits already existed for $\mathcal{C}$-sets, the addition of DPO to Catlab only required pushout complements. Like limits and colimits, pushout complements are computed component-wise for $\mathcal{C}$-sets, meaning that only basic code related to pushout complements of finite sets was required. More work was needed to implement SPO because no infrastructure for the category $\mathcal{C}$-**Par** existed at the time. However, with a specification of partial morphism pushouts in terms of pushouts and pullback complements of total morphisms [16, Theorem 3.2], the only engineering required for this feature was an efficient pullback complement for $\mathcal{C}$-sets. Lastly, for SqPO, an algorithm for final pullback complements for $\mathcal{C}$-sets was the only nontrivial component that needed to be implemented, based on [7, Theorem 1] and [2, Theorem 2]. This required generalizing examples of partial map classifiers from graphs to $\mathcal{C}$-sets. Because the partial map classifier can be infinite for even a finitely presented $\mathcal{C}$-set, this type of rewriting is restricted to acyclic schemas, which nevertheless includes graphs, Petri nets, semi-simplicial sets, and other useful examples.

As shown by the supplemental notebook, because Catlab is a library rather than a standalone application, users have a great deal of freedom in defining their own abstractions and automation techniques, using the full power of the Julia programming language. A great deal of convenience follows from having the scripting language and the implementation language be the same: we can specify the pattern of a rewrite rule via a pushout, or we can programmatically generate repetitive rewrite rules based on structural features of a particular graph. Providing libraries rather than standalone black-box software makes integration into other projects (in the same programming language) trivial, and in virtue of being open-source library, individuals can easily extend the functionality. By making these extensions publicly available, all members of the AlgebraicJulia ecosystem can mutually benefit from each other's efforts. As examples of this, the following additional features that have been contributed to Catlab.jl all serve to extend its utility as a general rewriting tool:

Computation of Homomorphisms and Isomorphisms of C-Sets. For rewriting algorithms to be of practical use, morphisms matching the left-hand-side of rules must somehow be supplied. The specification of a $\mathcal{C}$-set morphism requires a nontrivial amount of data that must satisfy the naturality condition. Furthermore, in confluent rewriting systems, manually finding matches is an unreasonable request to make of the end user, as the goal is to apply all rewrites possible until the term reaches a normal form. For this reason, DPO rewriting of $\mathcal{C}$-sets benefits from a generic algorithm to find homomorphisms, analogous to structural pattern matching in the tree term rewriting case.

The problem of finding a $\mathcal{C}$-set homomorphism $X \to Y$, given a finitely presented category $\mathcal{C}$ and two finite $\mathcal{C}$-sets X and Y, is generically at least as hard as the graph homomorphism problem, which is NP-complete. On the other hand, the $\mathcal{C}$-set homomorphism problem can be framed as a constraint satisfaction problem (CSP), a classic problem in computer science for which many algorithms are known [26, Chapter 6]. Since $\mathcal{C}$-sets are a mathematical model of relational databases [28], the connection between $\mathcal{C}$-set homomorphisms and constraint satisfaction is a facet of the better-known connection between databases and CSPs [30].

To make this connection precise, we introduce the slightly nonstandard notion of a typed CSP. Given a finite set T of *types*, the slice category $\mathbf{FinSet}/T$ is the category of *T-typed finite sets.* A *typed CSP* then consists of T-typed finite sets V and D, called the *variables* and the *domain*, and a finite set of *constraints* of form $(\mathbf{x}, R)$, where $\mathbf{x} = (x_1, \ldots, x_k)$ is a list of variables and $R \subseteq D^{-1}(V(x_1)) \times \cdots \times D^{-1}(V(x_k))$ is a compatibly typed k-ary relation. An *assignment* is a map $\phi : V \to D$ in $\mathbf{FinSet}/T$. The objective is to find a *solution* to the CSP, namely an assignment ϕ such that $(\phi(x_1), \ldots, \phi(x_k)) \in R$ for every constraint $(\mathbf{x}, R)$.

The problem of finding a $\mathcal{C}$-set morphism $X \to Y$ translates to a typed CSP by taking the elements of X and Y to be the variables and the domain of the CSP, respectively. To be precise, let the types T be the objects of $\mathcal{C}$. The variables $V : \{(c, x) : c \in \mathcal{C}, x \in X(c)\} \to \mathrm{Ob}\,\mathcal{C}$ are given by applying the objects functor $\mathrm{Ob} : \mathbf{Cat} \to \mathbf{Set}$ to $\int X \to \mathcal{C}$, the category of elements of X with its canonical projection. Similarly, the domain is $D := \mathrm{Ob}(\int Y \to \mathcal{C})$. Finally, for every generating morphism $f : c \to c'$ of $\mathcal{C}$ and every element $x \in X(c)$, introduce a constraint $((x, x'), R)$ where $x' := X(f)(x)$ and $R := \{(y, y') \in Y(c) \times Y(c') : Y(f)(y) = y'\}$ is the graph of $Y(f)$. By construction, an assignment $\phi : V \to D$ is the data of a $\mathcal{C}$-set transformation (not necessarily natural) and ϕ is a solution if and only if the transformation is natural. Thus, the solutions of the typed CSP are exactly the $\mathcal{C}$-set homomorphisms $X \to Y$.

With this reduction, CSP algorithms are straightforwardly ported to algorithms for finding $\mathcal{C}$-set morphisms, where the types and special structure permits optimizations, one example being the use of the discrete opfibration condition to accelerate the search. We only consider assignments that satisfy the typing relations. We have adapted backtracking search [26, Section 6.3], a simple but fundamental CSP algorithm, to find $\mathcal{C}$-set homomorphisms. By also maintaining a partial inverse assignment, this algorithm is easily extended to finding $\mathcal{C}$-set monomorphisms, an important constraint when matching for rewriting. Since a monomorphism between finite $\mathcal{C}$-sets X and Y is an isomorphism if and only if $X(c)$ and $Y(c)$ have the same cardinality for all $c \in \mathcal{C}$, this extension also yields an algorithm for isomorphism testing, which is useful for checking the correctness of rewrites.

Typed Graph Rewriting with Slice Categories. Slice categories offer a form of constraining $\mathcal{C}$-sets without altering the schema. Consider the example of rewriting string diagrams encoded as hypergraph cospans [3]. These can be used to represent terms in a symmetric monoidal theory, where it is important to restrict diagrams to only those which draw from a fixed set of boxes with particular arities, given by a monoidal signature Σ, which induces the unique hypergraph $H\Sigma$ which has all box types from Σ and a single vertex. Working within the slice category $\mathbf{Hyp}/H\Sigma$ prevents us from performing rewrites which violate the arities of the operations specified by Σ.

There are two ways to implement rewriting in $\mathcal{C}$-**Set**$/X$ for a particular $\mathcal{C}$: the computation can be performed with the objects L, I, R, G being $\mathcal{C}$-set morphisms, or it can be performed in $[\int X, \mathbf{Set}]$. Programming with generic categorical abstraction greatly lowered the barrier to implementing both of these: for the former, what was needed was to relate the pushout and pushout complement of $\mathcal{C}$-**Set**$/X$ to the corresponding computations in $\mathcal{C}$-**Set**. The barrier to the latter was to compute the category of elements and migrate data between the two representations, code which had already been implemented. As the former strategy requires less data transformation, it is preferred.

Open System Rewriting with Structured Cospans. The forms of rewriting discussed up to this point have concerned rewriting closed systems. Structured cospans are a general model for open systems, which formalize the notion of gluing together systems which have designated inputs and outputs. Open systems are modeled as cospans of form $La \rightarrow x \leftarrow Lb$, where the apex x represents the system itself and the feet La and Lb represent the inputs and outputs, typically discrete systems such as graphs without edges. Here, $L : A \rightarrow X$ is a functor that maps from the system category A to the system interface category X, and L must be a left adjoint between categories with finite colimits.[4] Larger systems are built up from smaller systems via pushouts in X, which glue systems together along a shared interface: $(La \rightarrow x \leftarrow Lb \rightarrow y \leftarrow Lc) \mapsto (La \rightarrow x +_{Lb} y \leftarrow Lc)$.

When L, I, and R are each structured cospans, there is extra data to consider when rewriting, as shown in Fig. 10. In ordinary DPO rewriting, if the R of one rewrite rule equals the L of another, a composite rewrite rule can be constructed, which could be called *vertical* composition. In the case of structured cospans, *horizontal* composition emerges from composing the L, I, and R of two structured cospan rules pairwise, visualized in Fig. 11. These two forms of composition together yield a double category of structured cospan rewrites, where horizontal arrows are in correspondence with structured cospans and squares are in correspondence with all possible rewrites [6].

While this compositional approach to building open systems can be an illuminating way to organize information about a complex system, there can also be computational benefits. When searching for a match in a large $\mathcal{C}$-set, the search space grows as $O(n^k)$ where k is the size of the pattern L and n is the size of G. However, after decomposing G into a composite of substructures and restricting matches to homomorphisms into a specific substructure, the search space is limited by $O(m^k)$ where $m < n$ is the size of the substructure. Not only does this accelerate the computation, but it can be semantically meaningful to restrict matches to those which do not cross borders.

[4] The L of structured cospans should not be confused with the L of the rewrite rule $L \leftarrow I \rightarrow R$.

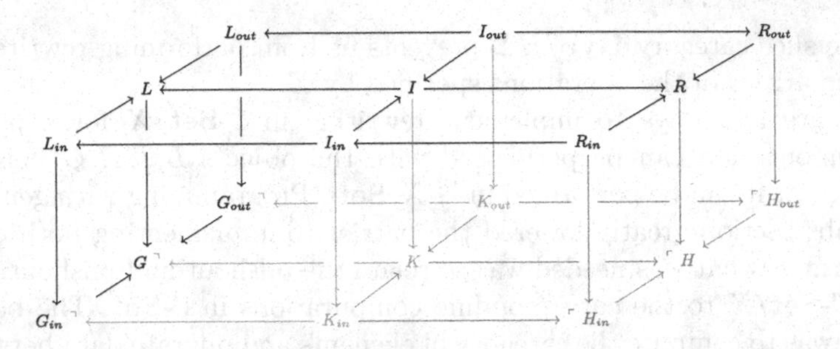

Fig. 10. Applying a structured cospan rewrite rule. $\mathcal{C}$-sets and morphisms in black are the initial data: the upper face represents the open rewrite rule, the upper left edge represents the open pattern to be matched, and the left face represents the matching. Green morphisms are computed by pushout complement in $\mathcal{C}$-**Set**. The purple morphisms are computed by the rewriting pushouts and red morphisms are computed by the structured cospan pushouts. Figure adapted from [6, Section 4.2]. (Color figure online)

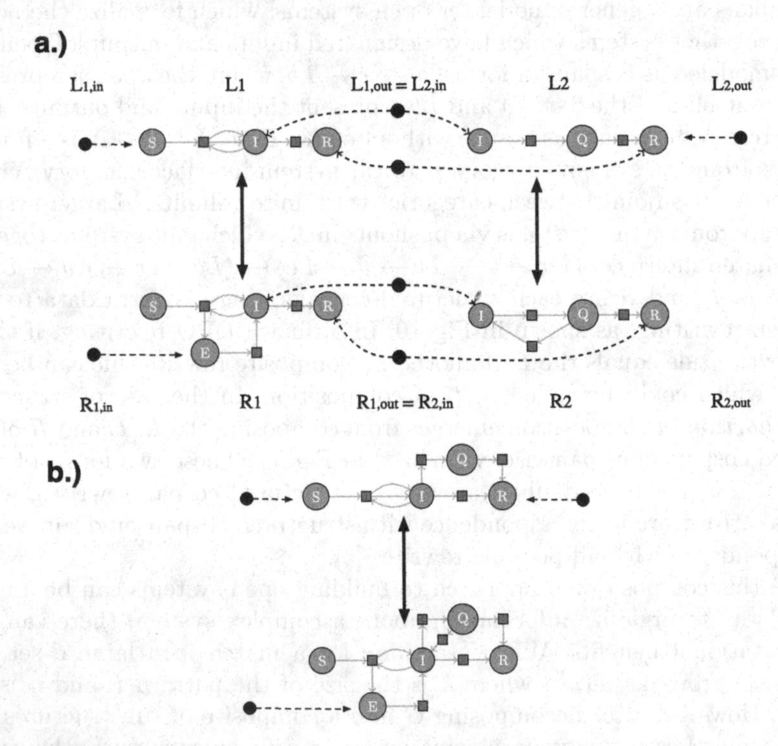

Fig. 11. a.) Example of horizontal composition of structured cospan rewrite rules. The L and R structured cospans are positioned on the top and bottom, respectively. For clarity, I cospans are omitted. **b.)** The result of composition.

Distributed Graph Rewriting. Distributed graphs offer an alternative formalism that allows one to decompose a large graph into smaller ones while maintaining consistency at the boundaries, and thus it is another strategy for parallelizing computations over graphs. The content of a distributed graph can be succinctly expressed in the language of category theory as a diagram in **Grph**. Because Catlab has sophisticated infrastructure in place for manipulating categories of diagrams, it merely takes specializing the codomain of the Diagram datatype to **Grph** to represent distributed graphs and their morphisms. Note that we can easily generalize to distributed semi-simplicial sets or other $\mathcal{C}$-sets (Fig. 12). A future plan will be to implement algorithms for colimits of diagrams as specified in [22], which would lead to a rewriting tool for distributed graphs.

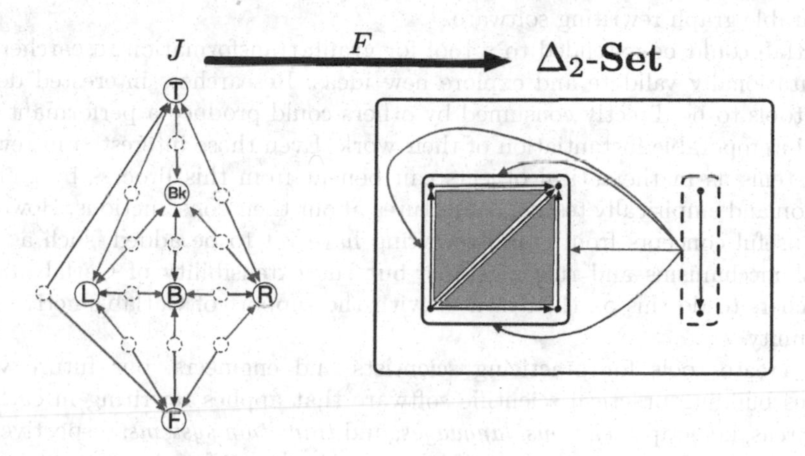

Fig. 12. Constructing the surface of a cube compositionally with a distributed graph. We construct the assembled cube as a $\mathcal{C}$-set simply by taking the colimit of the diagram.

Further Extensions. Examples of further features, such as negative application conditions, parallel rewriting, rewriting with functions applied to attributes, matching variables on attributes, (e.g. one rule which can identify any triangle that has exactly two edges with an equal length attribute and rewrite to make all three edges have that length) are found in the supplemental notebook.

5 Conclusions and Future Work

There are many desiderata for software development in academic and industrial settings alike, such as velocity of development, robustness to future changes in design, and correctness. We demonstrated how designing software with category-theoretic abstractions facilitates the achievement all three of these, using the mature field of graph rewriting software as a case study.

While current graph transformation software in use is often very specialized to particular domains, such as chemistry, we show that DPO, SPO, and SqPO rewriting can be efficiently performed on $\mathcal{C}$-sets, which are viewed as a subset of typed graphs (discrete opfibrations) with desirable theoretical and performance characteristics, and we have presented the first practical implementation for this.

This result allows generic rewrite operations to be used in a variety of contexts, when it would otherwise be time-consuming and error-prone to develop custom rewrite algorithms for such a multitude of data structures or to work with typed graphs and enforce the discrete opfibration condition by other means. We also extended these implementations to the first practical implementations of homomorphism search, structured cospan rewriting, and distributed graphs for arbitrary $\mathcal{C}$-sets. Our internal benchmark showed that $\mathcal{C}$-set rewriting can leverage the discrete opfibration condition to outperform typed graphs in memory and speed, and an external benchmark showed a significant speedup relative to comparable graph rewriting software.

Catlab could be extended to a tool for graph transformation researchers to computationally validate and explore new ideas. Researchers interested developing tools to be directly consumed by others could produce a performant and easily interoperable instantiation of their work. Even those interested in rewriting systems as mathematical objects can benefit from this process by gaining intuition and empirically testing conjectures about their constructions. However, many useful concepts from graph rewriting have yet to be added, such as rule control mechanisms and rule algebras, but the extensibility of Catlab allows researchers to do this on their own or with the support of Catlab's active user community.

To create tools for practicing scientists and engineers, our future work involves building practical scientific software that applies rewriting in each its main areas, i.e. graph *relations*, *languages*, and *transition systems*: respectively, a theorem prover for symmetric monoidal categories by performing e-graph equality saturation [31] with rewriting, a tool for defining and exploring a language of open epidemiological models, and a general agent-based model simulator.

References

1. Azzi, G.G., Bezerra, J.S., Ribeiro, L., Costa, A., Rodrigues, L.M., Machado, R.: The verigraph system for graph transformation. In: Heckel, R., Taentzer, G. (eds.) Graph Transformation, Specifications, and Nets. LNCS, vol. 10800, pp. 160–178. Springer, Cham (2018). https://doi.org/10.1007/978-3-319-75396-6_9
2. Behr, N., Harmer, R., Krivine, J.: Concurrency theorems for non-linear rewriting theories. In: Gadducci, F., Kehrer, T. (eds.) ICGT 2021. LNCS, vol. 12741, pp. 3–21. Springer, Cham (2021). https://doi.org/10.1007/978-3-030-78946-6_1
3. Bonchi, F., Gadducci, F., Kissinger, A., Sobocinski, P., Zanasi, F.: String diagram rewrite theory I: rewriting with Frobenius structure. arXiv preprint arXiv:2012.01847 (2020)
4. Braatz, B., Golas, U., Soboll, T.: How to delete categorically-two pushout complement constructions. J. Symb. Comput. **46**(3), 246–271 (2011)
5. Cheng, Y., Ding, P., Wang, T., Lu, W., Du, X.: Which category is better: benchmarking relational and graph database management systems. Data Sci. Eng. **4**(4), 309–322 (2019). https://doi.org/10.1007/s41019-019-00110-3
6. Cicala, D.: Rewriting structured cospans: A syntax for open systems. arXiv preprint arXiv:1906.05443 (2019)

7. Corradini, A., Duval, D., Echahed, R., Prost, F., Ribeiro, L.: AGREE – algebraic graph rewriting with controlled embedding. In: Parisi-Presicce, F., Westfechtel, B. (eds.) ICGT 2015. LNCS, vol. 9151, pp. 35–51. Springer, Cham (2015). https://doi.org/10.1007/978-3-319-21145-9_3
8. Corradini, A., Heindel, T., Hermann, F., König, B.: Sesqui-pushout rewriting. In: Corradini, A., Ehrig, H., Montanari, U., Ribeiro, L., Rozenberg, G. (eds.) ICGT 2006. LNCS, vol. 4178, pp. 30–45. Springer, Heidelberg (2006). https://doi.org/10.1007/11841883_4
9. Corradini, A., Montanari, U., Rossi, F.: Graph processes. Fund. Inform. **26**(3, 4), 241–265 (1996)
10. Ehrig, H., Pfender, M., Schneider, H.J.: Graph-grammars: an algebraic approach. In: 14th Annual Symposium on Switching and Automata Theory (SWAT 1973), pp. 167–180. IEEE (1973)
11. Halter, M., Patterson, E., Baas, A., Fairbanks, J.: Compositional scientific computing with Catlab and SemanticModels. arXiv preprint arXiv:2005.04831 (2020)
12. Harmer, R., Oshurko, E.: Reversibility and composition of rewriting in hierarchies. arXiv preprint arXiv:2012.01661 (2020)
13. Hayman, J., Heindel, T.: Pattern graphs and rule-based models: the semantics of kappa. In: Pfenning, F. (ed.) FoSSaCS 2013. LNCS, vol. 7794, pp. 1–16. Springer, Heidelberg (2013). https://doi.org/10.1007/978-3-642-37075-5_1
14. Heckel, R., Lambers, L., Saadat, M.G.: Analysis of graph transformation systems: native vs translation-based techniques. arXiv preprint arXiv:1912.09607 (2019)
15. Kashiwara, M., Schapira, P.: Categories and Sheaves. Springer, Heidelberg (2006). https://doi.org/10.1007/3-540-27950-4
16. Kennaway, R.: Graph rewriting in some categories of partial morphisms. In: Ehrig, H., Kreowski, H.-J., Rozenberg, G. (eds.) Graph Grammars 1990. LNCS, vol. 532, pp. 490–504. Springer, Heidelberg (1991). https://doi.org/10.1007/BFb0017408
17. Lack, S., Sobociński, P.: Adhesive categories. In: Walukiewicz, I. (ed.) FoSSaCS 2004. LNCS, vol. 2987, pp. 273–288. Springer, Heidelberg (2004). https://doi.org/10.1007/978-3-540-24727-2_20
18. Lack, S., Sobociński, P.: Adhesive and quasiadhesive categories. RAIRO-Theor. Inform. Appl. **39**(3), 511–545 (2005)
19. Löwe, M.: Algebraic approach to single-pushout graph transformation. Theor. Comput. Sci. **109**(1–2), 181–224 (1993)
20. Minas, M., Schneider, H.J.: Graph transformation by computational category theory. In: Engels, G., Lewerentz, C., Schäfer, W., Schürr, A., Westfechtel, B. (eds.) Graph Transformations and Model-Driven Engineering. LNCS, vol. 5765, pp. 33–58. Springer, Heidelberg (2010). https://doi.org/10.1007/978-3-642-17322-6_3
21. Patterson, E., Lynch, O., Fairbanks, J.: Categorical data structures for technical computing. arXiv preprint arXiv:2106.04703 (2021)
22. Peschke, G., Tholen, W.: Diagrams, fibrations, and the decomposition of colimits. arXiv preprint arXiv:2006.10890 (2020)
23. Rensink, A., Boneva, I., Kastenberg, H., Staijen, T.: User manual for the groove tool set. University of Twente, The Netherlands, Department of Computer Science (2010)
24. Ringer, T., Palmskog, K., Sergey, I., Gligoric, M., Tatlock, Z.: QED at large: a survey of engineering of formally verified software. arXiv preprint arXiv:2003.06458 (2020)
25. Rushby, J.: Automated test generation and verified software. In: Meyer, B., Woodcock, J. (eds.) VSTTE 2005. LNCS, vol. 4171, pp. 161–172. Springer, Heidelberg (2008). https://doi.org/10.1007/978-3-540-69149-5_18

26. Russell, S., Norvig, P.: Artificial Intelligence: A Modern Approach (2010)
27. Schultz, P., Spivak, D.I., Vasilakopoulou, C., Wisnesky, R.: Algebraic databases. arXiv preprint arXiv:1602.03501 (2016)
28. Spivak, D.I.: Functorial data migration. Inf. Comput. **217**, 31–51 (2012). https://doi.org/10.1016/j.ic.2012.05.001
29. Taentzer, G.: AGG: a graph transformation environment for modeling and validation of software. In: Pfaltz, J.L., Nagl, M., Böhlen, B. (eds.) AGTIVE 2003. LNCS, vol. 3062, pp. 446–453. Springer, Heidelberg (2004). https://doi.org/10.1007/978-3-540-25959-6_35
30. Vardi, M.Y.: Constraint satisfaction and database theory: a tutorial. In: Proceedings of the Nineteenth ACM SIGMOD-SIGACT-SIGART Symposium on Principles of Database Systems, pp. 76–85 (2000). https://doi.org/10.1145/335168.335209
31. Willsey, M., Nandi, C., Wang, Y.R., Flatt, O., Tatlock, Z., Panchekha, P.: EGG: fast and extensible equality saturation. Proc. ACM Program. Lang. **5**(POPL), 1–29 (2021)

Invariant Analysis for Multi-agent Graph Transformation Systems Using k-Induction

Sven Schneider(✉), Maria Maximova, and Holger Giese

Hasso Plattner Institute, University of Potsdam, Potsdam, Germany
{sven.schneider,maria.maximova,holger.giese}@hpi.de

Abstract. The analysis of behavioral models such as Graph Transformation Systems (GTSs) is of central importance in model-driven engineering. However, GTSs often result in intractably large or even infinite state spaces and may be equipped with multiple or even infinitely many start graphs. To mitigate these problems, static analysis techniques based on finite symbolic representations of sets of states or paths thereof have been devised. We focus on the technique of k-induction for establishing invariants specified using graph conditions. To this end, k-induction generates symbolic paths backwards from a symbolic state representing a violation of a candidate invariant to gather information on how that violation could have been reached possibly obtaining contradictions to assumed invariants. However, GTSs where multiple agents regularly perform actions independently from each other cannot be analyzed using this technique as of now as the independence among backward steps may prevent the gathering of relevant knowledge altogether.

In this paper, we extend k-induction to GTSs with multiple agents thereby supporting a wide range of additional GTSs. As a running example, we consider an unbounded number of shuttles driving on a large-scale track topology, which adjust their velocity to speed limits to avoid derailing. As central contribution, we develop pruning techniques based on causality and independence among backward steps and verify that k-induction remains sound under this adaptation as well as terminates in cases where it did not terminate before.

Keywords: k-inductive invariant checking · causality · parallel and sequential independence · symbolic analysis · bounded backward model checking

1 Introduction

The verification of formal models of dynamic systems featuring complex concurrent behavior w.r.t. formal specifications is one of the central problems in model driven engineering. However, the required expressiveness of modeling and specification formalisms that must be used for these complex dynamic systems often leads to undecidable analysis problems. For example, the formalism of GTSs

N. Behr and D. Strüber (Eds.): ICGT 2022, LNCS 13349, pp. 173–192, 2022.
https://doi.org/10.1007/978-3-031-09843-7_10

considered in this paper is known to be Turing complete. Hence, fully-automatic procedures for the analysis of meaningful properties on the behavior of such GTS-based systems returning definite correct judgements cannot always terminate. Analysis becomes even more intricate when the start graph is not precisely known or when the system behavior is to be verified for a large or even infinite number of start graphs.

The technique of (forward) model checking generates the entire state space and checks this state space against the given specification. However, this technique is inapplicable when the state space is intractably large or even infinite. To mitigate this problem, large or even infinite sets of concrete states that are equivalent w.r.t. the property to be analyzed may be aggregated into symbolic states. Model checking then generates symbolic state spaces consisting of symbolic states and symbolic steps between them. However, these symbolic state spaces may still be intractably large depending on the size of the models[1] and there is usually no adequate support for multiple symbolic start states.

In backward model checking, a backward state space is generated from a set of target states derived from the specification by incrementally adding all steps leading to states that are already contained in the backward state space. For invariant properties, the target states are given by the states not satisfying the candidate invariant. As for model checking, sets of concrete states may be aggregated into symbolic states, which may also lead to a single symbolic target state. Clearly, in backward model checking, only backward paths containing exclusively reachable states are significant but during the analysis also paths containing unreachable states may be generated requiring techniques to prune such paths as soon as possible.

The technique of k-induction is a variant of bounded backward model checking for establishing state invariants. In k-induction, generated backward paths are *(a)* limited to length k and *(b)* end in a state violating the candidate invariant. Definite judgements are derived in two cases. A backward path extended to a start state leads to candidate invariant refutation and the candidate invariant is confirmed when no backward path of length k is derivable.

In this paper, we extend earlier work on k-induction from [6,19] by solving the following open problem. When the system under analysis features concurrency such as in a multi-agent context, backward steps may be independent as k backward steps may be performed by k different agents that may be logically/spatially apart. In that case, the k backward steps do not accumulate knowledge on why the violating graph could be reached preventing the derivation of a definite judgement. This problem can even occur when every target state contains a single agent since backward steps can still introduce further agents. To solve this problem, we introduce several novel GTS-specific pruning techniques. Firstly, we prune backward paths in which the last added step does not depend on the already accumulated knowledge. This *causality pruning* avoids the inclusion of steps of unrelated agents in a backward path. Secondly, we prune states containing an agent that is permanently blocked from further backward

[1] Approaches such as CEGAR [4] also aim at minimizing symbolic state spaces.

steps. This *evolution pruning* (assuming that agents existed in the start graph or are created in some step) is required when all backward steps of a certain agent have been pruned by some other pruning technique (while other agents are still able to perform backward steps). Thirdly, when a state is removed in evolution pruning, we propagate this state prunability forward across backward steps until the blocked agent has an alternative backward step. This *evolution-dependency pruning* is, in conjunction with our explicit handling of independent steps, able to prune also other backward paths (with common suffix) where independent steps of other agents are interleaved differently. For these three novel pruning techniques, we ensure that they do not affect the correctness of derived judgements and that our approach presented here is a conservative extension in the sense that it terminates whenever the single-agent approach terminated before.[2]

As a running example, we consider an unbounded number of shuttles driving on a large-scale track topology, which avoid collisions with each other. As a candidate invariant to be confirmed, shuttles in fast driving mode should not drive across construction sites to avoid derailing. To ensure this candidate invariant, warnings are installed at a certain distance in front of construction sites. Agents in this running example are the shuttles and backward steps can be performed by different shuttles on the track topology. However, only the steps of the single shuttle violating the speed limit at a construction site as well as (possibly) the steps of shuttles that forced the shuttle to navigate to that construction site are in fact relevant to the analysis. Any other steps (possibly of shuttles far away on the considered track topology) should not be considered during analysis. Hence, the novel pruning techniques are designed to focus our attention on the relevant steps of relevant agents only.

Invariant analysis for GTSs has been intensively studied. Besides the approach from [19], which is restricted to single-agent GTSs, earlier approaches for establishing invariants for GTSs lack a formal foundation such as [2] or are restricted to k-induction for $k = 1$ such as [7] or to syntactically limited nested conditions such as [6]. Moreover, tools such as GROOVE [12], HENSHIN [11], and AUTOGRAPH [18] can be used for invariant analysis if the considered GTSs induce small finite state spaces. However, there are some approaches that also support invariant analysis for infinite state spaces. For example, the tool AUGUR2 [1] abstracts GTSs by Petri nets but imposes restrictions on graph transformation rules thereby limiting expressiveness. Moreover, static analysis of programs for GTSs w.r.t. pre/post conditions has been developed in [16] and [17]. Finally, an approach for the verification of invariants (similar to k-induction) is considered in [24] where graphs are abstracted by single so-called shape graphs, which have limited expressiveness compared to the nested graph conditions used in this work.

[2] Intuitively, GTSs have no built-in support for different agents as opposed to other non-flat formalisms (such as e.g. process calculi) where a multi-agent system is lazily constructed using a parallel composition operation where interaction steps between agents are then resolved at runtime. For such different formalisms, causality is much easier to analyze but it is one of the many strengths of GTSs that agents can interact in complex patterns not restricted by the formalism at hand.

The representation of causality and the focus on causally connected steps during analysis is important in various domains. For example, for Petri nets where tokens can be understood as agents, event structures and causal/occurrence nets have been used extensively to represent causality in a given run (see e.g. [15,22,23]). Similarly, causality-based analysis can also be understood as cone of influence-based analysis [3] where events are derived to be insignificant when they are logically/spatially disconnected from considered events.

This paper is structured as follows. In Sect. 2, we recapitulate the technique of k-induction based on labeled transition systems. In Sect. 3, we recall preliminaries on graph transformation and introduce our running example. In Sect. 4, we present an abstraction of GTSs to symbolic states and steps. In Sect. 5, we extend existing notions capturing causality and compatibility among steps to the employed symbolic representation. In Sect. 6, we discuss the k-induction procedure with the novel pruning techniques relying on causality and fairness among multiple agents in the GTS. Finally, in Sect. 7, we close the paper with a conclusion and an outlook on future work. Further details are given in a technical report [21].

2 Labeled Transition Systems and k-Induction

A *Labeled Transition System (LTS)* $\mathcal{L} = (Q, Z : Q \twoheadrightarrow \mathbf{B}, L, R \subseteq Q \times L \times Q)$ consists of a set of states Q, a state predicate Z identifying start states in Q, a set L of step labels, and a binary step relation R on Q where each step has a step label from L. An LTS $\mathcal{L}$ represents a state space and induces paths $\tilde{\pi} \in \Pi(\mathcal{L})$ traversing through its states. We write $\mathcal{L}_1 \subseteq \mathcal{L}_2$ and $\mathcal{L}_1 \cup \mathcal{L}_2$ for their componentwise containment and union, respectively.

A state predicate $P : Q \twoheadrightarrow \mathbf{B}$ is an *invariant* of $\mathcal{L}$ when P is satisfied by all states reachable from start states. A *shortest violation* of an invariant is given by a path $\tilde{\pi}$ of length n traversing through states s_i when *(a)* $\tilde{\pi}$ starts in a start state and never revisits a start state (i.e., $Z(s_i)$ iff $i = 0$) and *(b)* $\tilde{\pi}$ ends in a violating state and never traverses another violating state (i.e., $\neg P(s_i)$ iff $i = n$).

The k-induction procedure attempts to decide whether a shortest violation for a candidate invariant P exists. For shortest violations, in iteration $0 \leq i \leq k$ the paths of length i that may be suffixes of shortest violations are generated. That is, in iteration $i = 0$, all paths of length 0 consisting only of states q satisfying $\neg P(q)$ are generated. In iterations $i > 0$, each path $\tilde{\pi}$ of length $i - 1$ starting in state q is extended to paths $\tilde{\pi}'$ of length i by prepending all backward steps $(q', a, q) \in R$ such that $P(q')$ is satisfied. The k-induction procedure *(a)* rejects the candidate invariant P when in some iteration a path starting in a start state is generated, *(b)* confirms the candidate invariant P when in some iteration no path is derived, and *(c)* terminates without definite judgement when in the last iteration $i = k$ some path is generated.

Pruning techniques restrict the set of generated paths in each iteration to a relevant subset and only the retained paths are then considered for the abortion criteria *(a)–(c)*. While the additional computation that is required for pruning can be costly, pruning can speed up the subsequent iterations by reducing the number of paths to be considered in the next iteration. More importantly, pruning may prevent the generation of paths of length k, which lead to an indefinite judgement. For example, when $A:Q \rightarrow \mathbf{B}$ is an *assumed invariant* (either established in an earlier application of the same or another technique or assumed without verification), all paths in which some state q satisfies $\neg A(q)$ are pruned as in [6,19] attempting to limit constructed paths to reachable states. Further pruning techniques introduced later on are designed specifically for the case of GTSs taking the content of states and the nature of steps among them into account.[3]

3 Graph Transformation and Running Example

Our approach generalizes to the setting of $\mathcal{M}$-adhesive categories and $\mathcal{M}$-adhesive transformation systems with nested application conditions as introduced in [10]. Nevertheless, to simplify our presentation, we consider the $\mathcal{M}$-adhesive category of typed directed graphs (short graphs) using the fixed type graph TG from Fig. 1a (see [8–10] for a detailed introduction). In visualizations of graphs such as Fig. 1b, types of nodes are indicated by their names (i.e., S_i and T_i are nodes of type Shuttle and Track) whereas we only use the type names for edges. We denote the empty graph by $\emptyset$, monomorphisms (monos) by $f:H \hookrightarrow H'$, and the initial morphism for a graph H by $\mathrm{i}(H) : \emptyset \rightarrow H$. Moreover, a graph is finite when it has finitely many nodes and edges and a set S of morphisms with common codomain X is jointly epimorphic, if morphisms $g, h : X \rightarrow Y$ are equal when $\forall f \in S.\ g \circ f = h \circ f$ holds.

In our running example, we consider an unbounded number of shuttles driving on a large-scale track topology where subsequent tracks are connected using *next* edges (see again TG in Fig. 1a and the example graph in Fig. 1b). Each shuttle either drives fast or slow (as marked using *fast* or *slow* loops). Shuttles approaching track-forks (i.e., a track with two successor tracks) decide non-deterministically between the two successor tracks. Certain track-forks consist of a regular successor track and an emergency exit successor track (marked using an *ee* loop) to be used only to avoid collisions with shuttles on the regular successor track. Construction sites may be located on tracks (marked using *cs* loops) and, to inform shuttles about construction sites ahead, warnings are installed four tracks ahead of them (marked using *warn* edges instead of *next* edges). To exclude the possibility of shuttles derailing, analysis should confirm

[3] The computational trade-off between pruning costs and costs for continued analysis of retained paths will play out differently for each example but, due to the usually exponential number of paths of a certain length, already the rather simple pruning technique based on assumed invariants was highly successful in [6,19] where it was also required to establish a definite judgement at all.

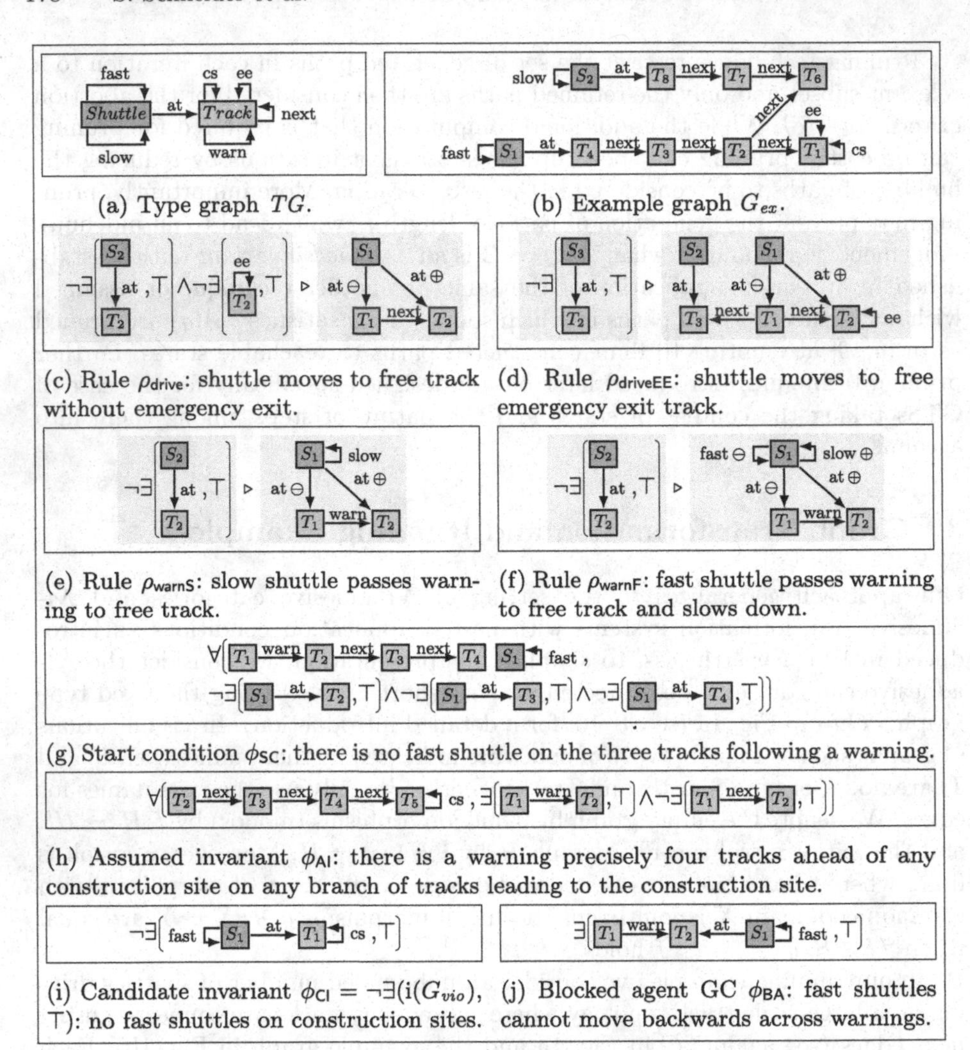

(a) Type graph TG.

(b) Example graph G_{ex}.

(c) Rule ρ_{drive}: shuttle moves to free track without emergency exit.

(d) Rule ρ_{driveEE}: shuttle moves to free emergency exit track.

(e) Rule ρ_{warnS}: slow shuttle passes warning to free track.

(f) Rule ρ_{warnF}: fast shuttle passes warning to free track and slows down.

(g) Start condition ϕ_{SC}: there is no fast shuttle on the three tracks following a warning.

(h) Assumed invariant ϕ_{AI}: there is a warning precisely four tracks ahead of any construction site on any branch of tracks leading to the construction site.

(i) Candidate invariant $\phi_{\mathsf{CI}} = \neg\exists(\mathrm{i}(G_{vio}), \top)$: no fast shuttles on construction sites.

(j) Blocked agent GC ϕ_{BA}: fast shuttles cannot move backwards across warnings.

Fig. 1. Running example.

the candidate invariant **P** stating that shuttles never drive fast on construction sites. Assumed invariants are used to rule out track topologies with undesired characteristics such as missing *warn* edges. We model this shuttle scenario using a GTS with rules featuring application conditions as well as assumed and candidate invariants all given by (nested) Graph Conditions (GCs). For this purpose, we now recall GCs and GTSs in our notation.

The graph logic GL from [10] allows for the specification of sets of graphs and monos using GCs. Intuitively, for a host graph G, a GC over a finite subgraph H of G given by a mono $m : H \hookrightarrow G$ states the presence (or absence) of graph elements in G based on m. In particular, the GC $\exists(f : H \hookrightarrow H', \phi')$ requires that m must be extendable to a match $m' : H' \hookrightarrow G$ of a larger subgraph H' where the nested sub-GC ϕ' restricts m'. The combination of propositional operators and the nesting of existential quantifications results in an expressiveness equivalent to first-order logic on graphs [5].

Definition 1 (Graph Conditions (GCs)). *If H is a finite graph, then ϕ is a graph condition (GC) over H, written $\phi \in \mathsf{GC}(H)$, if an item applies.*

- $\phi = \neg\phi'$ *and* $\phi' \in \mathsf{GC}(H)$.
- $\phi = \vee(\phi_1, \ldots, \phi_n)$ *and* $\{\phi_1, \ldots, \phi_n\} \subseteq \mathsf{GC}(H)$.
- $\phi = \exists(f : H \hookrightarrow H', \phi')$ *and* $\phi' \in \mathsf{GC}(H')$.

Note that the empty disjunction $\vee()$ serves as a base case not requiring the prior existence of GCs. We obtain the derived operators *false* $\bot$, *true* $\top$, *conjunction* $\wedge(\phi_1, \ldots, \phi_n)$, and *universal quantification* $\forall(f, \phi)$ in the expected way.

We now define the two satisfaction relations of GL capturing *(a)* when a mono $m : H \hookrightarrow G$ into a host graph G satisfies a GC over H and *(b)* when a graph G satisfies a GC over the empty graph $\emptyset$.

Definition 2 (Satisfaction of GCs). *A mono $m : H \hookrightarrow G$ satisfies a GC ϕ over H, written $m \models \phi$, if an item applies.*

- $\phi = \neg\phi'$ *and* $\neg(m \models \phi')$.
- $\phi = \vee(\phi_1, \ldots, \phi_n)$ *and* $\exists 1 \leq i \leq n.\ m \models \phi_i$.
- $\phi = \exists(f : H \hookrightarrow H', \phi')$ *and* $\exists m' : H' \hookrightarrow G.\ m' \circ f = m \wedge m' \models \phi'$.

A graph G satisfies a GC ϕ over the empty graph $\emptyset$, written $G \models \phi$, if the (unique) initial morphism $\mathrm{i}(G) : \emptyset \hookrightarrow G$ *satisfies ϕ.*

For our running example, *(a)* the GC ϕ_{AI} from Fig. 1h expresses the assumed invariant stating that there is always a warning preceding each construction site[4], *(b)* the GC ϕ_{CI} from Fig. 1i expresses the candidate invariant **P** stating that there is no fast shuttle at a track with a construction site, and *(c)* the GC ϕ_{SC} from Fig. 1g expresses that there is no fast shuttle already in the critical section between a warning and a construction site. Note that in visualizations of GCs, we represent monos $f : H \hookrightarrow H'$ in quantifications by only visualizing the smallest subgraph of H' containing $H' - f(H)$.

We rely on the operation shift from e.g. [10] for shifting a GC ϕ over a graph H across a mono $g : H \hookrightarrow H'$ resulting in a GC $\mathsf{shift}(g, \phi)$ over H'. The following fact states that GC shifting essentially expresses partial GC satisfaction checking for a morphism decomposition $f \circ g$.

[4] To ease the presentation, we omit further assumed invariants excluding graphs with duplicate *next* edges or tracks with more than two successor/predecessor tracks.

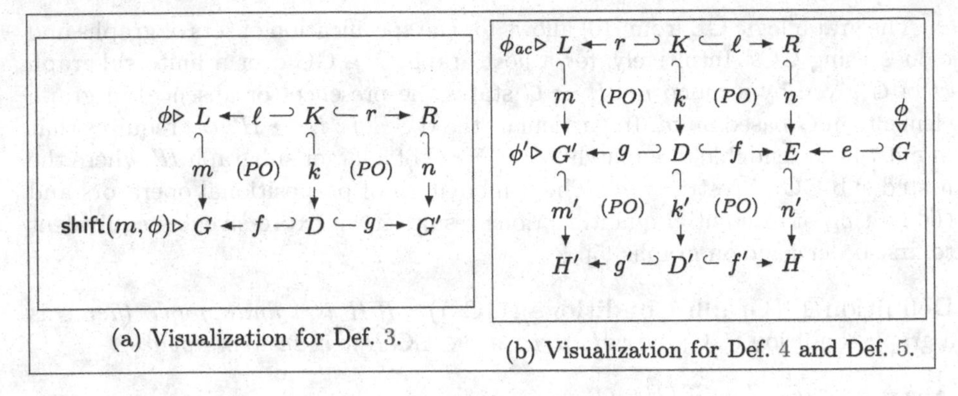

(a) Visualization for Def. 3.

(b) Visualization for Def. 4 and Def. 5.

Fig. 2. Visualizations for definitions.

Fact 1 (Operation shift [10]**).** $f \models \mathsf{shift}(g, \phi)$ *iff* $f \circ g \models \phi$

GTSs with multiple start graphs are now defined by specifying these start graphs using a GC over the empty graph. We employ the Double Pushout (DPO) approach to graph transformation with nested application conditions (see [8–10] for details) in which rules contain two morphisms $\ell : K \hookrightarrow L$ and $r : K \hookrightarrow R$ describing the removal of the elements in $L - \ell(K)$ and the addition of elements in $R - r(K)$ as well as a left-hand side (nested) application condition given by a GC over L to be satisfied by the match morphism.

Definition 3 (Graph Transformation System (GTS)). *A pair* $S = (\phi_0, P)$ *is a* graph transformation system (GTS)*, if* ϕ_0 *is a GC over the empty graph* $\emptyset$ *and* P *is a finite set of graph transformation rules (short rules) of the form* $\rho = (\ell : K \hookrightarrow L, r : K \hookrightarrow R, \phi)$ *where* L*,* K*, and* R *are finite and* ϕ *is a GC over* L*.*

If G*,* G' *are graphs,* $\sigma = (\rho, m : L \hookrightarrow G, n : R \hookrightarrow G')$ *is a step label containing a rule* $\rho = (\ell : K \hookrightarrow L, r : K \hookrightarrow R, \phi)$ *of* S*, a match* m*,*[5] *and a comatch* n*, the DPO diagram in Fig. 2a exists, and* $m \models \phi$*, then* $G \Rightarrow_\sigma G'$ *is a (GT) step of the LTS* $\mathcal{L}_{graphs}$ *induced by the GTS* S*. Also, the notion of derived rules* $\mathsf{drule}(\sigma) = (f, g, \mathsf{shift}(m, \phi))$ *captures the transformation span of the step and the instantiated application condition.*

For our running example, we employ the GTS $S = (\phi_{\mathsf{SC}} \wedge \phi_{\mathsf{CI}}, \{\rho_{\mathsf{drive}}, \rho_{\mathsf{driveEE}}, \rho_{\mathsf{warnS}}, \rho_{\mathsf{warnF}}\})$ using the GCs and rules from Fig. 1. For each rule, we use an integrated notation in which L, K, and R are given in a single graph where graph elements marked with $\ominus$ are from $L - \ell(K)$, graph elements marked with $\oplus$ are from $R - r(K)$, and where all other graph elements are in K. The application condition of each rule is given on the left side of the $\triangleright$ symbol. The rule ρ_{drive} states that a shuttle can advance to a next track T_2 when no other shuttle is on T_2 and when T_2 is not marked to be an emergency exit. The rule ρ_{driveEE} states

[5] Note that our approach extends to the usage of general match morphisms.

that a shuttle can advance to a next track T_2 marked to be an emergency exit when the regular successor track T_3 is occupied by another shuttle. The rule ρ_{warnS} states that a slow shuttle can advance to a next track T_2 passing by a warning when no other shuttle is on T_2. Finally, the rule ρ_{warnF} states that a fast shuttle can slow down and advance to a next track T_2 passing by a warning when no other shuttle is on T_2.

To accumulate the knowledge captured in application conditions and the candidate invariant over steps of a backward path, we employ the operation L from e.g. [10] for shifting a GC ϕ' over a graph R across a rule $\rho = (\ell : K \hookrightarrow L, r : K \hookrightarrow R, \phi)$ resulting in a GC $\mathsf{L}(\rho, \phi')$ over L. The following fact states that the operation L translates post-conditions of steps into equivalent pre-conditions.

Fact 2 (Operation L [10]). *$G \Rightarrow_{\rho,m,n} G'$ implies ($m \models \mathsf{L}(\rho, \phi')$ iff $n \models \phi'$).*

For our running example, we expect the k-induction procedure to confirm the candidate invariant ϕ_{CI} for $k \geq 4$ realizing that a fast shuttle at a construction site must have passed by a warning 4 steps earlier due to the assumed invariant ϕ_{AI}, which ensures that the shuttle drives slowly onto the construction site later on.[6] When applying the k-induction procedure, we start with the minimal graph G_{vio} representing a violation (see the graph used in ϕ_{CI} in Fig. 1i). To extend a given backward path from G to G_{vio} by prepending a backward step using a certain rule, we first extend G to a graph E by adding graph elements to be then able to apply the rule backwards to E (as discussed in more detail in the next section based on a symbolic representation of states and steps). Consider the graph G_{ex} in Fig. 1b, which can be reached using this iterative backward extension from G_{vio} by a path of length 5 (see Fig. 5a). Since the relevant shuttle S_1 has no further enabled *backward* step from G_{ex} according to the rules of the GTS (because fast shuttles cannot advance backwards over *warn* edges), any path leading to G_{ex} and any other path that varies by containing additional/fewer/differently ordered independent steps can be pruned (as discussed in more detail in Sect. 6). For example, the similar path (see Fig. 5b) where the shuttle S_2 has only been moved backwards to T_6 is pruned as well. Hence, with such additional pruning techniques, we mitigate the problem that the relevant shuttle S_1 does not move backwards in every backward step of every path. Instead, it is sufficient that S_1 is being moved backwards three times in some path. Still, all interleavings of backward steps must be generated (since, for arbitrary GTSs, it cannot be foreseen which interleaving results in a prunable path later on) but pruning one of these paths can result in the pruning of many further paths.

[6] The candidate invariant ϕ_{CI} could also be violated because *(a)* it is not satisfied by all start graphs (which is excluded since $\phi_{\mathsf{SC}} \wedge \phi_{\mathsf{CI}}$ captures the start graphs of the GTS), *(b)* a slow shuttle becomes a fast shuttle between a warning and a construction site (for which no rule exists in the GTS), and *(c)* a pair of a warning and a construction site could wrap a fast shuttle at runtime (for which no rule exists in the GTS).

4 Symbolic States and Steps

Following [19], concrete states of a GTS are given by graphs and its symbolic states are given by pairs (G, ϕ) of a graph G and a GC ϕ over G. A symbolic state (G, ϕ) represents all graphs H for which some $m : G \hookrightarrow H$ satisfies ϕ.

This symbolic representation extends to GT steps and symbolic steps and paths thereof. To obtain a backward step from a state (G, ϕ) (cf. Fig. 2b), *(i)* G is overlapped with the right-hand side graph R of some rule ρ where the overlapping consists of the comatch n of the backward step and the embedding morphism e and *(ii)* the GC ϕ and the application condition ϕ_{ac} of the rule are shifted to the resulting symbolic state (G', ϕ'). As discussed before, further graph elements are added using e as required for the k-induction procedure in which we start with (usually very small) graphs representing violations and then accumulate additional context *also* in terms of additional graph elements.

Definition 4 (Symbolic Step). *If (G', ϕ') and (G, ϕ) are symbolic states, $\rho = (\ell{:}K \hookrightarrow L, r{:}K \hookrightarrow R, \phi_{ac})$ is a rule, $\sigma = (\rho, m{:}L \hookrightarrow G', n{:}R \hookrightarrow E)$ is a step label, $G' \Rightarrow_{\sigma} E$ is a DPO step, $e{:}G \hookrightarrow E$ is a mono, e and n are jointly epimorphic, and $\phi' = \mathsf{L}(\mathsf{drule}(\sigma), \mathsf{shift}(e, \phi)) \wedge \mathsf{shift}(m, \phi_{ac})$, then $(G', \phi') \Rrightarrow_{\sigma,e} (G, \phi)$ is a symbolic step of the LTS $\mathcal{L}_{symb}$ induced by the GTS S (see Fig. 2b).*

To obtain concrete paths $\hat{\pi}$ represented by a symbolic path π, the implicit requirements given by the GCs in symbolic states and the incremental context extensions via monos e are resolved. This entails a forward propagation of additional graph elements resulting in a consistent perspective throughout all graphs traversed in $\hat{\pi}$. However, making these additional graph elements explicit may change satisfaction judgements for application conditions and assumed or candidate invariants implying that a symbolic path may represent no concrete path relevant in the context of k-induction or even no concrete path at all. Since some pruning techniques require that we are able to operate on the symbolic step relation, we only concretize symbolic paths using forward propagation that may represent concrete paths being shortest violations.

Definition 5 (Concretization of Symbolic Path). *A concrete path $\hat{\pi}$ is a concretization of a symbolic path π with first state (G', ϕ') for a mono $m' : G' \hookrightarrow H'$ satisfying ϕ', written $\hat{\pi} \in \mathsf{refine}(\pi, m')$, if an item applies.*

- *$\pi = (G', \phi')$ and $\hat{\pi} = H'$.*
- *$\pi = (G', \phi') \cdot \sigma \cdot e \cdot (G, \phi) \cdot \pi'$, $\sigma = (\rho, m, n)$, $\sigma' = (\rho, m' \circ m, n' \circ n)$, $H' \Rightarrow_{\sigma'} H$, $\hat{\pi}' \in \mathsf{refine}((G, \phi) \cdot \pi', n' \circ e)$, and $\hat{\pi} = H' \cdot \sigma' \cdot \hat{\pi}'$ (see Fig. 2b).*

The symbolic representation given by symbolic paths is complete in the sense of the following lemma stating that the concrete paths of a GTS correspond to the concretizations of all symbolic paths.

Lemma 1 (Full Coverage). $\Pi(\mathcal{L}_{graphs}) = \bigcup\{\mathsf{refine}(\pi, m') \mid \pi \in \Pi(\mathcal{L}_{symb})\}$

Proof (Sketch). By mutual inclusion of the sets and induction over the length of paths in both cases. Every concrete path of the GTS is represented by a symbolic

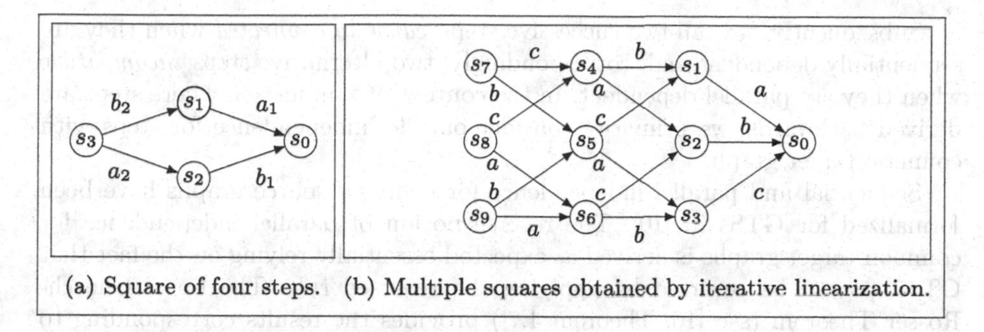

(a) Square of four steps. (b) Multiple squares obtained by iterative linearization.

Fig. 3. Linearization of parallel independent backward steps.

path where monos c are identities and GCs are $\top$. Every symbolic path only represents concrete paths of the GTS since the operation refine checks satisfaction of the GCs with the additional contexts accumulated via the e monos.

For our running example, the violating symbolic state used for the symbolic paths of length 0 during k-induction is $(G_{vio}, \top)$ (see Fig. 1i). Note that we implicitly rewrite symbolic states (G, ϕ) into symbolic states (G', ϕ') using the symbolic model generation technique from [20] to accumulate all positive requirements of G and ϕ in the graph G' and to store the remaining negative requirements (stating how G' cannot be extended) in ϕ'. Without this technique, k-induction would be limited to candidate invariants of the form $\neg\exists(\mathrm{i}(G), \top)$ and graph patterns required by positive application conditions would not be explicitly contained in the graph and could therefore not be overlapped leading to indefinite judgements in some cases. However, if multiple states (G', ϕ') are obtained using this rewriting, we would perform k-induction for each of these states separately. For the running example, $(G_{vio}, \top)$ is obtained by rewriting $(\emptyset, \neg\neg\exists(\mathrm{i}(G_{vio}), \top))$ using this technique.

5 Causality and Independence in GTS

According to [8, p. 8] in the context of GTSs, causal independence of rule applications allows for their execution in arbitrary order.

In the general setting of an LTS $\mathcal{L}$, considering Fig. 3a, *(a)* the two parallel steps with source s_3 (to s_1 and s_2), *(b)* the two parallel steps with target s_0 (from s_1 and s_2), *(c)* the two sequential steps traversing through s_1 (from s_3 and to s_0), or *(d)* the two sequential steps traversing through s_2 (from s_3 and to s_0) are independent iff the respective remaining two steps exist resulting in the square given in Fig. 3a (which we represent by $((s_3, b_2, s_1), (s_1, a_1, s_0), (s_3, a_2, s_2), (s_2, b_1, s_0)) \in \mathsf{SQ}(\mathcal{L})$) where, for $x \in \{a, b\}$, the labels x_1 and x_2 are required to be equivalent in an LTS specific sense in each case *(a)*–*(d)*. Clearly, in such an obtained square, each pair of sequential steps is sequentially independent and each pair of parallel steps (with common source/target) is parallel independent.

Subsequently, we call two successive steps *causally connected* when they are sequentially dependent and, correspondingly, two alternative steps *incompatible* when they are parallel dependent. In the context of k-induction where steps are derived backwards, we primarily consider parallel independence for steps with common target graph.

Sequential and parallel independence for common source graphs have been formalized for GTSs in [10]. The reverse notion of parallel independence for common target graphs is derived as expected essentially relying on the fact that GT steps can be reversed by applying the reversed rule. The Local Church-Rosser Theorem (see [10, Theorem 4.7]) provides the results corresponding to the discussion for LTSs from above. Technically, for concrete GT steps, for $x \in \{a, b\}$, two step labels σ_{x_1} and σ_{x_2} must then use the same rule and must match essentially the same graph elements.[7] Moreover, for symbolic steps, for $x \in \{a, b\}$, we additionally require that the step labels σ_{x_1}, e_{x_1} and σ_{x_2}, e_{x_2} state the same extensions using e_{x_1} and e_{x_2}.[8]

We use the operation $\mathsf{linearize}$ to obtain all linearizations for a given set of parallel steps with common target. For example, given the two parallel steps with target s_0 in Fig. 3a, $\mathsf{linearize}$ constructs the two further backward steps and the square given in Fig. 3a when the two steps are parallel independent and no further backward steps and no square otherwise. In general, for a given LTS $\mathcal{L}$ and a subset $\delta \subseteq Q \times L \times Q$ of size $n \geq 0$ of parallel steps of $\mathcal{L}$ with common target, $\mathsf{linearize}(\delta) = (\overline{sq}, \delta')$ generates the set $\overline{sq} \subseteq \mathsf{SQ}(\mathcal{L})$ of all squares that can be constructed by rearranging those parallel steps into corresponding sequences of length at most n and a set δ' of all generated steps including δ. More precisely, $\mathsf{linearize}$ iteratively constructs a square for each pair of distinct parallel independent steps with common target (considering for this the steps from δ and all steps generated already).[9] For the cases of $n = 0$ and $n = 1$ no additional steps are generated. For the cases of $n = 2$ and $n = 3$, Fig. 3a and Fig. 3b depict the maximal set δ' of resulting steps that may be generated when all pairs of distinct parallel steps are parallel independent throughout the application of $\mathsf{linearize}$ (note that we omit in Fig. 3b the differentiation between different a_i, b_i, and c_i steps for improved readability).

When some pair of steps with common target is not parallel independent (which is often the case), fewer squares and steps are generated.

---

[7] The considered GT steps must preserve the matched graph elements and thereby explain how one match is propagated over a GT step resulting in the other match.

[8] Similarly to the requirement on matches, which must essentially match the same graph elements, the extension monos must extend the graphs with the same graph elements up to the propagation along the considered symbolic steps.

[9] For concrete GT steps, we rely on [10, Theorem 4.7] to obtain a construction procedure for the operation $\mathsf{linearize}$. Also, this construction procedure extends to the case of symbolic steps as the additional GCs in symbolic states are extended precisely by the application conditions of the two involved rules in exchanged order only.

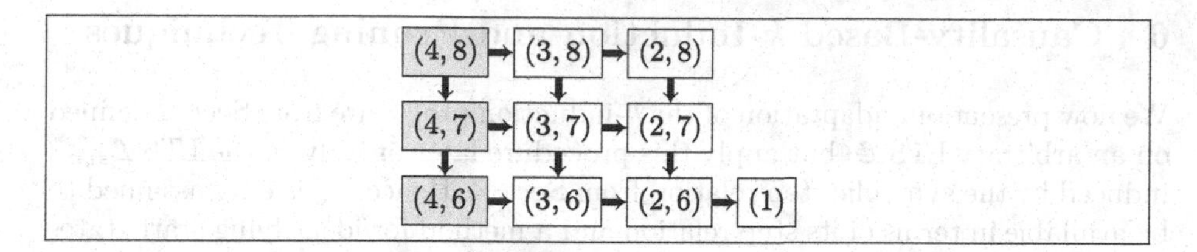

Fig. 4. Fragment of backward state space constructed for running example. We abbreviate symbolic states by only providing a tuple of the track numbers on which shuttles are located. See Fig. 1b or Fig. 5a for the graph part of state (4,8).

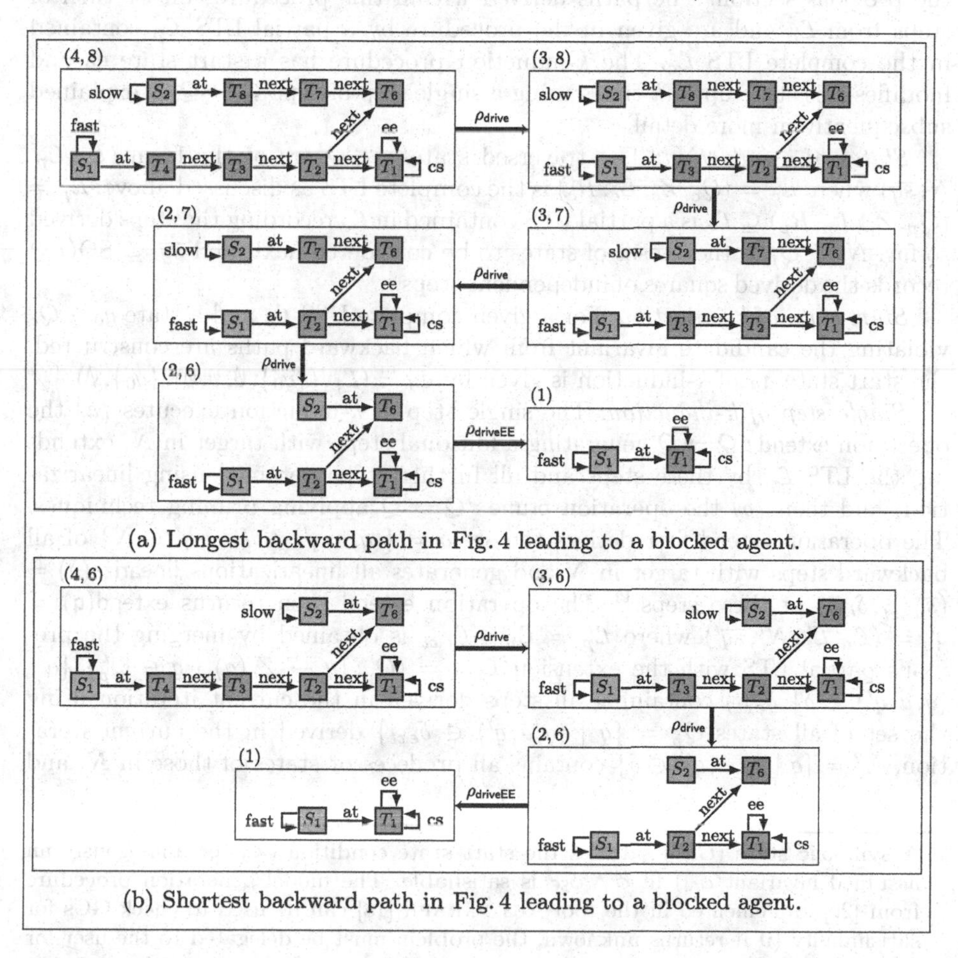

(a) Longest backward path in Fig. 4 leading to a blocked agent.

(b) Shortest backward path in Fig. 4 leading to a blocked agent.

Fig. 5. Two backward paths. We abbreviate symbolic states by providing a tuple of the track numbers on which shuttles are located as used in Fig. 4 and the graph part of the symbolic state.

6 Causality-Based k-Induction and Pruning Techniques

We now present our adaptation of the k-induction procedure from Sect. 2 defined on an arbitrary LTS $\mathcal{L}_c$ but apply this procedure later on only to the LTS $\mathcal{L}_{symb}$ induced by the symbolic step relation from Sect. 4. Hence, $\mathcal{L}_c$ is only assumed to be available in terms of its step relation and a method for identifying start states as well as states satisfying the assumed invariant.[10] Hereby, we rely on the notion of parallel independence of steps with common target and linearizations of such steps resulting in sequences of sequentially independent steps as introduced in the previous section. The paths derived within this procedure consist then of steps from $\mathcal{L}_c$ and are given in the procedure by a partial LTS $\mathcal{L}_p$ contained in the complete LTS $\mathcal{L}_c$. The k-induction procedure has a start state q_0 and modifies this state up to k times using a single step of type $\mathcal{Q} \twoheadrightarrow \mathcal{Q}$ as explained subsequently in more detail.

States of k-induction: The traversed states $\mathsf{q} \in \mathcal{Q}$ are of the form $(\mathcal{L}_c, \mathcal{L}_p, N, \overline{sq})$ where $\mathcal{L}_c = (Q_c, Z_c, L_c, R_c)$ is the complete LTS as discussed above, $\mathcal{L}_p = (Q_p, Z_p, L_p, R_p) \subseteq \mathcal{L}_c$ is a partial LTS contained in $\mathcal{L}_c$ recording the steps derived so far, $N \subseteq Q_p$ is the subset of states to be considered next, and $\overline{sq} \subseteq \mathsf{SQ}(\mathcal{L}_c)$ records the derived squares of independent steps.

Start state of k-induction: For a given complete LTS $\mathcal{L}_c$ and a state $q_0 \in Q_c$ violating the candidate invariant from which backward paths are constructed, the start state q_0 of k-induction is given by $\mathsf{q}_0 = (\mathcal{L}_c, (\{q_0\}, \emptyset, \emptyset, \emptyset), \{q_0\}, \emptyset)$.

Single step of k-induction: The single step of k-induction executes *(a)* the operation $\mathsf{extend} : \mathcal{Q} \twoheadrightarrow \mathcal{Q}$ generating additional steps with target in N, extending the LTS $\mathcal{L}_p$ by these steps and all further steps obtained using linearization, and then *(b)* the operation $\mathsf{prune} : \mathcal{Q} \twoheadrightarrow \mathcal{Q}$ applying pruning techniques. The operation extend first derives the set $\delta = \{(q, a, q') \in R_c \mid q' \in N\}$ of all backward steps with target in N and generates all linearizations $\mathsf{linearize}(\delta) = (\overline{sq}_{ext}, \delta_{ext})$ of these steps.[11] The operation extend then returns $\mathsf{extend}(\mathsf{q}) = \mathsf{q}' = (\mathcal{L}_c, \mathcal{L}'_p, N', \overline{sq}')$ where $\mathcal{L}'_p = \mathcal{L}_p \cup \mathcal{L}_{ext}$ is obtained by merging the previous partial LTS with the extension $\mathcal{L}_{ext} = (Q'_p, \{q \mapsto Z_c(q) \mid q \in Q'_p\}, \{a \mid (q, a, q') \in \delta\}, \delta_{ext})$ containing all steps derived in the current iteration using the set of all states $Q'_p = \{q \mid (q, a, q') \in \delta_{ext}\}$ derived in the current iteration, $N' = \{q \mid (q, a, q') \in \delta\}$ contains all predecessor states of those in N, and

[10] A symbolic state (G, ϕ) satisfies the start state condition ϕ_{SC} (or analogously an assumed invariant ϕ_{AI}) iff $\phi \wedge \phi_{\mathsf{SC}}$ is satisfiable. The model generation procedure from [20] implemented in the tool AutoGraph [18] can be used to check GCs for satisfiability (if it returns unknown, the problem must be delegated to the user for ϕ_{SC} and satisfiability may be assumed for ϕ_{AI}). If $\phi \wedge \neg\phi_{\mathsf{SC}}$ (or, analogously, $\phi \wedge \neg\phi_{\mathsf{AI}}$) is also satisfiable, not every concretization of paths $(G, \phi) \cdot \pi$ will be a violation. This source of overapproximation can be eliminated using splitting of states as in [19].

[11] Note that, due to linearization, R_p may already contain some of the steps derived here. By implicitly comparing steps derived here to those in R_p, we ensure to not derive isomorphic copies of steps. Also, two distinct parallel independent steps do not need to be linearized if not both steps are already contained in R_p.

$\overline{sq}' = \overline{sq} \cup \overline{sq}_{ext}$ additionally includes all squares derived in the current iteration. The operation prune is then applied to q' and discussed separately below.

For our running example, consider Fig. 4 where, initially in state (1), there is a single (fast) shuttle S_1 located on track T_1. A second shuttle S_2 is then added onto track T_6 in the first backward step to $(2, 6)$. When reaching the state $(4, 8)$, of which the graph part is given in Fig. 1b, the shuttles S_1 and S_2 moved backwards 3 and 2 times, respectively. See also Fig. 5a for this backward path from $(4, 8)$ to (1) and an additional backward path in Fig. 5b from $(4, 6)$ to (1), which is also included in abbreviated form in Fig. 4. The pruning of state $(4, 6)$ in Fig. 5b due to the blocked agent (given by the shuttle S_1) leads to the pruning of also the states $(3, 6)$, $(2, 6)$, and (1) in Fig. 5b and consequently also the path in Fig. 5a.

Termination condition of k-induction: The k-induction procedure applies the single step up to k times on the start state q_0. When a state is derived with $N = \emptyset$, the procedure concludes satisfaction of the candidate invariant. When a state is derived with Z_p mapping some state q to $\top$, the procedure concludes non-satisfaction of the candidate invariant and returns $(\mathcal{L}_p, q)$ as a counterexample. When the single step has been applied k times and none of the previous two cases applies, the procedure returns an indefinite judgement.

GTS-specific pruning: For the GTS setting where $\mathcal{L}_c = \mathcal{L}_{symb}$ as discussed above, we now present five pruning techniques (where the first two have been used already in [6,19]), which are used to remove certain states (and all steps depending on these states) recorded in the partial LTS $\mathcal{L}'_p$.

For *assumed invariant pruning*, we remove all states not satisfying the assumed invariant ϕ_{AI} as in prior work on GTS k-induction. For our running example, when moving the shuttle S_1 backwards from $(4, 6)$, a *next* edge is added leading to track T_4, which is forbidden by the assumed invariant ϕ_{AI} from Fig. 1h. Hence, this backward step of that shuttle is pruned.

For *realizability pruning*, we first determine states q that are identified to be start states via $Z_p(q) = \top$. Since each such state q represents a violating path leading to the refutation of the candidate invariant at the end of the iteration, we attempt to exclude false positives where each symbolic path π in $\mathcal{L}_p$ from q to the violating state q_0 cannot be concretized to a GTS path according to Definition 5. For this purpose, for $q = (G, \phi)$, we use the model generation procedure from [20] to generate extensions $m : G \hookrightarrow G'$ satisfying $\phi \wedge \phi_{\mathsf{SC}} \wedge \phi_{\mathsf{AI}}$. We then attempt to concretize some symbolic path π from q to q_0 to a concrete path $\hat{\pi}$ using m. If some $\hat{\pi}$ is obtained representing a shortest violation, the k-induction procedure terminates after this iteration refuting the candidate invariant. If the model generation procedure does not terminate, q may be a false positive and the k-induction procedure terminates with an indefinite judgement. However, if both cases do not apply, q is removed from $\mathcal{L}'_p$.

Certainly, *any* derived state q may not allow for a concretization along the same lines. However, not checking each such state for realizability along the same lines may only lead to indefinite judgements and there is a trade off between the cost for realizability pruning and the cost of exponentially more backward extensions leading to q to be generated and analyzed.

For *causality pruning*, a state q' is pruned when there is some symbolic backward step $q' \Rightarrow_{(\rho,m,n),e} q$ where n and e have non-overlapping images. We thereby ensure that the number of weakly connected components[12] of the graph under transformation does not increase over backward steps. For our running example, we prune states where further shuttles are added that are structurally not connected to the subgraph originating from the start state. Note that further shuttles can still be included as for the graph in Fig. 1b where the shuttle S_2 has been added according to the rule ρ_{driveEE} used in the first backward step.

For *evolution pruning*, a state q is pruned when it contains an agent (given in our running example by shuttles) for which permanent blockage is detected. Note that, as explained in Sect. 1, the inability of some agent to partake in a backward step does not preclude the ability of some other agent to partake in a backward step. Hence, when not removing such states, irrelevant steps of additional agents may prolong analysis or even prevent definite judgements. Also note that an agent is in general allowed to be blocked forever when it reaches its local configuration in a start graph of the GTS allowing other agents to perform backward steps to jointly reach a start graph. Since GTSs are Turing complete, no precise identification of such agents can be achieved and, to preclude the derivation of incorrect judgements, we must underapproximate the set of such agents. Technically, we attempt to identify all agents in states q that will unexpectedly never again be able to partake in a backward step using an additional blocked agent GC ϕ_{BA}. Such a blocked agent GC is (a finite disjunction of GCs) of the form $\exists(\mathrm{i}(H), \top)$ where H represents a minimal pattern containing a blocked agent. For our running example, see Fig. 1j for the GC ϕ_{BA} capturing a fast shuttle (i.e., an agent) that is blocked by not being able to move backwards across a *warn* edge. To maintain soundness of k-induction, we can verify the blocked agent GC ϕ_{BA} by checking that there is no symbolic backward step from $(H, \top)$ preventing that any further backward steps from q can reach a state where the matched agent can partake in a backward step. A state $q = (G, \phi)$ is then pruned using the blocked agent GC ϕ_{BA} when $\exists(\mathrm{i}(G), \phi) \wedge \phi_{\mathsf{AI}} \wedge \phi_{\mathsf{BA}}$ is satisfiable. For our running example, the shuttle S_1 is blocked according to the GC ϕ_{BA} in the states $(4, 6)$, $(4, 7)$, and $(4, 8)$ (marked blue in Fig. 4), which are therefore pruned.

For *evolution-dependency pruning*, we extend the state-based evolution pruning to a step- and square-based pruning technique propagating the information about blocked agents forward across steps. In particular, given a step (q', a, q) where q' was pruned (due to a blocked agent), q is also pruned unless there is a backward step (q'', b, q) to a non-pruned state q'' that is parallel dependent to (q', a, q). The step (q'', b, q) then potentially represents an alternative backward path not leading to a blocked agent.[13] However, only relying on the notion of

[12] Two nodes n_1 and n_2 of a graph G are in a common weakly connected component (given by a set of nodes of G) of G iff there is a sequence of the edges of G from n_1 to n_2 where edges may be traversed in either direction.

[13] Parallel independent backward steps are always performed by different agents.

parallel independence considering steps from a global perspective and not tracking which agents actually participated in the two backward steps can lead to an underapproximation of the steps that can be pruned potentially leading to avoidable indefinite returned judgements.[14] That is, backward steps of two distinct agents can be parallel dependent, which would then not allow to propagate the knowledge of one of them being blocked forwards. Constructing explicitly the squares in our backward state space generation procedure is essential for dissecting alternative backward steps. The forward propagation of prunability thereby allows to prune states and hence also all other paths traversing through these additionally pruned states where different step interleavings (of other agents) are executed (hence assuming that the blocked agent would be treated unfairly in all these other paths).

The usage of squares in k-induction supports evolution-dependency pruning since pruning a state also prunes all paths traversing through it, which would not be the case when we would construct a set of (disconnected) backward sequences or a tree (or forest) of backward steps. Moreover, minimizing the size of the state space representation using squares reduces the number of states for which blocked agents must be detected and from which evolution-dependency pruning must be performed. Also, when only constructing backward sequences instead, there would e.g. in our running example be a backward path not moving the initially given shuttle S_1 backwards to a situation where that shuttle would be blocked. Hence, employing a directed acyclic graph given by the square-based compressed backward state space, we can easily detect states occurring in different backward paths and thereby do not need to treat fairness among different agents beyond generating the backward state space using breadth-first search.

For our running example, the pruning of the state $(4, 6)$ and the non-existence of a backward step parallel dependent to the step from $(4, 6)$ to $(3, 6)$ leads to the pruning of the state $(3, 6)$ as well. Analogously, the states $(2, 6)$ and then (1) are also pruned leaving an empty state space, which leads to termination and candidate invariant confirmation at the end of the iteration.

Finally, we state that the presented k-induction procedure is sound and at least as complete as the previous variants from [6,19].

Theorem 1 (Soundness of k-Induction). *For a given GTS S, a candidate invariant ϕ_{CI}, an assumed invariant ϕ_{AI}, and a blocked agent GC ϕ_{BA}, the k-induction procedure confirms/refutes ϕ_{CI} only if ϕ_{CI} is an invariant/is no invariant. Also, it returns such a definite judgement whenever the k-induction procedure from [6, 19] without the novel pruning techniques and the use of causality and independence did.*

Proof (Sketch). Extending [6,19], we only need to ensure that the *novel* pruning techniques never prune states/paths that would otherwise be extended to shortest violations (the pre-existing assumed invariant pruning and realizability

[14] This pruning technique can be refined by attributing agents to steps to then determine prunable states with greater precision complicating forward propagation.

pruning do not need to be reexamined here). Causality pruning only removes steps where a disconnected agent is introduced: these steps can never help in gathering knowledge about the past of the actors involved in the violation and, moreover, the inclusion of such disconnected agents can always be delayed to later steps where they are then connected to a part of the current graph. The validity of the blocked agent GC ϕ_{BA} ensures that evolution pruning only prunes states containing an agent permanently blocked precluding the reachability of a start graph of the GTS. Evolution-dependency pruning then only prunes states/paths from which that agent unavoidably reaches such a blocking situation lacking alternative backward steps.

7 Conclusion and Future Work

We extended the k-induction procedure from [6,19] to support the verification of state invariants also for multi-agent GTSs. The presented extension relies on novel pruning techniques determining generated backward paths that cannot be extended to paths capturing a violation of the candidate invariant. It only returns sound judgements on candidate invariants, succeeds when the prior versions in [6,19] did, and succeeds for additional multi-agent GTSs.

In the future, we will extend our approach to Probabilistic Timed Graph Transformation Systems (PTGTSs) [13] in which dependencies among agents are also induced by the use of clocks (as in timed automata). This additional coupling among agents will complicate our analysis but will also reduce the number of possible backward paths to be constructed. Moreover, we will extend our prior implementations on k-induction to the presented approach and will evaluate the expected performance gain when restricting backward steps to a fixed underlying static topology fragment as in [14].

References

1. Augur 2. Universität Duisburg-Essen (2008). http://www.ti.inf.unidue.de/en/research/tools/augur2
2. Becker, B., Giese, H.: On safe service-oriented real-time coordination for autonomous vehicles. In: 11th IEEE International Symposium on Object-Oriented Real-Time Distributed Computing (ISORC 2008), 5–7 May 2008, Orlando, Florida, USA. pp. 203–210. IEEE Computer Society (2008). ISBN: 978-0-7695-3132-8. https://doi.org/10.1109/ISORC.2008.13, http://ieeexplore.ieee.org/xpl/mostRecentIssue.jsp?punumber=4519543
3. Berezin, S., Campos, S., Clarke, E.M.: Compositional reasoning in model checking. In: de Roever, W.-P., Langmaack, H., Pnueli, A. (eds.) COMPOS 1997. LNCS, vol. 1536, pp. 81–102. Springer, Heidelberg (1998). https://doi.org/10.1007/3-540-49213-5_4
4. Clarke, E., Grumberg, O., Jha, S., Lu, Y., Veith, H.: Counterexample-guided abstraction refinement. In: Emerson, E.A., Sistla, A.P. (eds.) CAV 2000. LNCS, vol. 1855, pp. 154–169. Springer, Heidelberg (2000). https://doi.org/10.1007/10722167_15

5. Courcelle, C.: The expression of graph properties and graph transformations in monadic second-order logic. In: Rozenberg, G. (ed.) Handbook of Graph Grammars and Computing by Graph Transformations, Volume 1: Foundations. World Scientific, pp. 313–400 (1997). ISBN: 9810228848
6. Dyck, J.: Verification of graph transformation systems with k-inductive invariants. Ph.D. thesis. University of Potsdam, Hasso Plattner Institute, Potsdam, Germany (2020). https://doi.org/10.25932/publishup-44274
7. Dyck, J., Giese, H.: k-inductive invariant checking for graph transformation systems. In: de Lara, J., Plump, D. (eds.) ICGT 2017. LNCS, vol. 10373, pp. 142–158. Springer, Cham (2017). https://doi.org/10.1007/978-3-319-61470-0_9
8. Ehrig, H., Ehrig, K., Prange, U., Taentzer, G.: Fundamentals of Algebraic Graph Transformation. Springer-Verlag, Berlin (2006). https://doi.org/10.1007/3-540-31188-2
9. Ehrig, H., Ermel, C., Golas, U., Hermann, F.: Graph and model transformation - general framework and applications. Monogr. Theoret. Comput. Sci. An EATCS Series. Springer Berlin, Heidelberg (2015). ISBN: 978-3-662-47979-7. https://doi.org/10.1007/978-3-662-47980-3
10. Ehrig, H., Golas, U., Habel, A., Lambers, L., Orejas. F.: $\mathcal{M}$-adhesive transformation systems with nested application conditions. Part 1: parallelism, concurrency and amalgamation. Math. Struct. Comput. Sci. **24**(4) (2014). https://doi.org/10.1017/S0960129512000357
11. EMF Henshin. The Eclipse Foundation (2013). http://www.eclipse.org/modeling/emft/henshin
12. Graphs for Object-Oriented Verification (GROOVE). University of Twente (2011). http://groove.cs.utwente.nl
13. Maximova, M., Giese, H., Krause, C.: Probabilistic timed graph transformation systems. J. Log. Algebr. Meth. Program. **101**, 110–131 (2018). https://doi.org/10.1016/j.jlamp.2018.09.003
14. Maximova, M., Schneider, S., Giese, H.: Compositional analysis of probabilistic timed graph transformation systems. In: FASE 2021. LNCS, vol. 12649, pp. 196–217. Springer, Cham (2021). https://doi.org/10.1007/978-3-030-71500-7_10
15. Nielsen, M., Plotkin, G.D., Winskel, G.: Petri nets, event structures and domains, Part I. Theor. Comput. Sci. **13**, 85–108 (1981). https://doi.org/10.1016/0304-3975(81)90112-2
16. Pennemann, K.-H.: Development of correct graph transformation systems. URN: urn:nbn:de:gbv:715-oops-9483. Ph.D. thesis. University of Oldenburg, Germany (2009). http://oops.uni-oldenburg.de/884/
17. Poskitt, C.M., Plump, D.: Verifying monadic second-order properties of graph programs. In: Giese, H., König, B. (eds.) ICGT 2014. LNCS, vol. 8571, pp. 33–48. Springer, Cham (2014). https://doi.org/10.1007/978-3-319-09108-2_3
18. Schneider, S.: AutoGraph. https://github.com/schneider-sven/AutoGraph
19. Schneider, S., Dyck, J., Giese, H.: Formal verification of invariants for attributed graph transformation systems based on nested attributed graph conditions. In: Gadducci, F., Kehrer, T. (eds.) ICGT 2020. LNCS, vol. 12150, pp. 257–275. Springer, Cham (2020). https://doi.org/10.1007/978-3-030-51372-6_15
20. Schneider, S., Lambers, L., Orejas, F.: Automated reasoning for attributed graph properties. Int. J. Softw. Tools Technol. Transfer **20**(6), 705–737 (2018). https://doi.org/10.1007/s10009-018-0496-3
21. Schneider, S., Maximova, M., Giese, H.: Invariant analysis for multi-agent graph transformation systems using k-induction. Tech. rep. 143. Hasso Plattner Institute, University of Potsdam (2022)

22. Smith, E.: On net systems generated by process foldings. In: Rozenberg, G. (ed.) ICATPN 1990. LNCS, vol. 524, pp. 253–276. Springer, Heidelberg (1991). https://doi.org/10.1007/BFb0019978
23. Smith, E.: On the border of causality: contact and confusion. Theor. Comput. Sci. **153**(1&2), 245–270 (1996). https://doi.org/10.1016/0304-3975(95)00123-9
24. Steenken, D.: Verification of infinite-state graph transformation systems via abstraction. Ph.D. thesis. University of Paderborn (2015). http://nbn-resolving.de/urn:nbn:de:hbz:466:2-15768

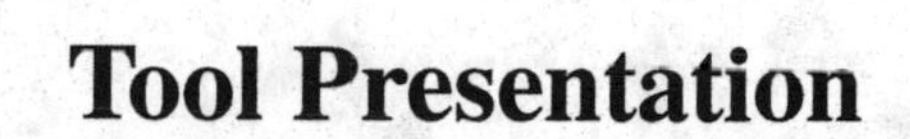

Tool Presentation

Tool Support for Functional Graph Rewriting with Persistent Data Structures - GrapeVine

Jens H. Weber(✉)

University of Victoria, Victoria, BC, Canada
jens@acm.org

Abstract. Existing graph transformation (GT) tools treat graphs as ephemeral data structures, i.e., the successful application of a GT rule to a graph G rewrites that graph to produce a modified graph G'. The original graph G is lost during that update. In contrast to ephemeral data structures, persistent data structures preserve access to all previous versions when data is modified and *fully persistent* data structures even allow all previous versions to be modified. In earlier work, we introduced the *Graph Rewriting and Persistence Engine* Grape as a tool for specifying and executing transformations on large-scale graphs and integrated it with a computational notebook platform (GrapePress). While the term "persistence" has been in the tool's acronym from the start, it was chosen to indicate that graphs were maintained in a database with transactional support. Until now, Grape (and GrapePress) treated graphs as ephemeral data structures, i.e., previous graph versions were not retained upon modification. This paper presents a major revision of the tool (called GrapeVine) to support *functional* graph rewriting based on a fully persistent data structure.

Keywords: Graph transformations · Grape · GrapeVine · persistent data structures · tools · graph processes · computational notebook · tools

1 Introduction

Graphs and graph transformation (GT) systems have been applied to a variety of problems with industrial relevance [12,17–19]. GT tools play an important role in making this feasible. A considerable number of such tools have been developed over the last decades [1,15]. Some of these tools aim to closely implement results from GT theory, e.g., results from algebraic graph rewriting [16,20]). Other tools are more concerned with incorporating a notion of GTs with practical applications, e.g., software design [10] and verification [13]. However, one aspect where all current GT tools fall short of reflecting mathematical theory is that they maintain graphs in *ephemeral* data structures, i.e., graphs are maintained in stateful objects that are updated when GT rules are applied and previous version of graphs

N. Behr and D. Strüber (Eds.): ICGT 2022, LNCS 13349, pp. 195–206, 2022.
https://doi.org/10.1007/978-3-031-09843-7_11

are lost. The use of ephemeral data structures for maintaining graphs is particularly limiting for example when an application needs to "look back" into graph derivation histories [8] or if it needs to perform some type of reasoning about graph processes [4]. Significant extensions of GT theory, such as the definition of transactional graph transformation systems as processes [2], cannot readily be implemented with current GT tools. From a more mundane perspective, it is well known from the programming languages and software engineering domains that stateless computation avoids complexity and is easier to understand and test.

All data in functional programming is immutable. Under the hood, this is implemented with *persistent* data structures which, in contrast to ephemeral data structures, retain all previous versions upon update [9]. A data structure is *partially persistent* if all versions can be accessed but only the latest version can be modified. Data structures where all versions can be modified are referred to as *fully persistent*.

In this paper, we present `GrapeVine`, a GT tool that maintains graphs in a fully persistent data structure and treats graphs as immutable data objects. `GrapeVine` is based on our earlier work on `Grape` (the *Graph Rewriting And Persistence Engine*) [21] and its integration with computational notebook technology (`GrapePress`) [22]. Like other tools, `Grape` (and `GrapePress`) uses an ephemeral data structure to store graphs, i.e., graphs are destructively rewritten "in-place". (The term "persistence" in the tool's name was originally chosen to indicate that data structure was persisted in a database with transactional support.) While `GrapeVine` is based on these earlier works, it constitutes a complete reimplementation of its core engine and data structures to achieve support for truly *functional* graph rewriting.

The rest of this paper is structured as follows: We discuss related work in the following section. Section 3 provides an overview of `GrapeVine` and its new persistent data structure. Section 4 comments on the tool demonstration provided in the appendix and Sect. 5 offers concluding remarks and an outlook to future work.

2 Related Work

Our current work on `GrapeVine` can be seen as a major revision of our earlier work on `Grape` [21] and `GrapePress` [22], with the fundamental difference that `GrapeVine` uses a "functional programming" paradigm and treats graphs as immutable objects. Historically, the name `Grape` referred to the GT engine and the domain-specific language for defining and controlling GTs, while the name `GrapePress` was used to refer to the computational notebook platform that integrates with `Grape`. While `GrapeVine` inherited the same architecture (i.e., its language and engine can be used without the computational notebook platform), we use a single name for the tool, going forward. Similarly, we will use the name `GrapePress` in this paper to also include the `Grape` engine.

`GrapeVine` has the ability to concurrently explore and compare many (all) possible derivations of arbitrarily many graphs. This feature is related to functionality provided by the `GROOVE` tool, which provides support for state-space exploration of GT systems [13]. However, GROOVE does not persist different versions of a graph but only fingerprints for the purpose of detecting collisions during model-checking (which is the primary function of graph process exploration in GROOVE).

The functionality provided by the fully persistent data structure implemented in GrapeVine resembles that of model versioning and indexing system like *Hawk* [3]. However, those approaches are primarily used for model management but they are not integrated with the computational model of the transformation tool. In contrast, GrapeVine realizes a truly functional computation model based on its fully-persistent graph data structure.

The data structure implemented in GrapeVine was inspired by the theoretical concept of a *graph process*, as defined by Corradini et al. [4], which is an a graph of graph transformation occurrences. Moreover, transaction-handling in GrapeVine is based on graph processes, as proposed by Baldan et al. [2].

3 GrapeVine Concepts

3.1 Overview

A core objective in the design of GrapeVine (and its predecessor GrapePress) has been to make it easy to integrate GT programming with common software engineering tools and processes. As such, GrapeVine was developed as an internal domain-specific language (DSL) to a general purpose programming language (Clojure). Regular tools like text editors, IDEs and configuration management systems can be used, without the need to install and learn a particular graphical tool for GT development and execution [21]. GrapeVine comes integrated with an optional computational notebook user interface, which provides graphical visualization of rules, graphs and graph history [22]. GrapeVine computational notebooks are an excellent way to quickly explore and document GT-based computations and systems. Notebook worksheets can be shared as "executable" papers, but they can also be saved as regular program code, to be used as part of larger software applications.

Under the hood, GrapeVine uses the Neo4J graph database management system, which makes it highly scalable to ultra-large graphs. GrapeVine is available as a Docker image and installs with a single command.

Graph Model. The tool uses directed, attributed, node- and edge-labeled (*danel*) graphs. Graphs do not need to be typed but it is possible to define constraints on graphs, which are enforced whenever rules are applied. GrapeVine comes with a number of predefined constraint types (e.g., for allowed node and edge types, cardinalities, and uniqueness of attribute values) but also allows users to define complex, user-defined constraints based on Oreja et al.'s logic of graph constraints [11].

Graph Transformation Rules. In the previous version of the tool (GrapePress), each GT rule could be defined with either single-pushout (SPO) or double-pushout (DPO) semantics [6]. The main difference concerned the way how any "dangling" edges would be handled during the execution of rules that delete nodes. SPO rules would delete such dangling edges, while the application of a DPO rule is simply not permitted in a context that would cause dangling edges to arise. GrapeVine has removed this choice and adopted SPO semantics throughout. We found that the complexities caused by providing this option (both in terms of tool implementation as well as in terms of reasoning about such "mixed" GT systems) are not justified, since the behaviour of DPO rules can be simulated in the SPO approach, given the availability of additional mechanisms, like negative application conditions.

The user can still choose between homomorphic and isomorphic matching semantics for each rule. Rules can be equipped with applications conditions (i.e., conditions that must be true for a rule to be applicable) [5] and negative application conditions (i.e., conditions that prevent rule application) [7].

Rules can be parameterized and rule parameters can be used to define or restrict the labels and attributes of graph elements. The previous version of the tool (GrapePress) treated attributes as variables and therefore also provided an *assignment* operator to change their values. The concept of variables no longer applies to the functional computation paradigm implemented in GrapeVine. Since graphs are immutable, attributes also have that property. This means that GTs that seek to "modify" attributes of a graph element (node or edge) need to replace that graph element with a new graph element of the same type, while copying the unchanged attributes and (re)defining the "changed" ones.

Control Structures. Since GrapeVine rules are defined with an internal DSL to a general purpose programming language (Clojure), the control structures of the host language are available for programming with GTs. The previous version of the tool (GrapePress) also provided a set of dedicated control structures (e.g., atomic blocks, loops, non-deterministic choice), which could be used to define composite transactions with ACID properties and backtracking for dealing with non-determinism during rule applications [21]. Given the paradigm shift to a purely functional model of computation in GrapeVine, these control structures have been revised completely. All GrapeVine control structures are defined as functions on graph sets rather than graphs. This allows the definition of deterministic operators for rule application. (Non-deterministic operators are also still available.) GrapeVine therefore no longer needs complex backtracking mechanisms to "undo" unsuccessful derivations. Similarly, there is no longer a need for a dedicated transaction manager to achieve ACID properties for programmed transformation units, since unsuccessful (partial) execution of such units can simply be "forgotten" [2].

Graph Queries. Graph queries are pattern-based matches for the purpose of returning data from the graph or testing the existence of conditions. The previous version of the tool (GrapePress) did not have a dedicated syntactic form for graph queries; graph queries were merely defined as rules that do not alter the graph. Instead, GrapeVine has a dedicated syntax for defining graph queries and enforces their read-only semantics. This is important because, in contrast to rule applications, the application of queries does not generate a new graph.

3.2 A Fully-Persistent Data Structure for Functional Graph Rewriting

There are several choices that can be made when designing a fully persistent data structure. Of course, there is always the naïve approach of copying a data item upon modification. However, that approach makes sense only for relatively small data items. Since in our application, each occurrence of a GT creates a new graph, one choice is to record only the differences from the "previous" graph in our data structure. We haven chosen that option and Fig. 1 presents the model that was implemented for this purpose.

Nodes of type *Graph* reference the graph elements (nodes and edges) that were created or deleted when the "last" transformation occurred. We also record which graph elements where read (preserved) when a transformation occurred. While the latter is not needed for implementing a versioned data structure, we capture this data for the purpose of using the tool for reasoning about the provenance of applied transformation rules.

Graphs are uniquely identified by a globally unique hash code (*id*). Graphs are arranged in a partial order, which is induced by the *previous* relationship. Computing the elements of a given graph G simply requires retrieving all elements that have been created by any graph in the history of G minus those elements that have been deleted in G's history. This query can efficiently be executed in the Neo4J graph database. The Neo4J query language is well understandable and we provide it below instead of creating our own mathematical formulation. The query looks up a graph (g) with the given id and collects all graphs gs in its history by repeatedly traversing *previous* edges. The query then collects all graph elements e that have a *created* edge from any node in gs without also having a *deleted* edge from any node in gs.

```
MATCH(g:Graph{id:".."})-[:previous*0..]->(gs)-[:created]->(e)
WHERE NOT (e)<-[:deleted]-(gs) RETURN e
```

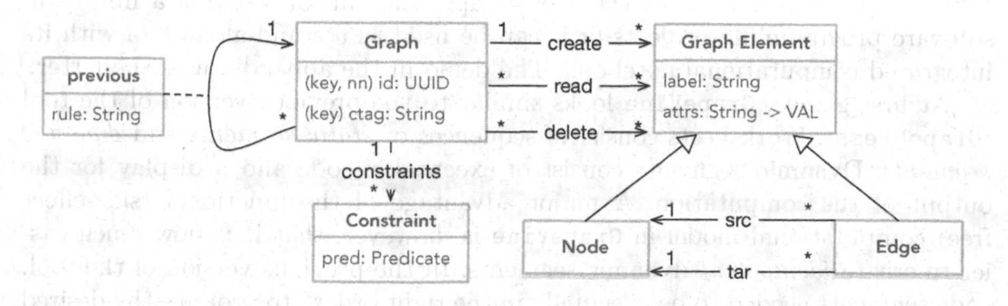

Fig. 1. Meta-model for persistent graph data structure in GrapeVine

Now that we have described the design chosen for implementing the persistent data structure for graphs in `GrapeVine`, we briefly want to comment on a disadvantage of the chosen approach, i.e., `GrapeVine` graphs are no longer represented "varbatim" in the Neo4J graph database. While the previous version of the tool (`GrapePress`) was able to operate on any *any* graph in a Neo4J database, independently of its origin, `GrapeVine` now requires graphs to conform to the meta-model in Fig. 1. For example, we note that `GrapeVine` now represents graph edges as Neo4J graph nodes, to allow incoming provenance edges from *Graph* nodes. (Neo4J does not support hypergraphs.)

We could have represented `GrapeVine` edges directly as edges in the Neo4J database if we replaced the provenance edges (*create, delete, read*) by "foreign key" attributes (*createdBy, deletedBy, readBy*) on edges. (Neo4J allows attributes on edges.) These attributes could store the IDs of the corresponding graphs (that created, deleted or read the edge), respectively. However, such a design would be inefficient as each graph lookup would then involve value-based "join" operations on potentially large sets of elements. Moreover, the benefit of such a design would be small, as other external tools accessing the graph database directly would still need to use a projection to assemble a concrete graph from its history. The only alternative that would avoid the need for such a projection would be the naïve approach of copying the graph upon each change, which does not appear scalable.

As mentioned in the previous section, `GrapeVine` no longer requires a sophisticated transaction manager like the one used on the previous (stateful) version of the tool (`GrapePress`). Rather, `GrapeVine` implements transactions based on the notion of graph processes. [2] Unsuccessful or incomplete executions of composite GT programs can simply be "forgotten". Of course, from a practical point of view, these graphs would still fill up the database over time. `GrapeVine` therefore provides a mechanism to purge them. Any graph of interest can be "*committed*" to the database by tagging it with a *commit tag* (`ctag`) in Fig. 1. Conversely, `GrapeVine` offers a *rollback* operation that deletes all graphs that are not in the history of any committed graph.

4 A Taste of Interacting with `GrapeVine`

Like the previous version of the tool, `GrapeVine` can be used as a library in software programming projects or it can be used as a stand-alone tool with its integrated computational notebook. The demo in the appendix uses the latter.

At first glance, `GrapeVine` looks similar to the previous version of the tool (`GrapePress`). Worksheets consist of sequences of *static segments* and *dynamic segments.* Dynamic segments consist of executable code and a display for the output of the computation. A major advantage of the functional (side-effect free) computational model in `GrapeVine` is, however, that it is now much easier to create *idempotent* dynamic segments. In the previous version of the tool, code segments needed to be executed "in the right order" to produce the desired output result. For example, if a graph was rewritten by a graph transformation once, applying that transformation another time would usually result in a different output.

The demo shows how using *sets* of graphs rather than single graphs as a data type for computation simplifies the composition of (GT) operations and allows for the definition of a deterministic operator for rule application (which simply return the set of all possible direct derivations). Another advantage of this choice of data type is that unsuccessful rule applications simply return an empty set rather than requiring special handling of failure.

Finally, the demo illustrates how `GrapeVine`'s data structure maintains the derivation history of all graphs as a graph of graphs. We demonstrate how such history graphs can be reflected as regular `GrapeVine` graphs and how the tool supports visualizing and reasoning about derivation histories.

5 Conclusions and Future Work

The fact that current GT tools maintain graphs in ephemeral data structures limits their usefulness in practice. Applications that require knowledge of the derivation history of graphs cannot readily be supported. Current tools do not support computing with and reasoning about graph derivation history. Even in applications where this is not needed, stateful computation is harder to understand and test. Stateless computation is also preferrable when working with `GrapeVine` as a computational notebook. Users of the predecessor tool had to remember to execute code segments in worksheets "in the right order" since GTs had side-effects on *the* shared graph and few code segments would be idempotent.

Finally, from a tool developer's point of view, the associated stateful computation model is complex and prone to problems. For example, the transaction manager developed for the (stateful) prior version of this tool (`GrapePress`) was by far the most complex module of the tool. Removing it from `GrapeVine` was liberating, but also painful because of all the work that had gone into it.

Our current plan for evolving `GrapeVine` has two near-term objectives. Firstly, we will evolve the persistent data structure to make it *confluently* persistent. A data structure is called confluently persistent if there is a "merge" operation to join two branches originating from a common version [9]. While we did not talk about it in this paper, `GrapeVine` already has an operator to filter out "duplicate" graphs in a graph set (up to isomorphism). That operator is implemented along a similar mechanism as described by Rensink [14]. We are working on using this operator as a basis for defining a "merge" operation for `GrapeVine`'s data structure. Our second objective is to integrate theoretical results for verifying properties of graph transformation systems, in particular with respect to the confluence of rule sets, i.e., critical pair analysis.

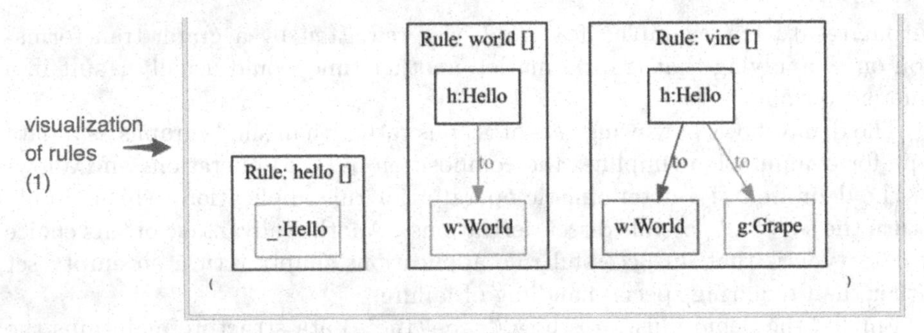

Above we have defined three simple GT rules.

static segment (2)

- Rule **hello** creates a node labelled "Hello".
- Rule **world** extends a "Hello" node with a "World" node and a "to" edge.
- Rule **vine** looks for a "Hello" node that is connected to a "World" node and replaces the latter with a "Grape" node.

In *GrapeVine*, the operation that creates a new (empty) graph is called `newgrape`. *(The reason why it is not called "newgraph" is because rule applications in GrapeVine actually operate on sets of graphs - dubbed* ***graph*** *enumerations (grapes)) rather than single graphs.*

Rule application is as simple as calling a function with the rule's name on a graph (*grape*). The statement below applies rule **hello** to an empty graph.

Graphs can be visualized in *GrapeVine*'s computation notebook using the `view` function, which also operates on *grapes*. To avoid too many nested parentheses when composing functions, we can use the Clojure threading macro. The following statement creates an empty graph, sequentially applies rule **hello** three times, and then calls `view` to visualize the result.

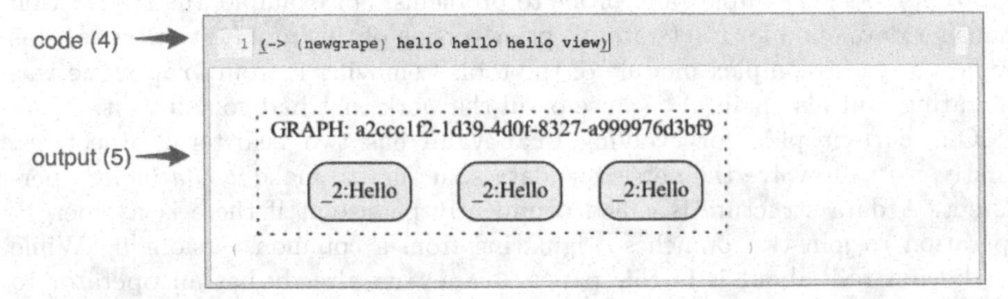

Nondeterministic vs. deterministic rule applications

The application of GT rules is generally non-deterministic. Applying rule `world` to a graph with three "Hello" nodes has three possible matches. In addition to the normal non-deterministic rule application operator, *GrapeVine* provides an operator for deterministic rule application. It simply returns a set of all possible derivations. That operator is invoked by adding an aterisk to the name of the rule (see below).

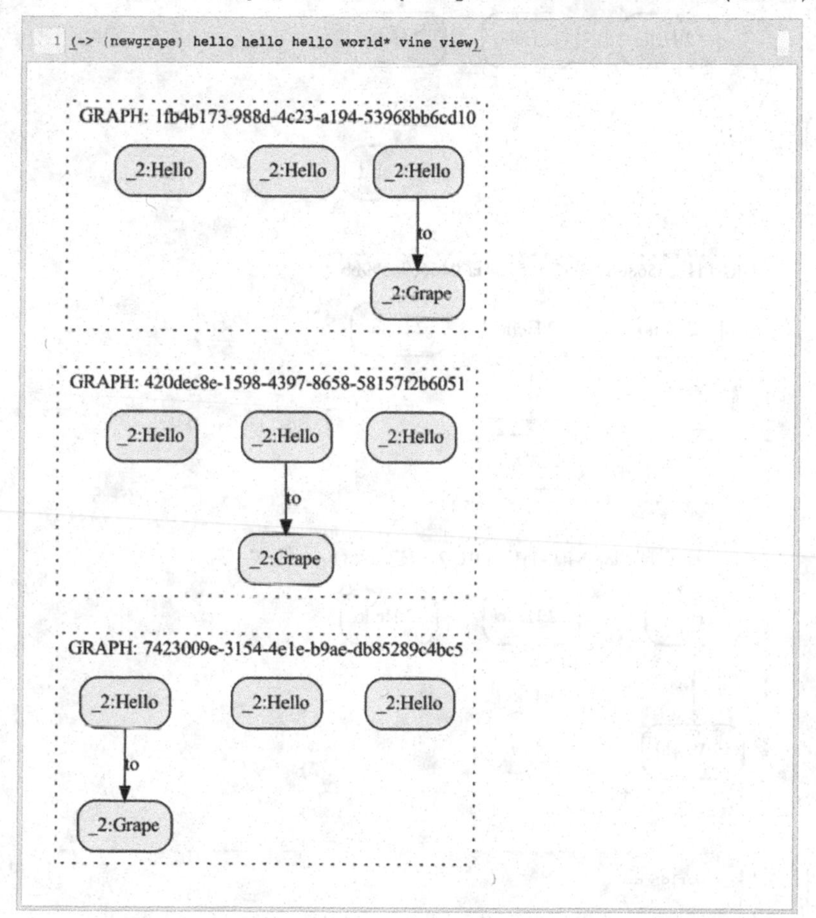

The above example also shows that rules can be applied to sets of graphs by simply applying them to all members in the set and producing the set of all results, i.e., `vine` is applied to three graphs.

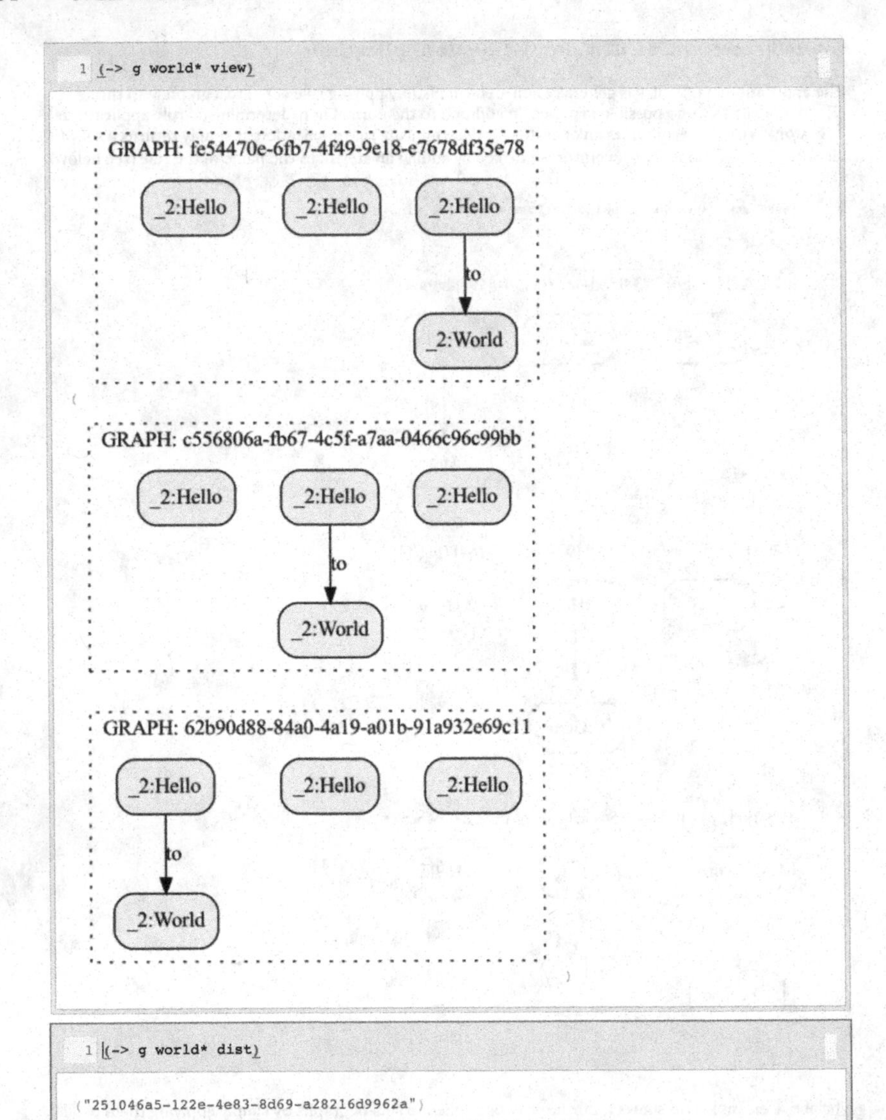
1 (-> g world* view)
GRAPH: fe54470e-6fb7-4f49-9e18-e7678df35e78
_2:Hello
_2:Hello
_2:Hello
to
_2:World
(
GRAPH: c556806a-fb67-4c5f-a7aa-0466c96c99bb
_2:Hello
_2:Hello
_2:Hello
to
_2:World
GRAPH: 62b90d88-84a0-4a19-a01b-91a932e69c11
_2:Hello
_2:Hello
_2:Hello
to
_2:World
)
1 (-> g world* dist)
("251046a5-122e-4e83-8d69-a28216d9962a")

Detailed occurrences

Of course we can define Clojure *vars* to refer to graph sets, as we do with *g* below. *GrapeVine* provides functions to query and visualize details of each occurrence in graph's history. The example below shows the `viewsteps` function which can be used to visualize the detailed changes made in a graph history (as defined by a sequence of steps).

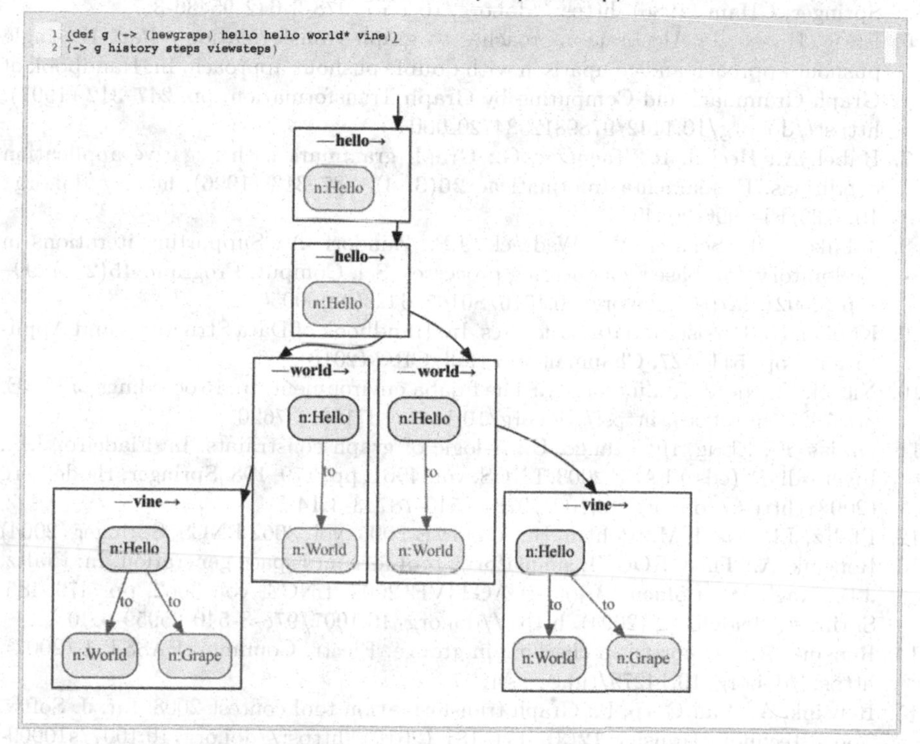

And we can simply get the graph sets produced by any of these steps in a graphs history:

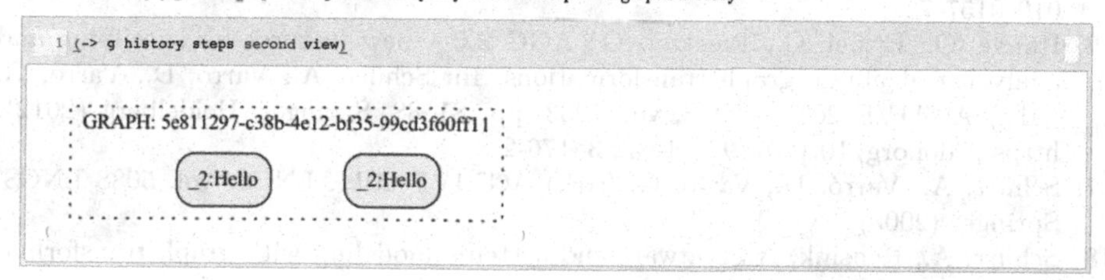

References

1. Aouat, A., Bendella, F., Deba, E.a.: Tools of model transformation by graph transformation. In: IEEE CSAE 2012, pp. 425–428 (2012). https://doi.org/10.1109/ICSESS.2012.6269495
2. Baldan, P., Corradini, A., Foss, L., Gadducci, F.: Graph transactions as processes. In: Corradini, A., Ehrig, H., Montanari, U., Ribeiro, L., Rozenberg, G. (eds.) ICGT 2006. LNCS, vol. 4178, pp. 199–214. Springer, Heidelberg (2006). https://doi.org/10.1007/11841883_15
3. Barmpis, K., Kolovos, D.: Hawk: towards a scalable model indexing architecture. In: Proceedings of the of ACM BigMDE 2013 (2013). https://doi.org/10.1145/2487766.2487771

4. Corradini, A., Montanari, U., Rossi, F.: Graph processes. Fundamenta Informaticae **26**(3, 4), 241–265 (1996). https://doi.org/10.3233/FI-1996-263402
5. Ehrig, H., Habel, A.: Graph Grammars with Application Conditions, pp. 87–100. Springer, CHam (1986). https://doi.org/10.1007/978-3-642-95486-3_7
6. Ehrig, H., et al.: Algebraic approaches to graph transformation - Part II: single pushout approach and comparison with double pushout approach. In: Handbook of Graph Grammars and Computing by Graph Transformation, pp. 247–312 (1997). https://doi.org/10.1142/9789812384720_0004
7. Habel, A., Heckel, R., Taentzer, G.: Graph grammars with negative application conditions. Fundamenta Informaticae **26**(3, 4), 287–313 (1996). https://doi.org/10.3233/FI-1996-263404
8. Jahnke, J.H., Schäfer, W., Wadsack, J.P., Zündorf, A.: Supporting iterations in exploratory database reengineering processes. Sci. Comput. Program. **45**(2–3), 99–136 (2002). https://doi.org/10.1016/S0167-6423(02)00056-4
9. Kaplan, H.: Persistent data structures. In: Handbook of Data Structures and Applications, pp. 511–527. Chapman and Hall/CRC (2018)
10. Nickel, U., Niere, J., Zündorf, A.: The fujaba environment. In: Proceedings of ICSE, pp. 742–745 (2000). https://doi.org/10.1145/337180.337620
11. Orejas, F., Ehrig, H., Prange, U.: A logic of graph constraints. In: Fiadeiro, J.L., Inverardi, P. (eds.) FASE 2008. LNCS, vol. 4961, pp. 179–198. Springer, Heidelberg (2008). https://doi.org/10.1007/978-3-540-78743-3_14
12. Pfaltz, J.L., Nagl, M., Bohlen, B.: AGTIVE 2003, vol. 3062 LNCS. Springer (2004)
13. Rensink, A.: The GROOVE simulator: a tool for state space generation. In: Pfaltz, J.L., Nagl, M., Böhlen, B. (eds.) AGTIVE 2003. LNCS, vol. 3062, pp. 479–485. Springer, Heidelberg (2004). https://doi.org/10.1007/978-3-540-25959-6_40
14. Rensink, A.: Isomorphism checking in groove. Electr. Commun. EASST **1** (2007). https://doi.org/10.14279/tuj.eceasst.1.77
15. Rensink, A., Van Gorp, P.: Graph transformation tool contest 2008. Int. J. Softw. Tools Technol. Transfer **12**(3), 171–181 (2010). https://doi.org/10.1007/s10009-010-0157-7
16. Runge, O., Ermel, C., Taentzer, G.: AGG 2.0 – new features for specifying and analyzing algebraic graph transformations. In: Schürr, A., Varró, D., Varró, G. (eds.) AGTIVE 2011. LNCS, vol. 7233, pp. 81–88. Springer, Heidelberg (2012). https://doi.org/10.1007/978-3-642-34176-2_8
17. Schürr, A., Varró, D., Varró, G. (eds.) AGTIVE 2011. LNCS, vol. 5088 LNCS. Springer (2008)
18. Schürr, A., Rensink, A.: Software and systems modeling with graph transformations theme issue of the Journal on Software and Systems Modeling. Softw. Syst. Model. **13**(1), 171–172 (2012). https://doi.org/10.1007/s10270-012-0254-z
19. Schürr, A., Varró, D., Varró, G.: AGTIVE 2011, vol. 7233 LNCS. Springer (2012)
20. Taentzer, G.: AGG: a graph transformation environment for modeling and validation of software. In: Pfaltz, J.L., Nagl, M., Böhlen, B. (eds.) AGTIVE 2003. LNCS, vol. 3062, pp. 446–453. Springer, Heidelberg (2004). https://doi.org/10.1007/978-3-540-25959-6_35
21. Weber, J.H.: GRAPE – a graph rewriting and persistence engine. In: de Lara, J., Plump, D. (eds.) ICGT 2017. LNCS, vol. 10373, pp. 209–220. Springer, Cham (2017). https://doi.org/10.1007/978-3-319-61470-0_13
22. Weber, J.H.: Grapepress - a computational notebook for graph transformations. In: Graph Transformation, LNCS, vol. 12741, pp. 294–302. Springer (2021). https://doi.org/10.1007/978-3-030-78946-6_16

Author Index

Printed in the United States
by Baker & Taylor Publisher Services